METHODS IN MOLECULAR BIOLOGY

Series Editor
John M. Walker
School of Life and Medical Sciences
University of Hertfordshire
Hatfield, Hertfordshire, UK

For further volumes:
http://www.springer.com/series/7651

For over 35 years, biological scientists have come to rely on the research protocols and methodologies in the critically acclaimed *Methods in Molecular Biology* series. The series was the first to introduce the step-by-step protocols approach that has become the standard in all biomedical protocol publishing. Each protocol is provided in readily-reproducible step-by-step fashion, opening with an introductory overview, a list of the materials and reagents needed to complete the experiment, and followed by a detailed procedure that is supported with a helpful notes section offering tips and tricks of the trade as well as troubleshooting advice. These hallmark features were introduced by series editor Dr. John Walker and constitute the key ingredient in each and every volume of the *Methods in Molecular Biology* series. Tested and trusted, comprehensive and reliable, all protocols from the series are indexed in PubMed.

Progenitor Cells

Methods and Protocols

Edited by

Mugdha V. Joglekar and Anandwardhan A. Hardikar

NHMRC Clinical Trials Centre, Faculty of Medicine and Health, The University of Sydney, Camperdown, NSW, Australia

Humana Press

Editors
Mugdha V. Joglekar
NHMRC Clinical Trials Centre
Faculty of Medicine and Health
The University of Sydney
Camperdown, NSW, Australia

Anandwardhan A. Hardikar
NHMRC Clinical Trials Centre
Faculty of Medicine and Health
The University of Sydney
Camperdown, NSW, Australia

ISSN 1064-3745 ISSN 1940-6029 (electronic)
Methods in Molecular Biology
ISBN 978-1-4939-9630-8 ISBN 978-1-4939-9631-5 (eBook)
https://doi.org/10.1007/978-1-4939-9631-5

This Humana imprint is published by the registered company Springer Science+Business Media, LLC, part of Springer Nature.
The registered company address is: 233 Spring Street, New York, NY 10013, U.S.A.

Dedication

"Aai," Mrs. Asha A. Hardikar—

So full of life, love, and kindness!

Preface

Progenitor cells have always been a topic of interest among researchers for their potential to differentiate into multiple lineages and ability to proliferate. They are considered to be the descendants of stem cells, and possess a limited capacity to self-renew and proliferate. It is rather fascinating that progenitor cells, which make up for a relatively minor proportion of cells in our body, retain the ability to proliferate and differentiate into multiple lineages, whereas most "specialized" cells seem to lose these properties during development. One question that progenitor cell biologists are intrigued with is the plasticity conferred on progenitor cells in retaining their capacity to differentiate. Progenitor cells reside in multiple tissues, and methodologies to isolate as well as to derive these from stem cells or terminally differentiated cells are available. These are also known to participate in regenerative and repair processes by migrating to damaged tissues or sites, apart from their role in routine maintenance of cells or tissue renewal in sites such as the skin, intestine, and blood. Understanding the cellular interactions and molecular mechanisms behind the function of progenitor cells can enlighten us about the differentiation process and aid in utilizing this knowledge toward regenerative biology applications in health and medicine.

This book captures the diversity in progenitor cell biology methods, through 21 chapters, from 21 different institutes/centers across the 4 continents of the world. Each chapter is unique in describing i) a technique that can be widely used by other fellow scientists in their respective cell systems or ii) a method of isolating/generating and characterizing progenitor cells either from a tissue or from embryonic stem cells.

The book highlights include chapters that describe isolating progenitor cells from different sources, including isolation from the adipose tissue, blood, bone marrow, ear (tympanic membrane), gut, heart, pancreatic islets, and the Wharton's jelly. Some of these describe the derivation of neural, pancreatic, and urothelial precursors from the embryonic stem cells, thus presenting the readers with methodologies associated with multiple tissues from the three germ layers in (wo)men and mice. The isolation of progenitors from the pancreas is described in two of the chapters, each with a different method. One of the protocols describes isolating islets followed by culturing them in vitro to generate progenitor cells via epithelial-to-mesenchymal transition, whereas another protocol describes colony-forming progenitors obtained from single-cell suspension of enzymatically dissociated pancreatic cells. It is very interesting to read how different labs answer a similar question with different approaches and obtain their own sets of methodologies.

Our book also showcases some unique and cutting-edge techniques that may be applicable to all researchers in the area of progenitor cell biology. These include digital droplet PCR, flow cytometry and cell sorting, mitochondrial assays, calcium ratiometric imaging, detecting replication using thymidine analogue, and 2D gel electrophoresis. Routine methods, including cell passaging, cell counting, cryofreezing, immunostaining, Western blotting, RNA isolation, and RT-PCR, are described in multiple chapters as per the need of progenitor cell characterization. Gene delivery to progenitor cells is described using lentiviral transduction methods as well as a novel method of using metal organic frameworks. Extracellular vesicles (EVs) have attracted much attention recently, and our book features a chapter that details isolation of EVs as well as their characterization using a nanoparticle tracking system and specialized microscopy methods. Other chapters of interest include protocols for encapsulation, transplantation in animal models, as well as T-cell proliferation assays. Single-cell assays to track the blood cell progeny and thereby progenitor cell divisions are of note. Overall, this book showcases several important and relevant chapters that most readers will find useful in their research.

We are fortunate and thankful to all our authors not only for their contribution in terms of scientific input with their chapter but also for keeping the timelines and for responding to reminders and all emails! This book could not have been possible without the encouragement from our department, students, and colleagues. The unconditional love and strong support from our daughters and parents is always a boost to life and work, including the long hours spent in editing this book.

We hope that our readers would feel enlightened and find in these methods the necessary means or tools to search for answers to their own questions that have been left unanswered. We also hope that our readers will find this book worthy of attention and an important addition to their laboratory protocols. Here's wishing to many more years of constructive research in progenitor cell biology, with multiple joys of discoveries, moments of getting awestruck by life under the microscope, and wondering at the marvel and magic of Mother Nature!

Camperdown, NSW, Australia

Mugdha V. Joglekar
Anandwardhan A. Hardikar

Contents

Dedication .... *v*
Preface .... *vii*
Contributors .... *xi*

1 Isolation and Molecular Characterization of Progenitor Cells from Human Umbilical Cord 1
Umesh Goyal, Ankita Sen, and Malancha Ta

2 Isolation and Characterization of Extracellular Vesicles from Mesenchymal Stromal Cells 15
Sally Yunsun Kim, Thanh Huyen Phan, Christina Limantoro, Bill Kalionis, and Wojciech Chrzanowski

3 Assessment of β-Cell Replication in Isolated Rat Islets of Langerhans 25
Louise T. Dalgaard

4 Droplet Digital PCR for Measuring Absolute Copies of Gene Transcripts in Human Islet-Derived Progenitor Cells 37
Cody-Lee Maynard, Wilson K. M. Wong, Anandwardhan A. Hardikar, and Mugdha V. Joglekar

5 Isolation and In Vitro Culture of Human Gut Progenitor Cells 49
Jessica Bruce, Gerard E. Kaiko, and Simon Keely

6 Isolation and Characterization of Colony-Forming Progenitor Cells from Adult Pancreas 63
Janine C. Quijano, Jacob R. Tremblay, Jeffrey Rawson, and Hsun Teresa Ku

7 A Novel Gene Delivery Approach Using Metal Organic Frameworks in Human Islet-Derived Progenitor Cells 81
Arpita Poddar, Mugdha V. Joglekar, Anandwardhan A. Hardikar, and Ravi Shukla

8 Encapsulation and Transplantation of Pancreatic Progenitor Cells 93
Luke Carroll, Auvro R. Mridha, and Bernard E. Tuch

9 Differentiation of Urothelium from Mouse Embryonic Stem Cells in Chemically Defined Conditions 103
Badwi B. Boumelhem, Stuart T. Fraser, and Stephen J. Assinder

10 Isolation and Characterization of Progenitor Cells from Human Adipose Tissue 117
Nitya Shree and Ramesh Bhonde

11 Identification and Analysis of Mouse Erythroid Progenitor Cells 125
Chanukya K. Colonne, Jia Hao Yeo, Campbell V. McKenzie, and Stuart T. Fraser

12 Single-Cell Assays Using Hematopoietic Stem and Progenitor Cells 147
Ashwini S. Hinge and Marie-Dominique Filippi

13 Isolation and Characterization of Cardiac Progenitor Cells 161
Parul Dixit and Rajesh G. Katare

14 Mitochondrial Assays Using Cardiac Stem Cells 175
Ayeshah A. Rosdah, Lea M. D. Delbridge, and Shiang Y. Lim

15 Two-Dimensional Electrophoresis and Mass Spectrometry for Protein Identification 185
Amaresh K. Ranjan and Anil Gulati

16 High-Efficiency Lentiviral Gene Modification of Primary Murine Bone-Marrow Mesenchymal Stem Cells 197
Dario Gerace, Binhai Ren, Rosetta Martiniello-Wilks, and Ann M. Simpson

17 Bone Marrow-Derived Progenitor Cells Mediate Immune Cell Regulation 215
Kisha N. Sivanathan and Patrick T. Coates

18 Flow Cytometry and Cell Sorting Using Hematopoietic Progenitor Cells 235
Sangeetha Vadakke-Madathil, Lalita S. Limaye, Vaijayanti P. Kale, and Hina W. Chaudhry

19 Isolation of Epidermal Progenitor Cells from Rat Tympanic Membrane 247
Lawrence J. Liew, Allen Y. Wang, and Rodney J. Dilley

20 Measurement of Store-Operated Calcium Entry in Human Neural Cells: From Precursors to Differentiated Neurons 257
Renjitha Gopurappilly, Bipan Kumar Deb, Pragnya Chakraborty, and Gaiti Hasan

21 Embryoid Body Differentiation of Mouse Embryonic Stem Cells into Neurectoderm and Neural Progenitors 273
Rachel A. Shparberg, Hannah J. Glover, and Michael B. Morris

Index *287*

Contributors

STEPHEN J. ASSINDER • *Disciplines of Physiology, Anatomy and Histology, School of Medical Science and Bosch Institute, University of Sydney, Camperdown, NSW, Australia*

RAMESH BHONDE • *School of Regenerative Medicine, Manipal Academy of Higher Education, Bangalore, India; Dr. D.Y. Patil Vidyapeeth, Pune, India*

BADWI B. BOUMELHEM • *Disciplines of Physiology, Anatomy and Histology, School of Medical Science and Bosch Institute, University of Sydney, Camperdown, NSW, Australia*

JESSICA BRUCE • *Viruses, Infections/Immunity, Vaccines and Asthma (VIVA) Research Program, Hunter Medical Research Institute, New Lambton Heights, NSW, Australia; School of Biomedical Sciences and Pharmacy, Faculty of Health and Medicine, University of Newcastle, Callaghan, NSW, Australia*

LUKE CARROLL • *School of Medical Sciences, Discipline Physiology, The University of Sydney, Sydney, NSW, Australia*

PRAGNYA CHAKRABORTY • *National Centre for Biological Sciences, Tata Institute of Fundamental Research, Bangalore, India*

HINA W. CHAUDHRY • *Icahn School of Medicine at Mount Sinai, New York, NY, USA*

WOJCIECH CHRZANOWSKI • *Faculty of Medicine and Health, The University of Sydney School of Pharmacy, Sydney, NSW, Australia; The University of Sydney Nano Institute, Sydney, NSW, Australia*

PATRICK T. COATES • *Faculty of Health and Medical Sciences, School of Medicine, University of Adelaide, Adelaide, SA, Australia; Central Northern Adelaide Renal Transplantation Service, Royal Adelaide Hospital, Adelaide, SA, Australia*

CHANUKYA K. COLONNE • *Disciplines of Physiology, Faculty of Medicine and Health, School of Medical Sciences, University of Sydney, Camperdown, NSW, Australia*

LOUISE T. DALGAARD • *Department of Science and Environment, Roskilde University, Roskilde, Denmark*

BIPAN KUMAR DEB • *National Centre for Biological Sciences, Tata Institute of Fundamental Research, Bangalore, India*

LEA M. D. DELBRIDGE • *Department of Physiology, University of Melbourne, Parkville, VIC, Australia*

RODNEY J. DILLEY • *Ear Science Institute Australia, Perth, WA, Australia; Ear Sciences Centre, School of Medicine and Centre for Cell Therapy and Regenerative Medicine, School of Biomedical Sciences, University of Western Australia, Perth, WA, Australia*

PARUL DIXIT • *Department of Physiology-HeartOtago, School of Biomedical Sciences, University of Otago, Dunedin, New Zealand*

MARIE-DOMINIQUE FILIPPI • *Division of Experimental Hematology and Cancer Biology, Cincinnati Children's Hospital Medical Center, Cincinnati, OH, USA*

STUART T. FRASER • *Disciplines of Physiology, Faculty of Medicine and Health, School of Medical Sciences, University of Sydney, Camperdown, NSW, Australia; Disciplines of Anatomy and Histology, Faculty of Medicine and Health, School of Medical Sciences, University of Sydney, Camperdown, NSW, Australia; Faculty of Medicine and Health, Bosch Institute, University of Sydney, Camperdown, NSW, Australia; The University of Sydney Nano Institute, University of Sydney, Camperdown, NSW, Australia*

Dario Gerace • *The School of Life Sciences and Centre for Health Technologies, University of Technology Sydney, Sydney, NSW, Australia; Department of Stem Cell and Regenerative Biology, Harvard Stem Cell Institute, Harvard University, Cambridge, MA, USA*

Hannah J. Glover • *Embryonic Stem Cell Lab, Bosch Institute and Discipline of Physiology, School of Medical Sciences, University of Sydney, Sydney, NSW, Australia*

Renjitha Gopurappilly • *National Centre for Biological Sciences, Tata Institute of Fundamental Research, Bangalore, India*

Umesh Goyal • *Indian Institute of Science Education and Research, Kolkata, West Bengal, India*

Anil Gulati • *Chicago College of Pharmacy, Midwestern University, Downers Grove, IL, USA*

Anandwardhan A. Hardikar • *NHMRC Clinical Trials Centre, Faculty of Medicine and Health, The University of Sydney, Camperdown, NSW, Australia*

Gaiti Hasan • *National Centre for Biological Sciences, Tata Institute of Fundamental Research, Bangalore, India*

Ashwini S. Hinge • *Division of Experimental Hematology and Cancer Biology, Cincinnati Children's Hospital Medical Center, Cincinnati, OH, USA*

Mugdha V. Joglekar • *NHMRC Clinical Trials Centre, Faculty of Medicine and Health, The University of Sydney, Camperdown, NSW, Australia*

Gerard E. Kaiko • *Viruses, Infections/Immunity, Vaccines and Asthma (VIVA) Research Program, Hunter Medical Research Institute, New Lambton Heights, NSW, Australia; School of Biomedical Sciences and Pharmacy, Faculty of Health and Medicine, University of Newcastle, Callaghan, NSW, Australia*

Vaijayanti P. Kale • *National Centre for Cell Sciences, Pune, India; Symbiosis Center for Stem Cell Research, Symbiosis School of Biological Sciences, Pune, India*

Bill Kalionis • *Department of Maternal-Fetal Medicine Pregnancy Research Centre, University of Melbourne, Melbourne, VIC, Australia; Department of Obstetrics and Gynaecology, Royal Women's Hospital, Parkville, VIC, Australia*

Rajesh G. Katare • *Department of Physiology-HeartOtago, School of Biomedical Sciences, University of Otago, Dunedin, New Zealand*

Simon Keely • *Viruses, Infections/Immunity, Vaccines and Asthma (VIVA) Research Program, Hunter Medical Research Institute, New Lambton Heights, NSW, Australia; School of Biomedical Sciences and Pharmacy, Faculty of Health and Medicine, University of Newcastle, Callaghan, NSW, Australia*

Sally Yunsun Kim • *Faculty of Medicine and Health, The University of Sydney School of Pharmacy, Sydney, NSW, Australia; The University of Sydney Nano Institute, Sydney, NSW, Australia*

Hsun Teresa Ku • *Department of Translational Research and Cellular Therapeutics, Diabetes and Metabolism Research Institute, Beckman Research Institute of City of Hope, Duarte, CA, USA; Irell & Manella Graduate School of Biological Sciences, Duarte, CA, USA*

Lawrence J. Liew • *Ear Science Institute Australia, Perth, WA, Australia; Ear Sciences Centre, School of Medicine and Centre for Cell Therapy and Regenerative Medicine, School of Biomedical Sciences, University of Western Australia, Perth, WA, Australia*

Shiang Y. Lim • *O'Brien Institute Department, St Vincent's Institute of Medical Research, Fitzroy, VIC, Australia; Department of Surgery, University of Melbourne, Parkville, VIC, Australia*

Christina Limantoro • *Faculty of Medicine and Health, The University of Sydney School of Pharmacy, Sydney, NSW, Australia; The University of Sydney Nano Institute, Sydney, NSW, Australia*

Lalita S. Limaye • *National Centre for Cell Sciences, Pune, India*

Rosetta Martiniello-Wilks • *The School of Life Sciences and Centre for Health Technologies, University of Technology Sydney, Sydney, NSW, Australia*

Cody-Lee Maynard • *Diabetes and Islet Biology Group, Faculty of Medicine and Health, NHMRC Clinical Trials Centre, The University of Sydney, Camperdown, NSW, Australia*

Campbell V. McKenzie • *Disciplines of Physiology, Faculty of Medicine and Health, School of Medical Sciences, University of Sydney, Camperdown, NSW, Australia*

Michael B. Morris • *Embryonic Stem Cell Lab, Bosch Institute and Discipline of Physiology, School of Medical Sciences, University of Sydney, Sydney, NSW, Australia*

Auvro R. Mridha • *School of Medical Sciences, Discipline Physiology, The University of Sydney, Sydney, NSW, Australia*

Thanh Huyen Phan • *Faculty of Medicine and Health, The University of Sydney School of Pharmacy, Sydney, NSW, Australia*

Arpita Poddar • *Ian Potter NanoBioSensing Facility and NanoBiotechnology Research Lab (NBRL), School of Science, RMIT University, Melbourne, VIC, Australia*

Janine C. Quijano • *Department of Translational Research and Cellular Therapeutics, Diabetes and Metabolism Research Institute, Beckman Research Institute of City of Hope, Duarte, CA, USA*

Amaresh K. Ranjan • *Chicago College of Pharmacy, Midwestern University, Downers Grove, IL, USA*

Jeffrey Rawson • *Department of Translational Research and Cellular Therapeutics, Diabetes and Metabolism Research Institute, Beckman Research Institute of City of Hope, Duarte, CA, USA*

Binhai Ren • *The School of Life Sciences and Centre for Health Technologies, University of Technology Sydney, Sydney, NSW, Australia*

Ayeshah A. Rosdah • *O'Brien Institute Department, St Vincent's Institute of Medical Research, Fitzroy, VIC, Australia; Department of Physiology, University of Melbourne, Parkville, VIC, Australia; Faculty of Medicine, Universitas Sriwijaya, Palembang, Indonesia*

Ankita Sen • *Indian Institute of Science Education and Research, Kolkata, West Bengal, India*

Rachel A. Shparberg • *Embryonic Stem Cell Lab, Bosch Institute and Discipline of Physiology, School of Medical Sciences, University of Sydney, Sydney, NSW, Australia*

Nitya Shree • *School of Regenerative Medicine, Manipal Academy of Higher Education, Bangalore, India*

Ravi Shukla • *Ian Potter NanoBioSensing Facility and NanoBiotechnology Research Lab (NBRL), School of Science, RMIT University, Melbourne, VIC, Australia*

Ann M. Simpson • *The School of Life Sciences and Centre for Health Technologies, University of Technology Sydney, Sydney, NSW, Australia*

Kisha N. Sivanathan • *Faculty of Health and Medical Sciences, School of Medicine, University of Adelaide, Adelaide, SA, Australia; Evergrande Center for Immunologic Diseases, Harvard Medical School and Brigham and Women's Hospital, Boston, MA, USA*

Malancha Ta • *Indian Institute of Science Education and Research, Kolkata, West Bengal, India*

Jacob R. Tremblay • *Department of Translational Research and Cellular Therapeutics, Diabetes and Metabolism Research Institute, Beckman Research Institute of City of Hope, Duarte, CA, USA*

Bernard E. Tuch • *School of Medical Sciences, Discipline Physiology, The University of Sydney, Sydney, NSW, Australia; Australian Foundation for Diabetes Research, Maroubra, NSW, Australia*

Sangeetha Vadakke-Madathil • *Icahn School of Medicine at Mount Sinai, New York, NY, USA*

Allen Y. Wang • *Ear Science Institute Australia, Perth, WA, Australia*

Wilson K. M. Wong • *Diabetes and Islet Biology Group, Faculty of Medicine and Health, NHMRC Clinical Trials Centre, The University of Sydney, Camperdown, NSW, Australia*

Jia Hao Yeo • *Disciplines of Anatomy and Histology, Faculty of Medicine and Health, School of Medical Sciences, University of Sydney, Camperdown, NSW, Australia*

Chapter 1

Isolation and Molecular Characterization of Progenitor Cells from Human Umbilical Cord

Umesh Goyal, Ankita Sen, and Malancha Ta

Abstract

Mesenchymal stem cells (MSCs) are multipotent precursor cells which have been isolated from different vascularized tissue sources. Due to their paracrine function of secreting trophic and immunomodulatory molecules, MSCs are successfully used in cell-based transplantations and provide an alternative medical paradigm for treating a variety of devastating disorders. Umbilical cord is a medical waste with a large, readily available donor pool. Since umbilical cord is a fetal tissue, MSCs derived from it are considered more primitive with proliferative and differentiation advantages over adult MSCs. We define here a simple, efficient, and reproducible protocol to isolate MSCs from WJ of human umbilical cord using a nonenzymatic procedure. Under the optimized culture conditions, the WJ-MSCs undergo robust proliferation, can be expanded up to 15–20 passages and express the characteristic MSC surface antigens. They can be differentiated into mesodermal lineages in vitro.

Key words Wharton's jelly, Mesenchymal stem cells, Explant culture, Umbilical cord, Arteries, Flow cytometry

1 Introduction

Mesenchymal stem cells (MSCs) offer great therapeutic promise in cell-based regenerative medicine and transplantation applications due to their immunomodulatory and trophic roles [1]. As perivascular cells, they have been identified and successfully isolated from many anatomical locations of the body such as bone marrow, adipose tissue, umbilical cord, and periodontal ligament [2, 3]. Being an ethically noncontroversial tissue with abundantly available donor pool, umbilical cord is an ideal and convenient source of MSCs. Moreover, as it is a by-product of pregnancy which is discarded after birth, its collection is painless and does not involve any invasive procedures or risk to the donor [4]. Since umbilical cord is a fetus-derived tissue, umbilical cord-derived MSCs are considered more primitive and exhibit certain advantages such as faster proliferation rate, extensive expansion capability,

Mugdha V. Joglekar and Anandwardhan A. Hardikar (eds.), *Progenitor Cells: Methods and Protocols*, Methods in Molecular Biology, vol. 2029, https://doi.org/10.1007/978-1-4939-9631-5_1, © Springer Science+Business Media, LLC, part of Springer Nature 2019

stronger immunosuppressive ability and no age-dependent variations, as compared to adult source derived-MSCs [5, 6].

MSCs have been isolated from different compartments of the umbilical cord. Wharton's jelly (WJ) is the cushioning, connective tissue matrix surrounding the blood vessels inside the umbilical cord and a rich source of MSCs [7]. Reports have shown WJ-MSCs to be therapeutic in several preclinical models of human diseases and also in clinical trials [6, 8]. As MSCs are present in low numbers in their niches, generating clinical quantity of MSCs for transplantations can be challenging. Here we describe isolation of MSCs from WJ of human umbilical cord by nonenzymatic, explant culture method which is simple, efficient, cost-effective and reproducible, and eliminates the use of any protease or enzymatic digestion. The isolation procedure involves removal of umbilical cord blood vessels, followed by chopping the perivascular tissue surrounding the blood vessels into small explants and placing them on a tissue culture dish. The WJ-MSCs isolated by this method can be subcultured and expanded till passage 15–20, while retaining all the basic characteristics defined for human MSCs by the International Society of Cellular Therapy (ISCT) [9]. They are adherent to plastic surfaces, positive for surface markers CD73, CD90 and CD105 while lacking CD34, CD45, HLA-DR, etc. and can undergo differentiation to osteogenic, chondrogenic and adipogenic lineages. In our protocol, for dissociation of cells, porcine trypsin has been replaced with TrypLE express which is an animal-origin free recombinant enzyme and gentle on cells. The WJ-MSCs isolated by this protocol could be cryopreserved for long-term storage, revived at high efficiency at any future time point and culture expanded again, without any compromises on their growth characteristics.

2 Materials

2.1 Umbilical Cord Explant Culture and WJ-MSC Expansion

1. MSC explant culture medium: KnockOut™ DMEM, 10% fetal bovine serum (FBS, mesenchymal stem cell qualified), 2 mM glutamine, 1× antibiotic–antimycotic. Take about 40 mL of KnockOut™ DMEM to a fresh sterile 50 mL conical tube, add 5 mL of FBS, 500 μL of 200 mM glutamine, 500 μL of antibiotic–antimycotic (100 ×) and adjust the volume to 50 mL with KnockOut™ DMEM. Filter through a 0.22 μm syringe filter (*see* **Note 1**). Store at 4 °C.
2. MSC expansion medium: KnockOut™ DMEM, 10% Fetal bovine serum (FBS, mesenchymal stem cell qualified), 2 mM glutamine, penicillin–streptomycin (100 U/mL). Take about 40 mL of KnockOut™ DMEM to a fresh sterile 50 mL conical tube, add 5 mL of FBS, 500 μL of 200 mM glutamine, 500 μL

of penicillin–streptomycin (10,000 U/mL) and adjust the volume to 50 mL with KnockOut™ DMEM. Filter through a 0.22 μm syringe filter (*see* **Note 1**). Store at 4 °C.

3. Freezing Mixture: 90% Fetal bovine serum (FBS, mesenchymal stem cell qualified), 10% dimethyl sulfoxide (DMSO, sterile-filtered) (*see* **Note 2**). Prepare 5–10 mL in sterile conical tube. Filter through a 0.22 μm syringe filter. Store at 4 °C.
4. 0.99% normal saline.
5. Saline–antibiotic–antimycotic (1×): Add 250 μL of antibiotic–antimycotic (100 ×) to 25 mL of 0.99% normal saline (*see* **Note 3**).
6. 1× Sterile Phosphate Buffer Saline (PBS): Add two tablets of PBS to 400 mL of distilled water and autoclave pH-7.4. Store at 4 °C.
7. Dulbecco's phosphate-buffered saline (DPBS) No calcium no magnesium.
8. TrypLE™ Express Enzyme (1×), no phenol red (*see* **Note 4**).
9. 70% ethanol.
10. Trypan Blue.
11. Hemocytometer.
12. Freezing vial.
13. Mr. Frosty™.
14. Isopropanol.
15. Sodium hypochlorite solution.
16. Plastic beaker.
17. 35 mm cell culture dishes.
18. Sterile surgical instruments: small and large-sized pointed forceps, scissors, blade, scalpel handle.
19. Biohazard disposable bags.

2.2 Immunophenotypic Characterization

1. Fluorochrome conjugated anti-human monoclonal CD 73 antibody and its respective isotype control.
2. Fluorochrome conjugated anti-human monoclonal CD 90 antibody and its respective isotype control.
3. Fluorochrome conjugated anti-human monoclonal CD 105 antibody and its respective isotype control.
4. Fluorochrome conjugated anti-human monoclonal CD 34 antibody and its respective isotype control.
5. 1× sterile PBS.
6. FACS tube.
7. Sheath fluid.
8. Flow cytometry instrument.

2.3 In Vitro Differentiation and Staining

1. Differentiation media (*see* **Note 5**).
2. 1× PBS solution.
3. 4% Paraformaldehyde solution.
4. Isopropanol.
5. 60% isopropanol.
6. Alizarin Red S stain.
7. Oil Red O stain.
8. Whatman No. 1 filter paper.
9. Funnel.
10. Oil Red O stock solution (0.5%): Weigh 0.05 g of Oil red O stain and add to 10 mL of isopropanol. Dissolve in warm water bath for 30 min.
11. Oil Red O working solution: Mix 3 parts of Oil Red O stock solution with 2 parts of distilled water and keep undisturbed for 10 min. Filter twice through Whatman No. 1 filter paper.
12. Alizarin Red S solution (1%): Weigh 0.05 g of Alizarin Red S and dissolve in 5 mL of double distilled water. Store at room temperature.
13. Microscope.

3 Methods

All the steps must be carried out with proper precautions and safety measures to maintain sterile conditions.

3.1 Isolation of WJ-MSCs from Human Umbilical Cord

1. Collect fresh human umbilical cord after full-term birth in a sterile 50 mL conical tube with proper consent from patient. Immediately transport it on ice to the laboratory for further processing (*see* **Note 6**) (Fig. 1a).
2. Rinse the collected umbilical cord sample with normal saline to remove blood and blood clots. Cut it into 2–3 cm pieces (Fig. 1b) and incubate them in a 50 mL conical tube containing normal saline with antibiotic–antimycotic (1×) for about 2 h at 4 °C (*see* **Note 3**) (Fig. 1c).
3. Following incubation, transfer the cord pieces to the laminar flow hood. Next, rinse them three times with sterile PBS followed by one quick wash with 70% ethanol for 30 s to disinfect them. Further, give three more washes with sterile PBS to remove any trace of ethanol (*see* **Note 7**).
4. Transfer one cord piece to a 10 cm culture dish (Fig. 1d). If twisted, straighten the cord piece using a forceps and then

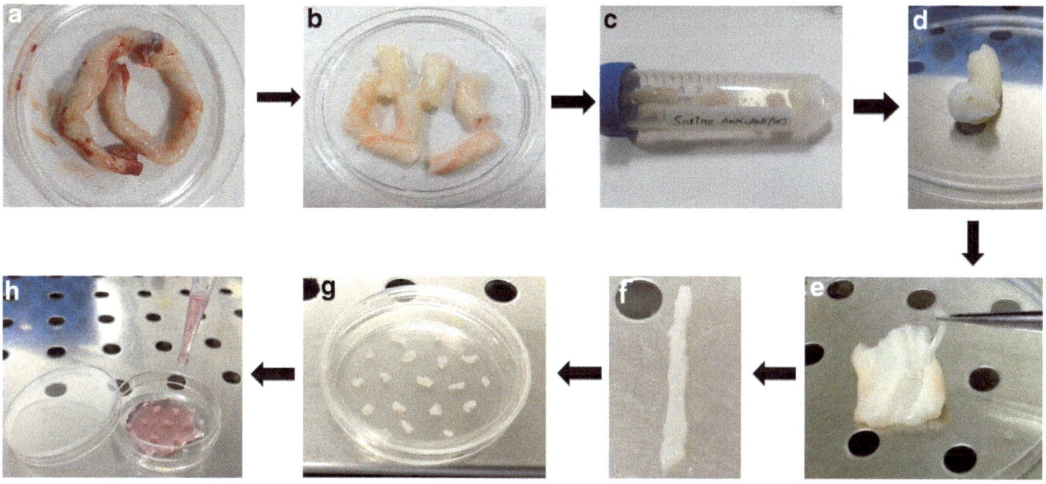

Fig. 1 Processing of human umbilical cord and plating of tissue explants for isolation of WJ-MSCs. Freshly collected human umbilical cord sample (**a**) was rinsed and the cord was cut into 2–3 cm long pieces (**b**). These pieces were further decontaminated in saline containing 1× antibiotic–antimycotic solution (**c**). An umbilical cord piece showing two arteries and one vein (**d**) was longitudinally slit along the length and an artery was traced and removed (**e**). The perivascular tissue surrounding the artery was cut into small explants and plated (**f**, **g**). Medium was added dropwise to the tissue explants pieces (**h**)

make a longitudinal slit along its length using a blade fixed to a scalpel handle (*see* **Note 8**).

5. Locate an artery and then cut open the flaps of tissue to trace and remove the artery, taking care that it does not break in between (Fig. 1e). After removing the artery completely, excise the underlying perivascular Wharton's jelly tissue (*see* **Note 9**) (Fig. 1f).
6. Chop the excised tissue into small 3–5 mm pieces using a scalpel and carefully place the pieces in a 35 mm cell culture dish using a fine forceps (*see* **Note 10**) (Fig. 1g).
7. To facilitate the attachment of the explant pieces, keep the dish for air drying for 5 min inside the laminar hood. Following this, gently and dropwise, add 2 mL of warm fresh MSC explant culture medium (Fig. 1h) (*see* **Note 11**).
8. In a similar manner, four to five 35 mm dishes can be plated from one umbilical cord sample.
9. Incubate the primary tissue culture dishes at 5% CO_2 and 37 °C temperature condition (*see* **Note 12**).
10. After 48 h, replace the medium with 2 mL of warm fresh MSC explant culture medium. Following this, medium change is given after every 72 h till sufficient number of cells come out from the tissue explants (*see* **Note 13**).
11. After 7–10 days, when sufficient number of cells have emerged out from the explant pieces (Fig. 2), remove the remaining

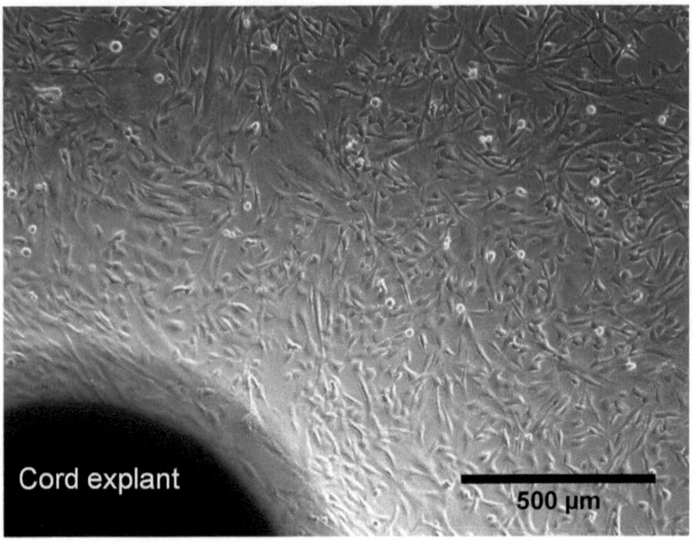

Fig. 2 WJ-MSCs emerging from explant tissue pieces after 7–10 days. Representative phase contrast image is shown. Bar is 500 μm as indicated

explant tissue along with the medium using 1 mL blunt end tips. Following this, add 2 mL warm fresh MSC explant culture medium to the dish and incubate at 37 °C, 5% CO_2 for another 24–48 h (*see* **Note 14**).

3.2 Establishment and Expansion of WJ-MSC Culture

1. After 24–48 h of explant removal, on reaching optimal confluency, carefully aspirate the medium and give two subsequent washes with DPBS (*see* **Note 15**).
2. Add 250 μL of TrypLE to each of the 35 mm cell culture dishes and incubate them for 4–6 min at 37 °C (*see* **Note 4**).
3. Add 1 mL of warm fresh MSC expansion medium to inactivate the TrypLE, collect the cells along with the medium in a 15 mL conical tube and centrifuge at 500 × *g* for 2 min.
4. After the spin, carefully aspirate the medium, add 500 μL of warm fresh MSC expansion medium and resuspend the cell pellet using a pipette tip. To count the viable number of cells, mix equal volumes of cell suspension and trypan blue and count the nonblue (live) cells in 10 μL mix using a hemocytometer (*see* **Note 16**).
5. Next, seed WJ-MSCs at a density of 5000 cells/cm^2 in 2 mL fresh warm MSC expansion medium in 35 mm dish. Incubate the dish at 37 °C for 72 h or until it becomes 70–80% confluent (*see* **Note 17**).
6. Again, detach cells from the dish using TrypLE, count and seed them by repeating **steps 2–5** (*see* **Note 15**).

3.3 Freezing and Reviving WJ-MSCs

1. After dislodging cells with TrypLE and counting them using a hemocytometer, centrifuge the left over, extra cells at 500 × *g* for 2 min at room temperature.
2. Aspirate the medium, add freezing mixture to the cell pellet and mix thoroughly using a pipette (*see* **Note 18**).
3. Aliquot cell suspension in cryogenic vials and keep them in a special container, Mr. Frosty™ containing isopropanol in −80 °C freezer for 24 h (*see* **Note 19**).
4. After 24 h, transfer the cryogenic vials from −80 °C to liquid nitrogen tank for long-term storage.
5. During cell revival, take out a cryogenic vial containing frozen WJ-MSCs from the liquid nitrogen tank. Immediately thaw it in a beaker containing prewarmed water at 37 °C.
6. As soon as the cell suspension thaws, add 1–2 mL of fresh, warm MSC expansion medium to the cryogenic vial inside the laminar flow hood. Mix thoroughly and transfer the content to a 15 mL conical tube containing 3–4 mL of fresh, warm MSC expansion medium.
7. Pellet down the cells at 500 × *g* for 2 min at room temperature to remove all traces of freezing mixture containing DMSO.
8. Aspirate the supernatant and resuspend the pellet in 500 μL of warm fresh MSC expansion medium. Count the number of live cells using a hemocytometer and plate 2–3 × 10^5 cells to each 35 mm cell culture dish with 2 mL of warm fresh MSC expansion medium (*see* **Note 20**).
9. Incubate the dish at 37 °C, 5% CO_2 for 48–72 h. After the first 24 h, remove the medium and add 2 mL of fresh, warm MSC expansion medium (*see* **Note 21**).
10. After 48–72 h of cell revival or at 80–90% confluency, detach the cells with TrypLE and seed cells for the next passage.

3.4 Surface Phenotype Characterization by Flow Cytometry

1. Dissociate WJ-MSCs grown on tissue culture vessels using TrypLE, neutralize with medium, count the number of cells and pellet by centrifugation as mentioned in Subheading 3.2, **steps 2–4** (*see* **Note 22**).
2. Wash the pelleted cells 1–2 times with ice cold PBS by centrifuging at 600 × *g* for 2 min.
3. Discard the supernatant and resuspend the cells in PBS at a concentration of 2 × 10^6 cells/mL (*see* **Note 23**).
4. Distribute ~1 × 10^5 cells or equivalent amount of cell suspension (50 μL) in each of the prelabeled FACS tubes.
5. Add the antibodies and their respective isotype controls to the labeled FACS tubes (*see* **Note 24**) and incubate on ice for 45–60 min. Analyze using flow cytometry (Fig. 3) (*see* **Note 25**).

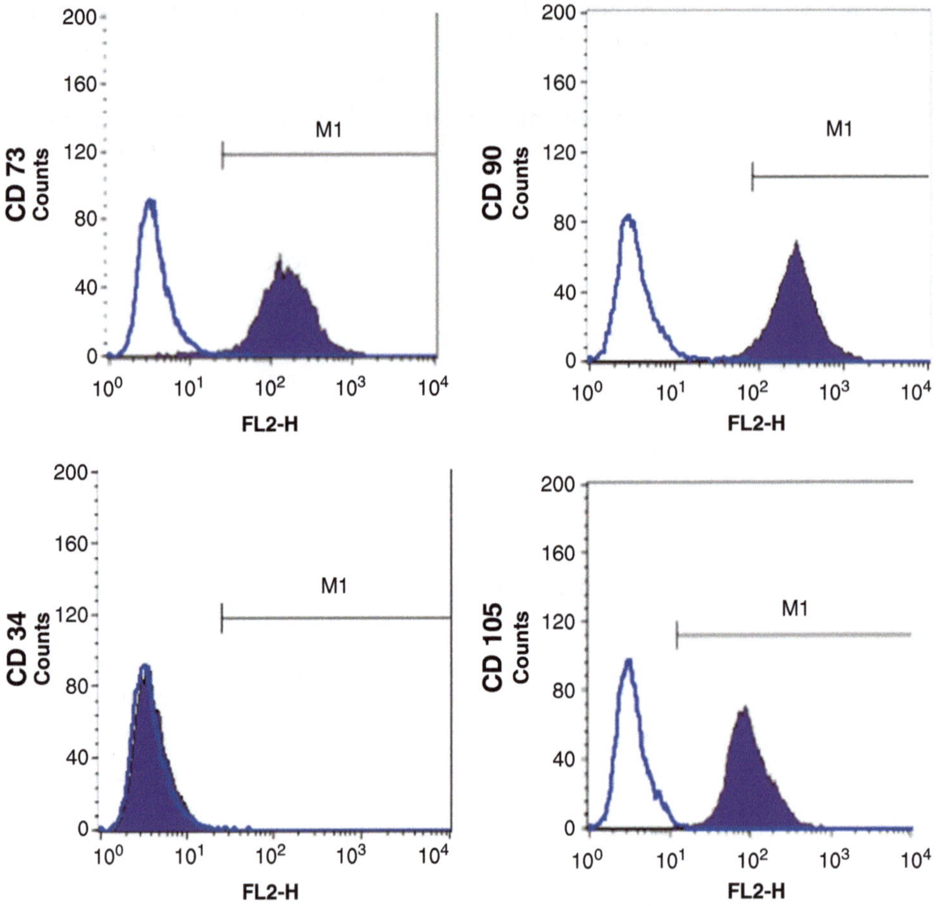

Fig. 3 Surface phenotype characterization by flow cytometry. WJ-MSCs at passage 4 were labeled with antibodies against human CD34, CD73, CD90, and CD105. Open histogram depicts background signal while shaded histogram represents positive reactivity with the specific antibody mentioned

3.5 In Vitro Multilineage Differentiation

1. Seed the isolated WJ-MSCs on 35 mm dishes at a density of 5000 cells/cm^2 in 2 mL of fresh, warm MSC expansion medium. Incubate the dishes at 37 °C and 5% CO_2 till they reach 85–90% confluency (*see* **Note 22**).
2. Next, remove expansion medium and initiate differentiation in vitro by adding 2 mL of prewarmed differentiation medium dropwise (*see* **Note 5**).
3. Continue the adipogenic differentiation for a total of ~14 days and osteogenic differentiation for 20–21 days, with media changes given every third day. Plate uninduced control cultures at an identical density and maintain in MSC expansion medium for similar duration (*see* **Note 26**).
4. Finally after completion of differentiation, stain the adipogenic differentiation dishes with Oil Red O and osteogenic differentiation dishes with Alizarin Red S, respectively and observe

them under bright field microscope. Stain the corresponding uninduced control cultures also, mainly for comparison purpose (*see* **Note 27**).

3.6 Staining

3.6.1 Alizarin Red S Staining

1. Aspirate medium from the osteogenic differentiation dish and give a quick wash with 1× PBS solution at room temperature (*see* **Note 28**).
2. Fix the cells with appropriate volume, required to cover the cell surface, of 4% paraformaldehyde solution and incubate for 20 min in dark at room temperature.
3. Give two quick washes with double distilled water and add 1–2 mL of 1% Alizarin Red S solution (*see* **Note 29**). Incubate in dark for 1 h at room temperature.
4. Following incubation give two brief washes with double distilled water to remove excess stain.
5. Add some double distilled water to the dish to prevent the cells from drying and then visualize under bright field microscope (Fig. 4a, b).

3.6.2 Oil Red O Staining

1. Carefully aspirate medium from the adipogenic differentiation dish and give a quick wash with 1× PBS solution at room temperature.
2. Fix the cells with appropriate volume, required to cover the cell surface, of 4% paraformaldehyde solution and incubate in dark for 30 min at room temperature.
3. Give a wash with 1× PBS, followed by 60% isopropanol for 3 min each at room temperature.
4. Aspirate 60% isopropanol and add 1–2 mL of Oil Red O staining solution (*see* **Note 29**). Incubate the dish in dark for 1 h at room temperature.
5. Aspirate the stain and give three washes with double distilled water to remove excess stain.
6. Cover the cells with double distilled water to prevent them from drying and observe under bright field microscope (Fig. 4c, d).

4 Notes

1. Constituted complete medium should be utilized within 2 weeks.
2. Concentration of FBS (90%) is critical for cryopreservation of mesenchymal stem cells to obtain maximum viability during cell revival. DMSO is used as cryoprotectant to reduce the

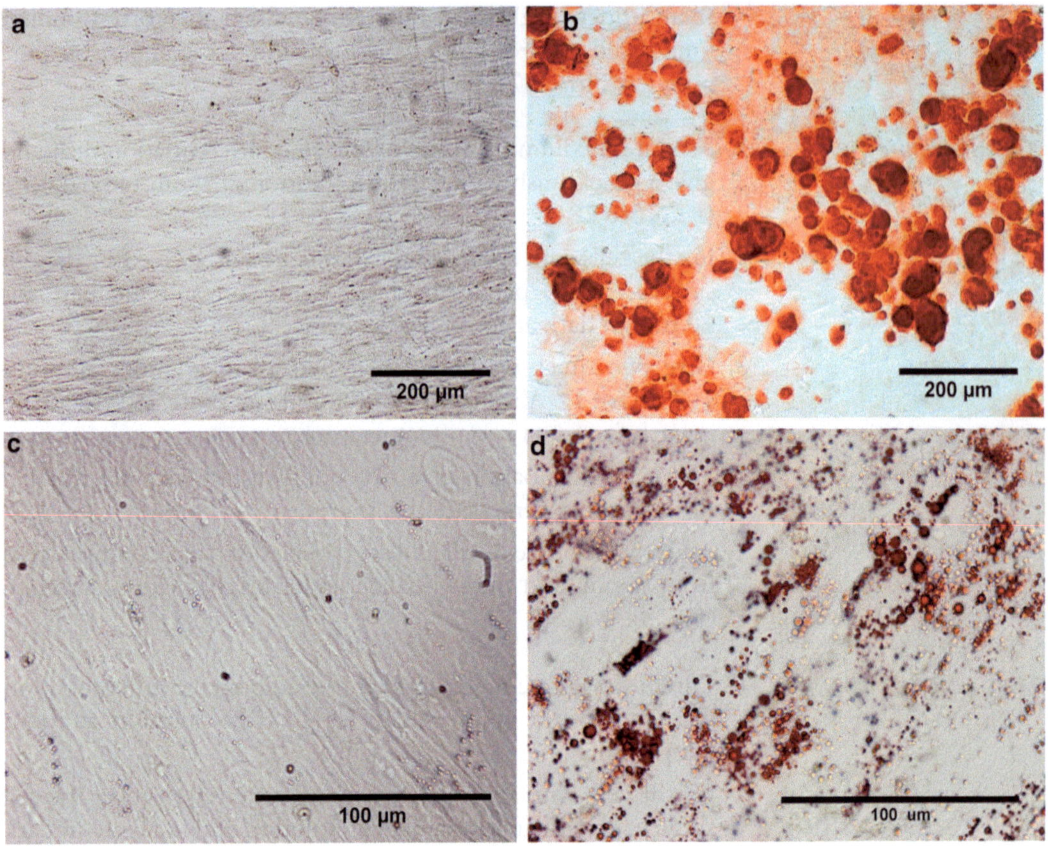

Fig. 4 WJ-MSCs, derived from human umbilical cord, were subjected to (**b**) osteogenic and (**d**) adipogenic differentiation and stained with Alizarin Red S and Oil Red O, respectively. Corresponding noninduced control cultures (**a**, **c**) without any differentiation stimuli are also shown. Scale bars = 200 μm (**a**, **b**) and 100 μm (**c**, **d**)

formation of intra and extracellular ice crystals which can render the cells to death.

3. The umbilical cord tissue pieces are incubated in saline containing antibiotic–antimycotic (1×) to disinfect them properly. 25–30 mL of saline containing antibiotic–antimycotic (1×) is enough for disinfection of cord pieces obtained from one cord sample.
4. Trypsin can be used in place of TrypLE. However, TrypLE is animal-origin free and gentle on cells. Avoid extending the time period of incubation with TrypLE or trypsin as that might affect the quality of cells.
5. Commercially available differentiation media have been used. Constitute the media as per manufacturer's instructions.
6. After collection of cord, it should be processed as soon as possible. Increasing the time between collection and processing may compromise the quality of the tissue/MSCs.

7. All washing steps are done using a 50 mL conical tube.
8. While processing one cord piece, place the conical tube containing the rest of the pieces on ice. The umbilical cord has two arteries and one vein, so the longitudinal slit must be made in a manner such as to avoid cutting through an artery.
9. Carefully remove the artery while tracing it such that it does not break as that would make further tracing difficult.
10. Place the explant tissue pieces such that they are evenly distributed but without overcrowding the dish. Approximately 10–15 pieces are usually placed in one 35 mm cell culture dish.
11. Attachment of the explant tissue pieces facilitates the MSCs to migrate out from them into the dish. However, do not air dry explant tissues for more than 5–7 min. After air drying step, medium should be added carefully and dropwise so as to prevent dislodging of the attached tissue pieces.
12. Discard the remaining tissue pieces in appropriate biohazard waste bag. All the liquid wastes obtained from the above steps should be collected in a beaker containing sodium hypochlorite solution and discarded properly.
13. Dishes should be handled gently to avoid dislodging of the explant tissue pieces. After 4–5 days, start observing the dishes under microscope to keep a track of cells emerging from tissue explants. Any possibility of contamination can be monitored by observing color and turbidity of the medium.
14. The tissue pieces should be removed very gently using blunt end tips, avoid scratching of the cell surface.
15. TrypLE, DPBS and MSC expansion medium should be warmed at 37 °C before use.
16. The total cell number can be calculated as total number of nonblue (live) cells in the four corner squares/4 × dilution factor × volume of cell suspension × 10,000.
17. Depending on the surface area of culture vessel used, volume of medium has to be added, but the cell seeding density should be maintained at 5000 cells/cm^2.
18. Based on the number of cells to be frozen, calculate volume of freezing mix to be added such that freezing concentration is maintained at 2×10^6 cells/mL.
19. Mr. Frosty™ containing isopropanol provides cooling rate of −1 °C/min, which is the optimal rate for cell preservation.
20. During cell revival from frozen condition, cells are seeded at higher density than usual to maximize recovery.
21. As cells are seeded at higher density, replenishment of nutrients helps in better recovery of cells.

22. For characterization of MSCs, cells from passage 4–6 are preferred as in earlier passages the cell population might be more heterogeneous.
23. Gently resuspend the cells in PBS to avoid disruption of cells.
24. Isotype controls are used as a negative control to measure and eliminate the nonspecific background signal caused by primary antibodies.
25. This protocol works for fluorochrome conjugated primary antibodies. For other primary antibodies that are not tagged, one may need to use fluorochrome conjugated secondary antibodies.
26. In the uninduced control cultures, cells might get overconfluent within 10–15 days and the monolayer might start to detach from the surface. Therefore, staining might have to be done before time.
27. Oil Red O stains intracellular lipid droplets, which is a characteristic of adipogenic differentiation. Alizarin Red S solution is used to identify and stain calcium depositions in osteogenic differentiated cells in vitro.
28. PBS solution should be free from calcium and magnesium ions.
29. A sufficient volume of solution should be added to completely cover the monolayer of cells.

Acknowledgments

This work was financially supported by SERB, DST, India. We are thankful to Dr. Jayanta Chatterjee, Aastha, Kalyani, for generously providing us with human umbilical cord samples. We thank CSIR and UGC India for the fellowship of Mr. Umesh Goyal and Ms. Ankita Sen, respectively.

References

1. Caplan AI (2017) Mesenchymal stem cells: time to change the name! Stem Cells Transl Med 6:1445–1451
2. da Silva Meirelles L, Chagastelles PC, Nardi NB (2006) Mesenchymal stem cells reside in virtually all post-natal organs and tissues. J Cell Sci 119:2204–2213
3. Hass R, Kasper C, Böhm S, Jacobs R (2011) Different populations and sources of human mesenchymal stem cells (MSC): a comparison of adult and neonatal tissue-derived MSC. Cell Commun Signal 9:12
4. Batsali AK, Kastrinaki MC, Papadaki HA, Pontikoglou C (2013) Mesenchymal stem cells derived from Wharton's jelly of the umbilical cord: biological properties and emerging clinical applications. Curr Stem Cell Res Ther 8:144–155
5. El Omar R, Beroud J, Stoltz JF, Menu P, Velot E, Decot V (2014) Umbilical cord mesenchymal stem cells: the new gold standard for mesenchymal stem cell-based therapies? Tissue Eng Part B Rev 20:523–544
6. Dalous J, Larghero J, Baud O (2012) Transplantation of umbilical cord-derived mesenchymal stem cells as a novel strategy to protect the central nervous system: technical aspects, preclinical

studies, and clinical perspectives. Pediatr Res 71:482–490

7. Troyer DL, Weiss ML (2008) Wharton's jelly-derived cells are a primitive stromal cell population. Stem Cells 26:591–599

8. Kalaszczynska I, Ferdyn K (2015) Wharton's jelly derived mesenchymal stem cells: future of regenerative medicine? Recent findings and clinical significance. Biomed Res Int 2015:430847

9. Dominici M, Le Blanc K, Mueller I, Slaper-Cortenbach I, Marini F, Krause D, Deans R, Keating A, Dj P, Horwitz E (2006) Minimal criteria for defining multipotent mesenchymal stromal cells. The International Society for Cellular Therapy position statement. Cytotherapy 8:315–317

Chapter 2

Isolation and Characterization of Extracellular Vesicles from Mesenchymal Stromal Cells

Sally Yunsun Kim, Thanh Huyen Phan, Christina Limantoro, Bill Kalionis, and Wojciech Chrzanowski

Abstract

Extracellular vesicles (EVs) have received immense attention in the past decade for their diverse use in diagnosis and therapeutics. Enhancing our understanding of EVs and increasing the reliability and reproducibility of EV research demands the use of standard isolation procedures and multiple characterization methods. Here we describe the most commonly used EV isolation method involving ultracentrifugation, and various characterization methods that include nanoparticle tracking analysis, atomic force microscopy and electron microscopy, which measure the size, concentration, and morphology of EVs.

Key words Extracellular vesicles, Ultracentrifugation, Nanoparticle tracking analysis, Atomic force microscopy, Electron microscopy

1 Introduction

Initially, the release of extracellular vesicles (EVs) was defined to be a part of removal mechanism of undesirable constituents from cells [1]. Scientific and technological advances in the last decade revealed that EVs have specialized abilities and play important roles in intercellular communication by modifying the behavior of recipient cells [2–4]. EVs deliver biological signals that direct a cascade of protective biological responses, which culminate in the determination of cellular function, particularly during postinjury cellular repair or inflammation [5, 6]. However, EVs can also carry "undesirable messages" that contribute to disease pathogenesis and severity, cancer progression and metastasis, or spread of infection [7]. Consequently, EVs can be used as biomarkers for the detection of diseases such as cancer [8, 9].

Because different molecular cargos of EVs are highly specialized and deterministic [10], unlocking their true potential as therapeutics or biomarkers, requires strict adherence to

Mugdha V. Joglekar and Anandwardhan A. Hardikar (eds.), *Progenitor Cells: Methods and Protocols*, Methods in Molecular Biology, vol. 2029, https://doi.org/10.1007/978-1-4939-9631-5_2, © Springer Science+Business Media, LLC, part of Springer Nature 2019

standardized, widely used isolation and characterization protocols. In this chapter, the commonly used EV isolation method of ultracentrifugation [11] is presented, followed by techniques for analysis of EVs. Specifically, methodologies to determine size, concentration, morphology, and heterogeneity of EVs are presented. These parameters are the key determinants of EV functionality.

The ability to reproducibly isolate EVs and to probe their physicochemical characteristics and molecular cargos is essential for identifying biological function of EVs [12]. Advanced characterization methods such as high-resolution flow cytometry, correlative atomic force microscopy, and nanoflow cytometry of EVs provide important information on EV biogenesis and potential release routes [13, 14]. Moreover, these methods allow us to elucidate the effect of environment and stressors such as hypoxia, infection or drug treatment on EV production and molecular composition [15]. Furthermore, rigorous standardized protocols and new innovative methodologies such as atomic force microscopy with infrared spectroscopy (AFM-IR), can answer the global challenge of resolving the heterogeneity between EV populations [16]. Advanced characterization and isolation tools allow for discrimination between different subclasses of EVs (microvesicles, oncosomes and exosomes), protein aggregates, and cell debris. Thus, these tools can drive the development of next-generation personalized healthcare products and technologies, including effective biosensors for rapid disease detection, EV-based drug delivery systems, and EV-based immunomodulators.

2 Materials

Prepare all solutions in nuclease-free conditions using molecular biology grade reagents. All reagents should be kept on ice to preserve EV integrity.

2.1 EV Isolation

1. Bovine serum albumin (BSA; trusted source and quality of BSA must be assured) 35% stock solution: Weigh 3.5 g BSA and transfer to a glass beaker. Add water and mix using a magnetic stirring plate. Make volume up to 10 mL with water. Sterilize by passing the solution through 0.2 μm syringe filter and aliquot in small volumes. Store at −20 °C until used.
2. Hank's Balanced Salt Solution without calcium and magnesium (HBSS(−)); 1×, pH 7.0. Store at room temperature.
3. EV isolation media: Mix 714 μL of BSA 35% stock solution, 500 μL of penicillin–streptomycin (100×), and 500 μL of L-glutamine (100×). Make volume up to 50 mL with serum-free basal media. Store at 4 °C.

4. Nuclease-free water or phosphate buffered saline (PBS; 1×): Depending on the downstream application; water is preferred for characterization by microscopy to avoid interference by PBS crystals.

2.2 Ultra-centrifugation

1. Tubes for ultracentrifugation: Rinse polycarbonate ultracentrifuge tubes using molecular biology grade ethanol (70%) and leave inside laminar flow hood to dry.
2. Ultracentrifuge: Set the temperature to 4 °C and equilibrate the temperature of the rotor to 4 °C.
3. Nuclease-free water or phosphate buffered saline (PBS; 1×): Depending on the downstream application; water is preferred for characterization by microscopy to avoid interference by PBS crystals.

2.3 Nanoparticle Tracking Analysis (NTA)

1. Nuclease-free water for diluting EV samples.
2. Filtered ultrapure water (e.g., Milli-Q® water), filtered and aliquoted.
3. Syringe pump fitted with 1 mL syringe driver.
4. 1 mL syringes.

2.4 Electron Microscopy

2.4.1 Scanning Electron Microscopy (SEM)

1. Polyethylenimine (PEI) 1% solution: Prepare the solution by using 0.5 mL of the commercially available 50% (w/v) PEI stock solution to dilute in 25 mL of ultrapure water in a 50 mL tube.
2. Thermanox: cut into approximately 1 cm × 1 cm pieces. Rinse briefly in 90% acetone, then in ultrapure water and incubate in 1% PEI solution for 60 min. Rinse twice with ultrapure water and air-dry.
3. Phosphate buffer (PB) solution (0.2 M, pH 7.2): Add 1.785 g sodium phosphate dibasic (MW: 268 g/mol) and 0.461 g sodium phosphate monobasic (MW: 138 g/mol) to 40 mL water, mix well using a magnetic stirrer and adjust solution to pH 7.2 using HCl or NaOH, then add distilled water to 50 mL.
4. 2.5% glutaraldehyde in 0.1 M PB solution: Take 1 mL of 25% stock solution (commercially available) and dilute 10× using 5 mL of 0.2 M PB solution and 4 mL of ultrapure water.
5. Osmium tetroxide 1% solution: Take 2.5 mL 4% stock solution (commercially available) and dilute to 10 mL using 7.5 mL ultrapure water.
6. Ethanol at 30%, 50%, 70%, 90%, 95%, 100% concentrations: Using 100% ethanol, make serial dilutions in ultrapure water (*see* **Note 1**).

2.4.2 Transmission Electron Microscopy (TEM)

1. Filtered PBS.
2. Carbon and Formvar coated copper grids.
3. 2.5% glutaraldehyde solution.
4. 2% (w/v) ammonium molybdate: Weigh 200 mg ammonium molybdate and dissolve in 10 mL water.

2.5 Atomic Force Microscopy (AFM)

1. Freshly cleaved mica: Fix mica onto a metal stub using glue and dry completely. Using sticky tape, cleave mica 2–3 times until an even layer of mica is cleaved.
2. Poly-L-lysine 0.1% (w/v) solution.
3. AFM probe for tapping mode (Resonance frequency 200–400 kHz, spring constant 13–77 N/m).

3 Methods

Carry out EV isolation and handling procedures with sterile techniques at room temperature with EV samples placed on ice, unless otherwise specified.

3.1 Isolation of EVs by Ultracentrifugation

1. Culture mesenchymal stromal cells until 80% confluence in complete growth media.
2. Wash cells twice using HBSS(−) buffer.
3. Incubate cells in EV isolation media (serum-free media containing 0.5% BSA, 1% penicillin-streptomycin, 1% glutamine) for 48 h at 37 °C and 5% CO_2.
4. After 48 h incubation, collect media and transfer into nuclease-free centrifuge tube.
5. Centrifuge media at 500 × *g* for 5 min to remove cells and debris.
6. Centrifuge supernatant at 2000 × *g* for 10 min.
7. Filter supernatant through a 0.22 μm filter to remove larger particles.
8. Transfer media to a thick-wall polycarbonate ultracentrifuge tube and label the tube on the side of the tube.
9. Place tubes inside the rotor with labels facing outside to ensure that the pellet is collected on the labelled side of each tube (*see* **Note 2**).
10. Centrifuge at 100,000 × *g* for 60 min at 4 °C.
11. Remove supernatant and resuspend the EV pellet in 1 mL PBS; (*see* **Note 2**).
12. Centrifuge at 100,000 × *g* for 60 min at 4 °C.
13. Remove supernatant and resuspend the EV pellet in 200 μL PBS or water.

3.2 Determining Size and Concentration of EVs Using NTA

1. Launch the NTA software (e.g., NanoSight NS300) and ensure that the hardware status information is displayed correctly (*see* **Note 3**).
2. Remove any liquids in the tubing by hand (*see* **Note 4**).
3. Using the syringe pump at maximum speed, infuse filtered ultrapure water through the tubing for cleaning. Repeat to ensure that the tubing is clean (*see* **Note 5**).
4. Withdraw all liquid from the tubing by hand.
5. Take 25 μL EV sample and dilute in 975 μL nuclease-free water, vortex to mix thoroughly.
6. Transfer diluted sample to 1 mL syringe and infuse the sample at maximum speed using the syringe pump while the camera is on. Stop infusing when EVs are visible.
7. Adjust focus so that the particles appear smooth and have a ring around each particle (Fig. 1, arrows).
8. Using the settings on the software, capture 3 × 60 s videos, with the speed of the syringe pump set to have approximately 50 particles per view (Fig. 1).
9. Process the videos using the data analysis features on the software to obtain information on the particle size and concentration.
10. For cleaning, withdraw all liquids by hand and infuse 1 mL ultrapure water using the maximum speed on the syringe pump, 2–3 times or until no particles are visible on the camera (*see* **Note 6**).
11. Infuse 20% ethanol using the syringe pump to fill the tubing and flow cell then switch off the system until the next use.

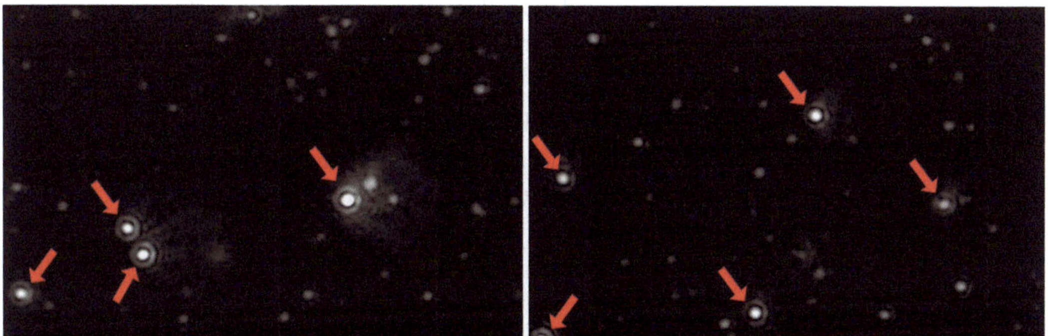

Fig. 1 Snapshots from EVs captured using NanoSight NS300 NTA software, which display the rings around EVs in the way it should be focused (arrows) and showing the approximate number of particles per view expected when following this protocol

3.3 Electron Microscopy for EVs

3.3.1 SEM

1. Incubate EV sample in PBS/water on PEI-coated coverslip for 1 h (*see* **Note 7**).
2. Gently remove and discard PBS/water and immediately add glutaraldehyde fixative.
3. Incubate samples in fixative (2.5% glutaraldehyde in 0.1 M PBS) for 30 min at room temperature.
4. Gently remove glutaraldehyde and discard to waste container. Rinse samples in 0.1 M PB solution three times, 5 min each.
5. Add secondary fixative 1% osmium tetroxide in water. Incubate samples at room temperature for 1 h.
6. Remove osmium to waste container and rinse gently using ultrapure water three times, 5 min each.
7. Incubate samples in series of ethanol 5 min per incubation as follows: 30% ethanol, 50% ethanol, 70% ethanol (twice), 90% ethanol (twice), 95% ethanol (twice), and 100% ethanol (thrice).
8. Retain samples in 100% ethanol during transportation to a critical freeze dryer.
9. Using the manufacturer's protocols, set up a 2 h gentle cycle (50% stirrer, slow CO_2 intake, delay 120 s, exchange speed 5, cycles 14, slow heat, and slow speed of gas).
10. Image EVs using the standard procedures of SEM.

3.3.2 TEM

1. Glow discharge copper grids coated with carbon and Formvar.
2. Place 5 μL EV sample on a piece of Parafilm.
3. Float grid on the sample and leave covered for 30 min.
4. Place fixative solution (2.5 μL of 2.5% glutaraldehyde) on another piece of Parafilm.
5. Transfer grid and float it on the fixative solution, leaving it covered for 30 min.
6. Wash grid with PBS three times by floating it on a PBS droplet, leaving it for 5 min during each wash.
7. Without letting the sample dry after the wash, negatively stain the sample using 2% (w/v) ammonium molybdate in water by floating the grid on the stain droplet for 5 min.
8. Remove excess stain by carefully blotting the grid on a filter paper.
9. Leave grid to air-dry before imaging.
10. Image EVs using the standard procedures of TEM.

3.4 AFM Analysis of EVs

1. Place freshly cleaved mica into poly-L-lysine solution and incubate for 1 h. Remove mica and air-dry.
2. Place a drop of EVs redispersed in a solvent (i.e., PBS, water) on functionalized mica and leave overnight in a partly closed container to allow slow and controlled evaporation of the solvent.
3. Attach mica to a sample holder.
4. Place the sample holder on the AFM scanner operating in tapping, soft tapping or noncontact mode (instrument dependent).
5. Approach the sample with AFM tip as per instrument producer instructions, and scan the samples (e.g., 10 × 10 μm) to localize EVs.
6. Zoom in to the area where vesicles were identified and scan smaller scan areas (e.g., 1 × 1 μm) using following parameters: 512 dpi; scan rate 0.5–0.8 Hz; setpoint 1.0–20 V (instrument dependent); integral gain 1–5; proportional gain 5–20. Zoom in further if single vesicle image is required.
7. Process images using AFM software (e.g., SPIP, Gwydion). Preferably only flattening should be done.
8. Present images as required in 2D or 3D. To obtain quantitative measurements of EV sizes use the line profiles and measure the height of the vesicle which is the measure of the diameter (Fig. 2).

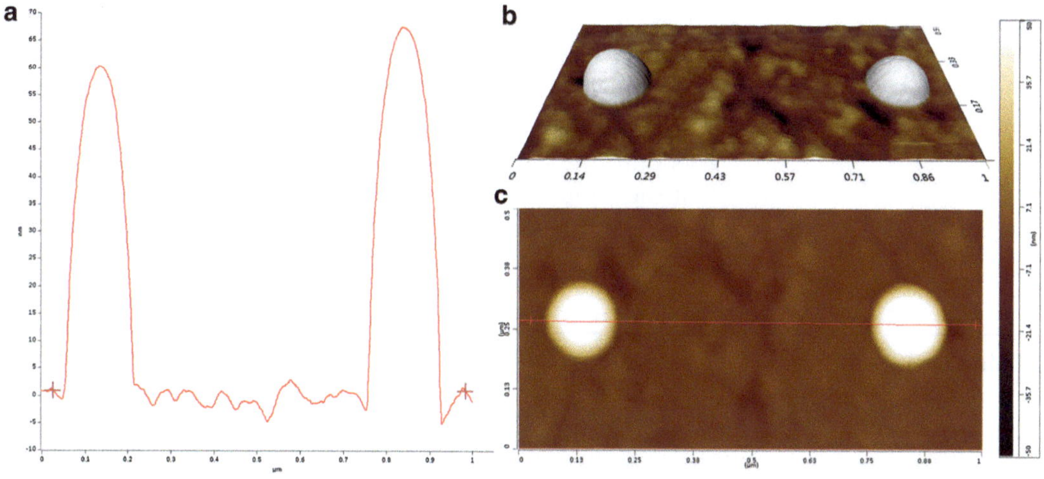

Fig. 2 Atomic force microscopy (AFM) of placenta stem cell-derived EVs, using the Analysis Studio™ software. (**a**) Profile analysis that displays the height/diameter of EVs. (**b**) Three-dimensional AFM image of EVs. (**c**) Two-dimensional AFM image of EVs with lines to indicate where the profile analysis in (**a**) was done

4 Notes

1. For ethanol solutions, the use of transfer pipettes over squeeze bottles (a popular choice for dehydration of biological samples) is recommended as it is less forceful and easier to rinse small EV samples.
2. EV pellet will not be visible so ensure the ultracentrifuge tube is labelled and place the tubes inside the rotor with label facing outward. When removing the tube from the rotor, recheck that the label is in the correct orientation and note where the pellet should be. Take care not to disturb the pellet during transportation, and removal of the supernatant.
3. When the NTA software is launched, check the connection status to ensure that the hardware is correctly detected by the software: the syringe pump, filter wheel, stage/pumps, and temperature. If any of the hardware is shown as "not found," the connections and power supply need to be checked.
4. The tubing should contain 20% ethanol, which is the suggested solvent that should remain inside the tubing after each use of the equipment.
5. Prior to infusing ultrapure water for cleaning, remove any residual liquid from the tubing by hand. While infusing water for cleaning, turn on the camera on the software to view the quality of the ultrapure water and to ensure there are no contaminating particles.
6. At this stage, other samples can be processed using the same procedures. Ensure that there are no visible particles while infusing ultrapure water for cleaning, to avoid cross-contamination and invalid results.
7. PEI coating on coverslips is done to promote the attachment of EVs onto the coverslips. To further assist this process, ensure that the sample is placed with a thin layer of solvent to reduce the void distance between the EVs and the surface of the PEI-coated coverslip.

Acknowledgments

The authors acknowledge The University of Sydney for the SOAR Fellowship for W.Ch. The authors acknowledge the facilities and the scientific and technical assistance of the Bosch Molecular Biology Facility and Australian Microscopy & Microanalysis Research Facility at the Australian Centre for Microscopy & Microanalysis, The University of Sydney.

References

1. Pan BT, Johnstone RM (1983) Fate of the transferrin receptor during maturation of sheep reticulocytes in vitro: selective externalization of the receptor. Cell 33(3):967–978
2. Marcilla A, Trelis M, Cortes A, Sotillo J, Cantalapiedra F, Minguez MT, Valero ML, Sanchez del Pino MM, Munoz-Antoli C, Toledo R, Bernal D (2012) Extracellular vesicles from parasitic helminths contain specific excretory/secretory proteins and are internalized in intestinal host cells. PLoS One 7(9): e45974. https://doi.org/10.1371/journal.pone.0045974
3. Luga V, Zhang L, Viloria-Petit AM, Ogunjimi AA, Inanlou MR, Chiu E, Buchanan M, Hosein AN, Basik M, Wrana JL (2012) Exosomes mediate stromal mobilization of autocrine Wnt-PCP signaling in breast cancer cell migration. Cell 151(7):1542–1556. https://doi.org/10.1016/j.cell.2012.11.024
4. Bjorge IM, Kim SY, Mano JF, Kalionis B, Chrzanowski W (2017) Extracellular vesicles, exosomes and shedding vesicles in regenerative medicine—a new paradigm for tissue repair. Biomater Sci 6(1):60–78. https://doi.org/10.1039/c7bm00479f
5. Su SA, Xie Y, Fu Z, Wang Y, Wang JA, Xiang M (2017) Emerging role of exosome-mediated intercellular communication in vascular remodeling. Oncotarget 8(15):25700–25712. https://doi.org/10.18632/oncotarget.14878
6. Bui TM, Mascarenhas LA, Sumagin R (2018) Extracellular vesicles regulate immune responses and cellular function in intestinal inflammation and repair. Tissue Barriers 6(2): e1431038. https://doi.org/10.1080/21688370.2018.1431038
7. Lane RE, Korbie D, Hill MM, Trau M (2018) Extracellular vesicles as circulating cancer biomarkers: opportunities and challenges. Clin Transl Med 7(1):14. https://doi.org/10.1186/s40169-018-0192-7
8. Urabe F, Kosaka N, Kimura T, Egawa S, Ochiya T (2018) Extracellular vesicles: toward a clinical application in urological cancer treatment. Int J Urol 25(6):533–543. https://doi.org/10.1111/iju.13594
9. Xu R, Rai A, Chen M, Suwakulsiri W, Greening DW, Simpson RJ (2018) Extracellular vesicles in cancer—implications for future improvements in cancer care. Nat Rev Clin Oncol 15 (10):617–638. https://doi.org/10.1038/s41571-018-0036-9
10. Keerthikumar S, Chisanga D, Ariyaratne D, Al Saffar H, Anand S, Zhao K, Samuel M, Pathan M, Jois M, Chilamkurti N, Gangoda L, Mathivanan S (2016) ExoCarta: a web-based compendium of Exosomal cargo. J Mol Biol 428(4):688–692. https://doi.org/10.1016/j.jmb.2015.09.019
11. Gardiner C, Di Vizio D, Sahoo S, Thery C, Witwer KW, Wauben M, Hill AF (2016) Techniques used for the isolation and characterization of extracellular vesicles: results of a worldwide survey. J Extracell Vesicles 5:32945. https://doi.org/10.3402/jev.v5.32945
12. Garcia-Manrique P, Matos M, Gutierrez G, Pazos C, Blanco-Lopez MC (2018) Therapeutic biomaterials based on extracellular vesicles: classification of bio-engineering and mimetic preparation routes. J Extracell Vesicles 7 (1):1422676. https://doi.org/10.1080/20013078.2017.1422676
13. Tian Y, Ma L, Gong M, Su G, Zhu S, Zhang W, Wang S, Li Z, Chen C, Li L, Wu L, Yan X (2018) Protein profiling and sizing of extracellular vesicles from colorectal cancer patients via flow cytometry. ACS Nano 12(1):671–680. https://doi.org/10.1021/acsnano.7b07782
14. Dipesh K, Bokai Z, Iqbal R, Curtis M, Quan L, Wojciech C (2018) Probing chemical and mechanical nanodomains in copolymer nanorods with correlative atomic force microscopy—nano-correscopy. Part Part Syst Charact 35 (6):1700409. https://doi.org/10.1002/ppsc.201700409
15. Szatanek R, Baj-Krzyworzeka M, Zimoch J, Lekka M, Siedlar M, Baran J (2017) The methods of choice for extracellular vesicles (EVs) characterization. Int J Mol Sci 18(6). https://doi.org/10.3390/ijms18061153
16. Kim SY, Khanal D, Tharkar P, Kalionis B, Chrzanowski W (2018) None of us is the same as all of us: resolving the heterogeneity of extracellular vesicles using single-vesicle, nanoscale characterization with resonance enhanced atomic force microscope infrared spectroscopy (AFM-IR). Nanoscale Horizons 3(4):430–438. https://doi.org/10.1039/C8NH00048D

References

Chapter 3

Assessment of β-Cell Replication in Isolated Rat Islets of Langerhans

Louise T. Dalgaard

Abstract

Pancreatic β-cells in the islets of Langerhans secrete insulin in response to the rise in glucose levels following food intake. The hypoglycemic action of insulin applies a strong evolutionary brake on β-cell division. However, under some conditions β-cells can be stimulated to enter cell cycle progression and divide, for example following exposure to increased glucose levels or during pregnancy. Here, a protocol is described for the isolation of rat adult islets of Langerhans, followed by culture of intact islets in Matrigel and measurement of β-cell replication by the incorporation of ethynyldeoxyuridine (EdU). EdU positive cells are revealed by a click reaction, nuclei are visualized using a DNA-binding fluorophore (Hoechst 33342), and β-cells are identified using immunofluorescence detection.

Key words Islet of Langerhans, β-Cell, Proliferation, Mitosis, Ethynyldeoxyuridine (EdU), Glucose, Insulin, Hoechst 33342

1 Introduction

In adults, proliferation of the pancreatic β-cells occurs at a slow rate under steady-state conditions. However, the β-cell mass has been shown to increase in the face of insulin resistance during obesity and pregnancy [1–7]. The production of adult β-cells predominantly occurs through replication rather than differentiation, although this is still debated [8, 9]. The total β-cell mass of the pancreas is regulated by at least five different mechanisms: proliferation, neogenesis, hypertrophy, apoptosis, and atrophy [7].

During progression into type 2 diabetes (T2D) an initial increase in β-cell mass occurs, likely driven by increases in postprandial glycaemia [10], but after a while the apoptosis rate increases and the β-cell mass decreases. Diabetes may arise as a consequence of failure to properly increase the β-cell mass; analyses show that the pancreatic mass donated from obese patients diagnosed with T2D is lower compared to obese individuals without diabetes [11]. Therefore, understanding the mechanisms of beta-cell

Mugdha V. Joglekar and Anandwardhan A. Hardikar (eds.), *Progenitor Cells: Methods and Protocols*, Methods in Molecular Biology, vol. 2029, https://doi.org/10.1007/978-1-4939-9631-5_3,

replication would allow for novel therapeutic interventions in preventing or treating diabetes.

Here, I present a detailed protocol for the isolation of islets from adult rats, followed by culture of intact islets on Matrigel-coated glass coverslips with incorporation of Ethynyldeoxyuridine (EdU) followed by its detection using click chemistry and identification of β-cells by insulin immunofluorescent staining.

The specific protocol outlined here identifies replicating β-cells. However, the protocol may easily be modified to identify other cell populations of interest: other mature endocrine cells of the islet (α-cells, δ-cells, PP-cells, which are glucagon, somatostatin or pancreatic polypeptide positive, respectively) or progenitor cells (Neurogenin 3 and/or Nestin positive) using suitable antibodies.

2 Materials

2.1 Isolation of Adult Rat Islets of Langerhans

1. All media and buffers should be sterile, but the actual procedure of islet isolation may be carried out at the bench.
2. Collagenase IV (*see* **Note 1**).
3. Instruments for dissection: A small pean, two small serrafine Dieffenbach clamps, one pair of fine serrated forceps, one pair of hooked forceps, one pair of large scissors for the opening of the abdomen, one pair of fine scissors for dissecting out the distended pancreas. Sterilize instruments (*see* **Note 2**).
4. Magnifier LED lamp with magnifying lens (10×) (well-stocked office supplies).
5. Two adult rats, weight 201–225 g, Wistar Hannover GALAS.
6. Hanks' buffered saline solution (HBSS).
7. Deoxyribonuclease I (DNase) (5 mg/mL dissolved in HBSS, store at −20 °C) in 500 μL aliquots.
8. Fetal bovine serum (FBS). Heat-inactivate at 56 °C for 10 min and aliquot aseptically into 15 mL and 37.5 mL aliquots and store at −20 °C.
9. Penicillin–streptomycin solution (10,000 U/mL). Aliquot aseptically into 5 mL aliquots and store at −20 °C.
10. Collagenase mix: For 500 mL collagenase mix add 0.7 g Collagenase IV (540 U/mg) and 500 μL DNase. Sterile filter using a cellulose acetate filter and store at −20 °C in 30–50 mL aliquots.
11. Quenching medium: 500 mL HBSS, 15 mL FBS, 5 mL penicillin–streptomycin. Keep on wet ice. The small amount of FBS in the Quenching medium inhibits trypsin activity from the exocrine pancreas and impurities in the collagenase.

12. Histopaque 1.077. Allow to equilibrate to room temperature before use.
13. Sterile 50 mL conical glass bottle, preferably with screw lids.
14. Cover for a 14 G Venflon needle and 10 mL plastic syringes.
15. Metal sieve CD-1 with Mesh 40 (Sigma S1145 and S0770).
16. RPMI 1640 with GlutaMAX 1 M HEPES pH 7.4 Complete islet medium: 500 mL RPMI1640 with GlutaMAX, 37.5 mL FBS (7.5%), 10 mL 1 M HEPES, 5 mL penicillin–streptomycin.

2.2 Culture of Isolated Islets, Free Floating

1. Stereomicroscope such as Olympus SZ51 or similar preferably with illumination from below.
2. Sterile polystyrene bacterial petri dishes (10 cm diameter).
3. Complete islet medium (*see* above).

2.3 Matrigel Culture of Islets and EdU Incorporation

1. Complete human serum (HS) islet medium: 500 mL RPMI 1640 with GlutaMAX, 10 mL (2%) pooled HS, 10 mL 1 M HEPES, 5 mL penicillin–streptomycin.
2. RPMI1640.
3. Matrigel Matrix Growth Factor Reduced.
4. Pipette tips, 200 μL, stored at −20 °C for use with Matrigel, which is liquid at temperatures below 0 °C.
5. 12-well cell culture plate.
6. Glass coverslips (16 mm diameter) Click-iT EdU Alexa Fluor 555 Imaging Kit.
7. Phosphate buffered saline (PBS).
8. Bovine serum albumin (BSA).
9. DMSO.
10. Paraformaldehyde (PFA).
11. PBT buffer: PBS with 0.2% BSA, 0.5% Triton X-100. Store at −20 °C.
12. Animal Free Blocker (Vector Labs).
13. Microscope slides.
14. Guinea pig anti-insulin (Abcam ab7842).
15. Goat Anti-Guinea pig Dylight 488 (Abcam ab102374).
16. Nail polish for sealing coverslips.
17. Prolong Gold Antifade Mountant.
18. Confocal imaging facility.

3 Methods

3.1 Perfusion of the Pancreas Via the Common Bile Duct

1. On the day of islet isolation: Thaw collagenase mix and keep on ice. Calculate 12.5 mL collagenase mix solution for each rat pancreas.
2. Histopaque 1.077 is removed from the refrigerator and must not be used before it has reached room temperature.
3. Add 4 mL collagenase mix to each 50 mL conical glass bottle, one per pancreas, and place on wet ice.
4. Items to bring for the animal facility/perfusion bench: Instruments, syringe (10 mL) and needles (27 G), wet ice, 50 mL conical flasks to place the perfused pancreata in, ethanol for disinfection, gloves, Benchkote and wipes, and a bag for disposal of waste.
5. Fill up the 10 mL syringe with diluted collagenase mix, bend the needle 90° and attach it and leave on the wet ice (Fig. 1).
6. The euthanization of the rats should follow local institutional animal care and utilization guidelines (*see* **Note 3**).

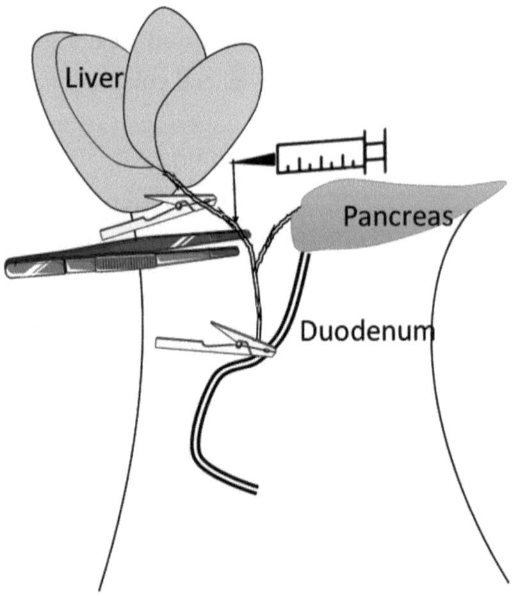

Fig. 1 Perfusion of the rat pancreas through the common bile duct. The stomach is cut across the abdomen and the skin pulled away. The intestines and the liver lobes are pushed aside. Two clamps are placed: one across the ampulla of Vater, where the common bile duct enters the duodenum, and the other just below the liver. These clamps will ensure that the pancreatic duct is the only route available for the collagenase perfusion. The needle is bend 90° and inserted into the duct, which is held up by serrated fine forceps

7. Immediately following euthanization, rinse the abdomen with 70% ethanol and cut open the abdomen using the large scissors. First cut the skin and then the peritoneum. Open the abdomen from side to side and push the intestines and the top liver lobes to the side in order to have access to the common bile duct (ductus choledochus communis) and the duodenum. Place the rat with its tail toward you and the head pointing away (Fig. 1).
8. Clamp the entrance of the common bile duct into the duodenum (ampulla of Vater) using a small Dieffenbach clamp.
9. Clamp the upper bile duct before the entrance to the liver using a second Dieffenbach clamp (Fig. 1).
10. Perfusion of the pancreas: Use the fine forceps held in the left hand to lift up the common bile duct. Under the lamp with magnifying lens use the right hand to insert the needle into the duct pointing the needle toward you. Avoid rolling of the duct by lifting the duct slightly with the left hand forceps, while simultaneously piercing the duct with the needle. Gently inject collagenase mix into the duct. The needle may, if necessary, be kept in place using a clamp. The signs of a successful perfusion are as follows: (1) a slight resistance in the syringe, (2) no fluid leaking, and (3) the pancreas swells slowly and entirely including the distal part (the tail of the pancreas).
11. One pancreas will use 5–10 mL collagenase mix. If the perfusion is not complete, then collagenase may be injected directly at different positions in the pancreas, especially in the distal part, which is most rich in islets. However, this will result in a much lower islet yield compared to a successful perfusion.
12. The pancreas is carefully dissected from the intestines using the fine scissors. Both the ventral and the dorsal part of the pancreas should be distended and dissected out. Place the pancreas in a petri dish on ice and remove fat and obvious connective tissue.
13. Repeat perfusion with the next rat (*see* **Note 4**).

3.2 Digestion of Pancreas and Isolation of Islets

1. Each perfused pancreas is transferred to a sterile 50 mL conical glass bottle on ice. Digest the pancreas at 37 °C in a shaking water bath (200 rotations/min) for 4 min.
2. Add 75 mL cold Quenching medium, mix carefully and let the conical glass bottle stand tilted at a 45° angle in ice for 3 min.
3. Pour off the supernatant one glass bottle at a time, until 25 mL is left.
4. Add an extra 2.5 mL collagenase mix solution per pancreas.
5. Further incubate in the shaking water bath at 37 °C and 200 rotation/min for 2 min.

6. Transfer to ice bath for 1 min.
7. Vigorously shake the tissue for 1 min to enhance dissociation.
8. Pour the entire content of each glass bottle into a 50 mL 'blue cap' (plastic)conical tube, rinse the glass bottle with ice-cold Quenching medium and add to the tube. Fill up the tube to the 50 mL mark. Keep the islets cold from now on.
9. Incubate for 10 min on ice.
10. Centrifuge the 50 mL tube at 4 °C, 1 min at 900 × *g* using a swinging bucket rotor.
11. Pour off the supernatant (take care to remove floating fat) and resuspend the pellet in 20 mL Quenching medium by pipetting up and down using a serological pipette.
12. Filter the tissue blend through a metal sieve with mesh to remove undigested connective tissue or fat into a new 50 mL centrifuge tube. Wash the previous tube with 2 × 15 mL Quenching medium.
13. Centrifuge the 50 mL tube at 4 °C, 1 min at 900 × *g* using a swinging bucket rotor.
14. Pour off the supernatant until 10 mL remains in the conical tube.
15. Homogenize tissue by pipetting 6–10 times using the plastic cover for a 14 G Venflon needle and a 10 mL plastic syringe. Avoid foaming.
16. From now, work at room temperature: Carefully add 10 mL Histopaque 1.077 below the homogenized tissue.
17. Centrifuge at 900 × *g*, 20 °C, *without* brakes and slow acceleration and deceleration, 22 min.
18. Hold tube to light source, locating the main part of islets in the intersection between the top and bottom phases.
19. Transfer islets to a new 50 mL conical tube and fill up with Quenching medium.
20. Then transfer the remaining top and middle phase to a new 50 mL conical tube and fill to the 50 mL mark with Quenching medium and mix well through inversion.
21. Centrifuge both 50 mL conical tubes at 200 × *g*, 20 °C, with brake, 7 min.
22. Remove supernatant through decanting, flick and add 5 mL Quenching medium.
23. Pour the contents of the 50 mL conical tubes into sterile 10 cm bacterial petri dishes, wash the conical tube 2 times with new Quenching medium to make certain all islets are transferred into the petri dish.

3.3 Culture of Free Floating Islets

1. Hand-pick the islets under a stereomicroscope into complete islet medium. Expected yield of islets: 200–300 per pancreas.
2. Pick islets into new petri dish with fresh complete islet medium until all exocrine tissue has been removed.
3. When clean, move islets to a clean petri dish with complete islet medium and culture at 37 °C in CO_2 incubator. Circling movements with the petri dish will collect the islets in the center, from where they can be easily transferred. Do not culture more than 150 islets in the same 10 cm petri dish as they tend to stick together—be sure to gently move plate sideways when placing it in the incubator in order to move islets away from the center during incubation.
4. Pick islets clean every day and transfer to fresh islet medium. After 2–3 days, all exocrine tissue is degraded and islets may be seeded on to Matrigel-coated glass coverslips or used for other experiments.

3.4 Matrigel Culture of Islets and EdU Incorporation

1. Prepare for Matrigel coating the following day: (a) Thaw Matrigel overnight at 4 °C on ice, (b) Store pipette tips (200 μL) at −20 °C overnight.
2. Coat coverslips (16 mm diameter) with Matrigel (Thin Gel Method): (a) Put slides into 12-well plate on ice. (b) Thaw Matrigel Matrix Growth Factor Reduced as recommended on ice. (c) Using cooled pipette tips, mix the Matrigel to homogeneity on ice and while keeping culture plates on ice, add 50 μL per square centimeter of growth surface (Ø16 mm ~2 cm^2 ~100 μL). (d) Incubate plates at 37 °C for 30 min (*see* **Note 5**) (e) Remove excess Matrigel by washing in serum-free RPMI 1640.
3. Seed islets onto Matrigel: Pick islets clean onto new petri dish, fill up with fresh RPMI. Pick 10–20 islets of similar sizes/coverslip. Do not scrape the bottom of the coverslip. Make sure islets do not touch each other on the slide. Let islets attach in small droplets on the coverslip for 30 min in the incubator while ensuring that the droplets do not dry out. Carefully add 1000 μL complete HS islet medium (37 °C) with 2% Human Serum (hinders fibroblast formation).
4. Incubate at 37 °C CO_2 for 48–96 h, during which time the islets will be treated with stimulators or inhibitors as per experimental conditions and pulsed with EdU for 24–48 h.
5. EdU labeling: Prepare the stock solutions for the Click-iT EdU imaging kit according to manufacturer's instructions.
6. Carefully remove 500 μL RPMI from each well on the 12-well plate and add to respective sterile 1.5 mL centrifuge tube (this is to ensure that the treatment initiated earlier continues).

7. Add 2.0 μL EdU (10 mM stock solution, component A of MP10338) to each 1.5 mL centrifuge tube and pipette carefully, avoid foaming.
8. Restore respective 500 μL EdU-mix into each well of the 12-well plate (final conc. 10 μM EdU/well).
9. Incubate at 37 °C CO_2 for 24–48 h (*see* **Note 6**).

3.5 Detection of EdU and Staining for Insulin

1. Islets fixation: Remove the media and wash once in 500 μL PBS. Add 200 μL 4% paraformaldehyde in PBS (PFA) per well and incubate at room temperature for 2 h.
2. Wash in PBS 200 μL 3 × 10 min.
3. Continue with **step 4** (below) or keep wrapped in multiwall plate at 4 °C in 500 μL PBS until later use.
4. Islet permeabilization: Add 200 μL 17% DMSO vol/vol (e.g., for total volume 3 mL; 510 μL DMSO + 2490 μL PBS). Incubate in the hood for 2 h.
5. Wash in 200 μL PBT 3 × 15 min.
6. Add 200 μL PBT and incubate for 1 h RT.
7. Wash in 3% BSA/PBS 2 × 10 min.
8. EdU detection: Prepare the click-iT reaction cocktail, 200 μL per well as per Table 1.
9. Reaction buffer additive—Solution F: Dilute the stock (10×) to 1:10 in milliQ water (or deionized water).
10. Add 200 μL Click-iT reaction cocktail per coverslip and incubate at RT in the dark for 1 h.
11. Wash in 200 μL 3% BSA/PBS 1 × 5 min.
12. Wash in 200 μL PBT 3 × 15 min.

Table 1
Overview of the EdU reactions for 1, 2, 5, and 10 coverslips for the Click-iT EdU Imaging kit from Thermo Fisher Scientific

	Number of coverslips			
Reaction components	**1**	**2**	**5**	**10**
1 × Click-iT reaction buffer (component D, MP10338, step 1.4)	430 μL	860 μL	2.2 mL	4.3 mL
$CuSO_4$ (component H)	20 μL	40 μL	100 μL	200 μL
Alexa Fluor azide (MP10338, step 1.3)	1.2 μL	2.5 μL	6 μL	12.5 μL
Reaction buffer additive (component F, MP10338, step 4.1)	50 μL	100 μL	250 μL	1.25 mL
Total volume	500 μL	1 mL μL	2.5 mL	5 mL

Note: Add components in the order listed

13. Block with 200 μL Animal Free Blocker (1:5) for 30 min (e.g., for 3000 μL animal-free blocker use 600 μL blocker and 2400 μL deionized H_2O).
14. Wash in 200 μL PBT 2 × 5 min.
15. Add 250 μL 1° antibody: Guinea pig anti-insulin diluted 1:500 in PBT, incubate overnight at 4 °C.
16. Wash in 200 μL PBT, 3 × 20 min.
17. Add 250 μL 2° antibody: Goat α-guinea pig, Alexa Fluor 488 labeled (1:200) and Hoechst (5 μg/mL) in PBT, incubate for 3 h at room temperature in the dark (*see* **Note 7**).
18. Wash 200 μL PBS, 3 × 20 min.
19. View under the fluorescent microscope and if the staining is satisfactory, proceed with mounting of the coverslip.
20. Mount with Prolong Gold antifade mounting medium: (a) Place one drop of Prolong Gold on a microscope slide, (b) Pick up the coverslip from the well using the blunt side of a broad scalpel, (c) Turn it down (islets facing downward) onto the droplet of mounting medium on the microscope slide. Dry away surplus mounting medium.
21. Let the mounting medium solidify overnight at room temperature on a level surface.
22. Next day, seal the edges of the coverslip using nail polish.
23. Image the islets on the coverslip by confocal microscopy.
24. The fluorescence excitation/emission wavelength maxima for the fluorophores are (in nm): EdU stain: 555/565, Dylight 488: 495/519, Hoechst 33342: 350/461.
25. Whole islets are scanned using the 20× objective on the *z*-axis with acquisition for every 10 μm. For exemplary images *see* Fig. 2.

4 Notes

1. Collagenase mix for rodent islet isolation: Collagenase IV isolated from *Clostridium histolyticum* is available from Sigma-Aldrich (C6079), ThermoFisher (17104019) and other suppliers: Various preparations of collagenases of different purity grades are available for islet isolation. Liberase TL has replaced Liberase RI (Roche) as a more refined and pure enzyme product, which is reported to result in 28% increased islet yield [12]. Optimization of the digestion time should be anticipated.
2. Fine surgical instruments may be purchased at Roboz.

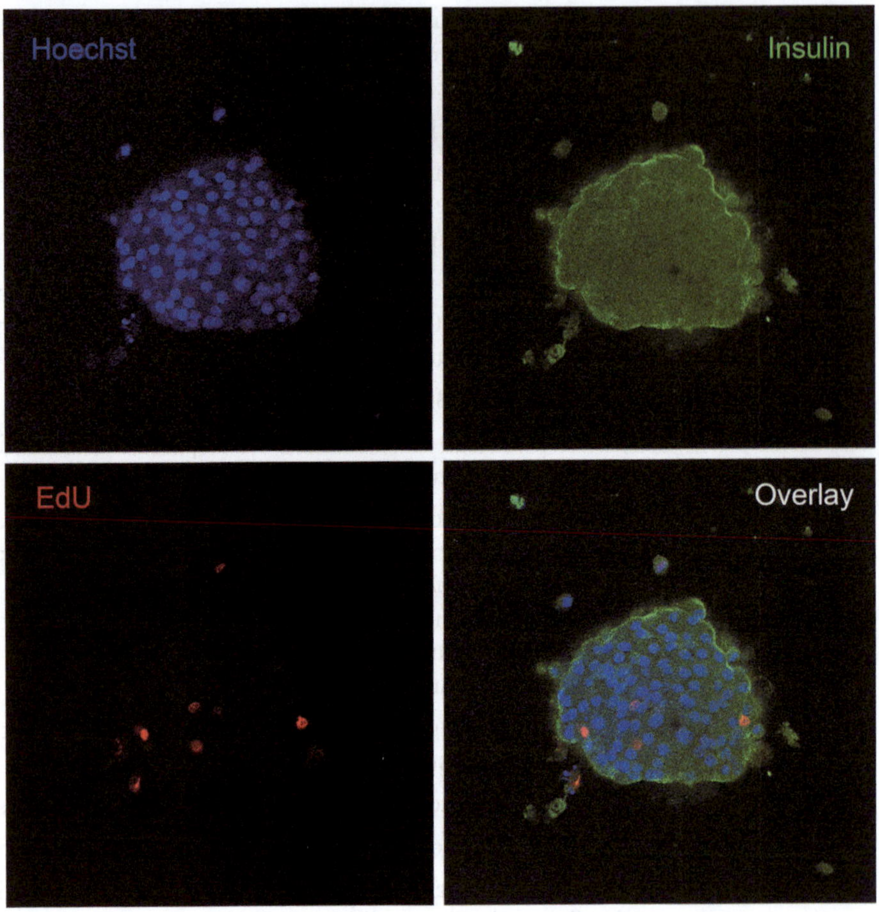

Fig. 2 Images of an adult rat islet of Langerhans cultured on Matrigel, stained for Hoechst (blue), insulin (green), EdU incorporation (red) and a composite image of all three stains (overlay)

3. Euthanization using CO_2 followed by cervical dislocation is recommended. The rat must be perfused immediately following euthanization. If this is not possible, the recommendation is to anesthetize the animal using Hypnorm–Dormicum (Fentanyl–fluanisone–midazolam) according to local guidelines.
4. Up to six rats may be perfused at the same time, but all perfusions must be done within 1 h as the pancreata will start to digest although they are kept on ice.
5. Matrigel contains chloroform, known to cause cancer. This procedure should be done in a ventilated hood.
6. Forty-eight hours of pulse with EdU should yield 2–3% incorporation, though not all of the labeled cells will be β-cells. We never observe labeled clusters and therefore consider this pulse interval suitable. Extended pulse periods are discouraged as EdU incorporates into the DNA and causes DNA damage, which will limit the viability of incorporated cells.

7. Suitable controls for the antibody staining could be prepared by leaving out either the primary or the secondary antibody to ensure that the expected specificity of the signal is obtained.

References

1. Meier JJ, Butler AE, Saisho Y, Monchamp T, Galasso R, Bhushan A, Rizza RA, Butler PC (2008) Beta-cell replication is the primary mechanism subserving the postnatal expansion of beta-cell mass in humans. Diabetes 57 (6):1584–1594
2. Butler PC, Meier JJ, Butler AE, Bhushan A (2007) The replication of beta cells in normal physiology, in disease and for therapy. Nat Clin Pract Endocrinol Metab 3(11):758–768
3. Van Assche FA, Aerts L, de PF (1978) A morphological study of the endocrine pancreas in human pregnancy. Br J Obstet Gynaecol 85 (11):818–820
4. Butler AE, Cao-Minh L, Galasso R, Rizza RA, Corradin A, Cobelli C, Butler PC (2010) Adaptive changes in pancreatic beta cell fractional area and beta cell turnover in human pregnancy. Diabetologia 53(10):2167–2176. https://doi.org/10.1007/s00125-010-1809-6
5. Saisho Y, Butler AE, Meier JJ, Monchamp T, Allen-Auerbach M, Rizza RA, Butler PC (2007) Pancreas volumes in humans from birth to age one hundred taking into account sex, obesity, and presence of type-2 diabetes. Clin Anat 20(8):933–942. https://doi.org/10.1002/ca.20543
6. Rhodes CJ (2005) Type 2 diabetes-a matter of beta-cell life and death? Science 307 (5708):380–384
7. Ackermann AM, Gannon M (2007) Molecular regulation of pancreatic beta-cell mass development, maintenance, and expansion. J Mol Endocrinol 38(1-2):193–206
8. Dor Y, Brown J, Martinez OI, Melton DA (2004) Adult pancreatic beta-cells are formed by self-duplication rather than stem-cell differentiation. Nature 429(6987):41–46
9. Georgia S, Bhushan A (2004) Beta cell replication is the primary mechanism for maintaining postnatal beta cell mass. J Clin Invest 114 (7):963–968. https://doi.org/10.1172/JCI22098
10. Schmidt SF, Madsen JG, Frafjord KO, Poulsen L, Salo S, Boergesen M, Loft A, Larsen BD, Madsen MS, Holst JJ, Maechler P, Dalgaard LT, Mandrup S (2016) Integrative genomics outlines a biphasic glucose response and a ChREBP-RORgamma axis regulating proliferation in beta cells. Cell Rep 16(9):2359–2372. https://doi.org/10.1016/j.celrep.2016.07.063
11. Donath MY, Halban PA (2004) Decreased beta-cell mass in diabetes: significance, mechanisms and therapeutic implications. Diabetologia 47(3):581–589
12. Yesil P, Michel M, Chwalek K, Pedack S, Jany C, Ludwig B, Bornstein SR, Lammert E (2009) A new collagenase blend increases the number of islets isolated from mouse pancreas. Islets 1(3):185–190. https://doi.org/10.4161/isl.1.3.9556

Chapter 4

Droplet Digital PCR for Measuring Absolute Copies of Gene Transcripts in Human Islet-Derived Progenitor Cells

Cody-Lee Maynard, Wilson K. M. Wong, Anandwardhan A. Hardikar, and Mugdha V. Joglekar

Abstract

Transcript analysis is a routinely used method to assess the expression profile of progenitor cells at different stages starting from their isolation to differentiation into specific lineages. It is a powerful way to understand similarities and differences between different cell types as well to estimate successful differentiation process. Transcript measurement is most commonly done using polymerase chain reaction (PCR) but other methods such as in situ hybridization, RNA sequencing are available. The quantitative PCR using TaqMan chemistry is a highly sensitive and reproducible method that measures gene transcripts as a relative abundance. With recent advances in technology, absolute quantitation of genes to single copy level is possible using digital PCR platforms.

Digital PCR is an improved method of PCR in which a single reaction is partitioned into multiple mini reactions. Gene transcripts are measured in each of these mini reactions thereby improving assay sensitivity and making absolute quantitation possible. Here we describe the generation of human islet-derived progenitor cells and measuring gene transcripts in these cells at different passages using digital droplet PCR.

Key words Digital PCR, Gene expression, Human islets, Islet-derived progenitor cells, RNA transcripts

1 Introduction

Digital PCR (dPCR) is a refined approach of the conventional quantitative (q) PCR method. The concept of dPCR was initially described and developed in between the late 1980s to 1990s [1–4]. dPCR was developed to offer the quantification of "absolute" copy numbers from a target sequence present in a sample in contrast to the standard qPCR which provides only a relative quantitative value of the target's abundance [5–7].

Cody-Lee Maynard and Wilson K. M. Wong are equal first authors and Anandwardhan A. Hardikar and Mugdha V. Joglekar are equal senior authors.

Mugdha V. Joglekar and Anandwardhan A. Hardikar (eds.), *Progenitor Cells: Methods and Protocols*, Methods in Molecular Biology, vol. 2029, https://doi.org/10.1007/978-1-4939-9631-5_4,

The dPCR components and amplification procedures are similar to qPCR, whereby amplicons are hybridized with fluorescence probes and the fluorescence is then monitored either in real time or cumulatively quantified at the end [7]. The major difference of dPCR to qPCR is their respective approach used to measure the sample target sequence [8]. dPCR uses the method of limiting DNA dilution with nanofluidic and emulsion chemistries to generate individual PCR partitions (i.e., creating multiple individual PCR reactions) and capture the random distribution of the DNA template. Poisson distribution statistics can then be used to measure the quantities of DNA present [1, 4, 8]. In comparison to the standard qPCR; dPCR provides more precision and sensitivity in allowing for detection of low-copy number genes of DNA (or RNA) template [6, 9–11], while have also seen to be less susceptible to inhibitors [12].

Human Islet-derived Progenitor cells (hIPCs) are proposed to be a potent alternative cell source for islet transplantation to treat Type 1 diabetes (T1D) [13–16]. In vitro lineage analyses on human islets revealed that β cells proliferate as well as undergo epithelial to mesenchymal transition (EMT), leading to the generation of a lineage-committed hIPC population [17–21]. The hIPCs being derived from human β cells have shown to retain chromatin conformation that favours key β-cell gene (including insulin) transcription [15–19, 22–24]. However, over multiple expansions and exposure to different conditions, hIPCs lose their characteristic epigenetic marks that define active β-cell transcriptome along with the overgrowth of non β-cell progenies which may further dilute the lineage commitment potential [23]. dPCR serves as a precise and sensitive method to detect the absolute copy numbers of key β-cell genes (which maybe expressed at low levels) within long-term cultures of hIPCs (containing β-cell and non β-cell progenies). We present in this chapter the generation of hIPCs from islets, RNA isolation, cDNA synthesis followed by digital droplet (dd) PCR used for measuring absolute copies of gene transcripts in hIPCs.

2 Materials

2.1 Generation of Human Islet-Derived Progenitor Cells (hIPCs)

1. CMRL serum-containing medium (CMRL-SCM): 450 mL of CMRL medium (no glutamine), 50 mL of heat-inactivated FBS, 5 mL of GlutaMAX (100×), 5 mL of Penicillin-Streptomycin (5000 U/mL) solution, 10 ng/mL Epithelial Growth Factor (EGF) (*see* **Note 1**). Store CMRL-SCM) in 4 °C for long-term storage.
2. Trypsin EDTA: Dissolve 4 g NaCl, 0.2 g KCl, 0.5 g D-glucose, 0.18 g $NaHCO_3$, 0.03 of KH_2PO_4, 0.05 g of Na_2HPO_4, 0.19 g EDTA, and 1.25 g bovine pancreatic trypsin in

300 mL of Milli-Q water. Make up the volume to 500 mL and filter the solution through 0.2 μm filters. Prepare aliquots of trypsin–EDTA and store in −20 °C for long-term storage. In-use aliquots can be kept at 4 °C.

2.2 Cellular RNA Isolation and cDNA Synthesis

1. TRIzol© reagent.
2. Chloroform.
3. Isopropyl alcohol (IPA).
4. 100% ethanol.
5. Nuclease-free water.
6. RNase AWAY® (*see* **Note 2**).
7. High capacity cDNA kit containing 10× RT buffer, 10× RT random primers, 25× dNTP mix with dTTP, and 50 U/μL MultiScribe reverse transcriptase enzyme (rMoMuLV).

2.3 Droplet Digital PCR

1. TaqMan®Assay(s) for the genes of interest.
2. ddPCR Supermix for Probes (no dUTP) (*see* **Note 3**).
3. ddPCR™ 96-well plates (Bio-Rad) (*see* **Note 4**).
4. Pierceable foil heat seal (Bio-Rad).
5. Plate Support Block (Bio-Rad).
6. PX1™ PCR Plate Sealer instrument (Bio-Rad).
7. Automated droplet generator oil (Bio-Rad).
8. Automated droplet generator tips (Bio-Rad).
9. Automated droplet generator cartridges (Bio-Rad).
10. Automated droplet generator cold block (store at −20 °C).
11. Automated Droplet Generator instrument (Bio-Rad).
12. QX200™ Droplet Reader instrument (Bio-Rad).
13. Droplet reader oil (Bio-Rad).

3 Methods

Prepare all cell culture media and reagents in a sterile biosafety class II cabinet. All reagents used are molecular biology grade and all plasticware nuclease-free. Maintain aseptic conditions by using biosafety class II cabinet during all cell culture procedures.

3.1 Generating Islet-Derived Progenitor Cells from Islets

1. Leave isolated human islets (*see* **Note 5**) in a conical tube to settle down for 10 min.
2. Discard excess storage media (*see* **Note 6**) if needed (leaving approximately only 1000 IEQ per mL) and gently resuspend the islets in the storage medium.

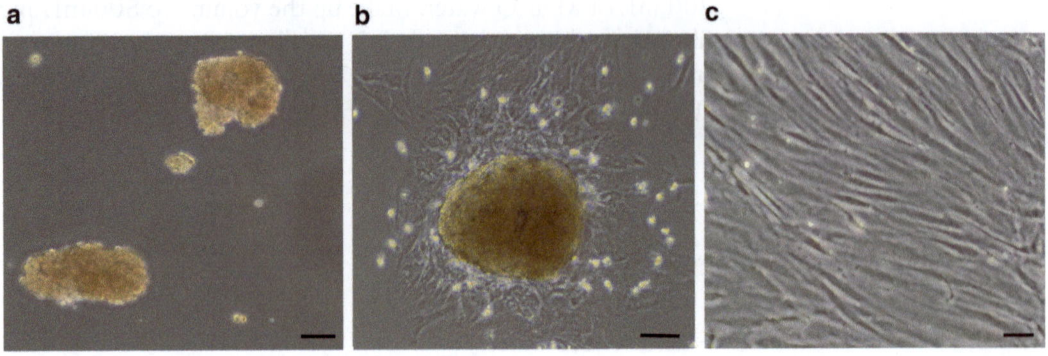

Fig. 1 Freshly isolated human islets (**a**) are allowed to attach to cell culture flasks in serum-containing medium, where cells within the islets migrate out (**b**) and eventually form highly proliferative monolayers of islet-derived cells (**c**). Images are captured with 10× objective and scale bar presents 20 μm

3. Transfer 2.5 mL (~2500 IEQ) of islets into a separate 15 mL tube.
4. Centrifuge the islets at 200 × *g* for 2 min.
5. Discard as much supernatant as possible without disturbing the islets.
6. Resuspend the islet pellet in 10 mL of CMRL-SCM and transfer into a T75 flask (*see* **Note** 7).
7. Observe islets in T75 flask under a microscope (Fig. 1a) and keep the T75 flask in a CO_2 incubator (5% CO_2) at 37 °C.
8. Observe the T75 flask every other day to check the attachment of islets to the surface of the flask.
9. Change media in a flask with fresh 10 mL of CMRL-SCM, when all islets have attached and the cells started to migrate out (Fig. 1b).
10. Afterward, keep changing medium every 3–4 days until all the cells have migrated out of islets and have formed a confluent monolayer of epithelial cells.

3.2 Culturing and Passaging hIPCs

1. Remove media from the confluent T75 flask of islet-derived cells.
2. Add 4 mL of warm (kept in 37 °C water bath for at least 15 min) trypsin–EDTA into T75 flask.
3. Place T75 flask into CO_2 incubator (for 2–3 min) until cells dislodge. Cells should dislodge in this time (*see* **Note 8**).
4. Confirm that cells have rounded up and are floating by observing the flask under a microscope.
5. Immediately add 1 mL of CMRL-SCM into the flask to stop the trypsin activity.
6. Resuspend cells thoroughly and then transfer into a 15 mL conical tube.

7. Centrifuge tube at 450 × *g* for 2 min.
8. Gently discard all media in the tube, leaving only the cell pellet.
9. Resuspend the cell pellet in 2 mL of CMRL-SCM.
10. Transfer 1 mL of cell suspension into a fresh (new) T75 flask. Make two such flasks.
11. Add 9 mL of CMRL-SCM into these two new flasks and label them as P1.
12. Incubate the T75 flasks in a CO_2 incubator (CO_2 at 5%) at 37 °C.
13. Change media when necessary to provide sufficient nutrient for cells and continue to passage the cells whenever the flasks are confluent (usually 3–4 days). In 2–3 passages, all cells attain typical mesenchymal fibroblast-like morphology (Fig. 1c). At this stage, these highly proliferative cultures are termed as hIPCs.
14. When the cells are in cell suspension (**step 9**, Subheading 3.2), one can perform cell counting using hemocytometer or can keep around 50,000 to one million cells for RNA isolation depending upon the downstream experiments.
15. Transfer the required cells in the suspension to a 1.7 mL microcentrifuge tube and spin down at 3000 × *g* for 2 min.
16. Remove the supernatant completely and store the dry cell pellet in −80 °C until RNA isolation is carried out.

3.3 Cellular RNA Isolation

1. Lyse the cells by adding 1 mL of TRIzol© to each sample in individual 1.7 mL microcentrifuge tubes (*see* **Note 9**). If a different volume of TRIzol© is to be used, the volume of all other reagents must be adjusted accordingly to maintain the correct ratios (*see* **Note 10**).
2. Add 200 μL of chloroform to each samples and mix by shaking vigorously for 1 min (*see* **Note 11**).
3. Immediately centrifuge the tubes at 12,000 × *g* for 15 min at 4 °C. Following centrifugation, the liquid components will now be separated into three distinct layers. The upper aqueous (clear) phase containing RNA, the middle interphase (white) containing DNA, and the lower organic phase (pink) containing proteins and lipids.
4. Carefully transfer the upper aqueous (clear) phase into a new 1.7 mL microcentrifuge tube. We recommend performing this step with a smaller pipette to remove this layer gradually. This minimises the risk of disturbing lower phases and contaminating the aqueous phase.
5. Add 500 μL of IPA to the transferred upper aqueous phase, and then mix gently by inverting 7–10 times. Do not vortex or shake tubes.
6. Incubate at room temperature for 10 min.

7. Centrifuge the tubes at 12,000 × *g* for 15 min at 4 °C. This centrifugation step is intended to pellet RNA. For better visibility and handling of the RNA pellet, we recommend orienting the tubes with the hinge facing outward (outer rim of the centrifuge). The RNA pellet will be observed at the bottom of the tube toward the hinge.
8. While samples are spinning, prepare a fresh volume of 75% ethanol. The amount will depend on the number of samples being processed (1 mL of 75% ethanol per sample).
9. Carefully aspirate and discard the supernatant. To reduce disturbance of the pellet, aspirate along the wall of the tube opposite to the hinge. Also, utilise appropriate small-volume pipettes as you begin to remove smaller volumes of the supernatant that are closer to the pellet.
10. Add 1 mL of the fresh 75% ethanol to each tube and briefly mix by vortexing.
11. Centrifuge the tubes at 12,000 × *g* for 15 min at 4 °C. Again, orient the tubes with the hinge facing outward to enable estimation of pellet location.
12. Carefully aspirate and discard the supernatant.
13. Allow the tubes to dry at room temperature for 5 min. We lay the tubes on the side with their lids open. If there are any large droplets, gently twist the tube to spread out the liquid and permit faster drying. However, be careful not to over dry the samples as this will decrease the solubility of the RNA.
14. Add 15 μL of nuclease-free water to each tube and resuspend the RNA pellet (*see* **Note 12**). Always store RNA on ice after this step.
15. Measure the concentration of RNA using a NanoDrop, if proceeding with downstream processing immediately. If you are not planning to continue processing the samples, you can skip this step and store the RNA at −80 °C.

3.4 cDNA Synthesis

1. Thaw all reagents except the reverse transcriptase enzyme and (if necessary) RNA on ice. The reverse transcriptase enzyme should always be stored at −20 °C and removed only on a −20 °C cold block just before addition.
2. Gently vortex all reagents, except the enzyme, and briefly centrifuge them at 10,000 × *g* for 10 s.
3. Measure the concentration of RNA. This is necessary as the reaction requires a total RNA input of 1 μg.
4. Aliquot 1 μg of each RNA sample into an individual 0.2 mL PCR tube and add an appropriate volume of nuclease-free water to make a total of 7.1 μL. Keep the diluted RNA on ice.

Table 1
Example calculations of reagents required for reverse transcription mix

Reagent	Volume per reaction (μL)	Volume per 96 samples (including 5% excess) (μL)
10× RT buffer	1	100.8
25× dNTPs	0.4	40.32
10× random primers	1	100.8
rMoMuLV reverse transcriptase (50 U/μL)	0.5	50.4
Total	2.9	292.32

5. To prepare the reverse transcription master mix for one reaction/sample, combine 1 μL of RT Buffer, 0.4 μL of dNTPs, 1 μL of Random Primer, and 0.5 μL of MultiScribe reverse transcriptase enzyme (rMoMuLV) for a total volume of 2.9 μL RT reagent mix per sample (Table 1). It is recommended to calculate for and prepare at least 5% excess volume to compensate for any pipetting error.
6. Add 2.9 μL of the reverse transcription master mix into each PCR tube containing diluted RNA to make a total reaction volume of 10 μL.
7. Invert the PCR tubes to mix, and briefly centrifuge at 10,000 × *g* for 10 s.
8. Place the tubes into a thermocycler with the following cycling conditions: 25 °C for 10 min, 37 °C for 2 h, 70 °C for 10 min, and hold at 4 °C.
9. Store cDNA at −20 °C or use immediately for downstream processing.

3.5 Droplet Digital PCR (ddPCR) for Gene Targets

1. If necessary, allow cDNA samples to thaw thoroughly on ice prior to commencing.
2. Prepare the ddPCR mastermix for one reaction/sample by combining 12.5 μL of ddPCR Supermix for Probes (no dUTP), 1.25 μL of the TaqMan®Assay Primer/Probe mix and 10 μL of nuclease-free water. It is recommended to calculate for and prepare at least 5% excess volume to compensate for any pipetting error. (Table 2).
3. Briefly vortex and centrifuge the mastermix.
4. Add 23.75 μL of the mastermix to the required amount of wells in the ddPCR™ 96-well plate. This plate is called sample plate that will contain PCR reactions. Plates must be prepared so that the columns that are to be used are completely filled (i.e., multiples of 8). This is necessary for the automatic droplet

Table 2
Example calculations of reagents required for ddPCR master mix

Reagent	Volume per reaction (μL)	Volume per 96 samples (including 5% excess) (μL)
2× ddPCR Supermix for probes (no dUTP)	12.5	1260
20× TaqMan®assay primer/probe mix	1.25	126
Nuclease-free water	10	1008
Total	23.75	2394

generator (**step** 7) to work. Add water in case you do not have enough samples.

5. Mix cDNA samples and then add 1.25 μL of the cDNA to the respective well. Mix by pipetting 20 times. Always include no template control (NTC) well containing water instead of cDNA to set up thresholds.
6. Place a pierceable heat seal onto the sample plate. Ensure the pierceable heat seal has the side with the red stripe visible/face-up.
7. Place the sample plate in the plate support block and seal the pierceable heat seal to the sample plate by keeping the plate support block in the PX1™ PCR Plate Sealer for 5 s at 180 °C.
8. Centrifuge at 450 × *g* for 1 min at room temperature.
9. Set up the automated droplet generator by first configuring the number of samples. The required droplet generation cartridges and tips will be displayed (2 tips are required per sample). Insert the required amount of consumables and check that the tip waste is empty.
10. Place the automated droplet generator cold block (*see* **Note 13**) into the machine with a new ddPCR™ 96-Well Plate. This plate is called as a droplet plate that will contain the newly generated droplets.
11. Place the sample plate (containing PCR reactions) into the automated droplet generator.
12. Run the droplet generation program and remove the droplet plate within 30 min of completion.
13. Turn on the PX1™ PCR Plate Sealer and configure the settings to 180 °C for 5 s.
14. Insert the droplet plate into the plate support block.
15. Place a pierceable heat-seal onto the droplet plate. Ensure that the pierceable heat-seal has the side with the red stripe visible/face-up.

16. Seal the pierceable heat-seal to the droplet plate in the PX1™ PCR Plate Sealer.
17. Perform the end stage PCR on the sealed droplet plate containing thousands of droplets of PCR reactions with the following cycling conditions: 95 °C for 10 min, 40 cycles (94 °C for 30 s, 60 °C for 1 min), 98 °C for 10 min, and hold at 12 °C. Ensure that the ramp rate is set to 2 °C/s, the lid is set to 105 °C and the sample volume is set to 40 μL.
18. Following completion of the cycling protocol, allow the droplet plate to sit in the thermocycler for at least 20 min before proceeding to the next step.
19. Turn on the droplet reader at least 10 min prior to use to allow for the reader to assess reader oil and waste level. Input the sample details in the "Setup" section of the QuantaSoft droplet reader software. Specify reader settings as follows: Experiment (Absolute Quantification ABS); Supermix (ddPCR Supermix for Probes (no dUTP)); Target 1(Channel 1 (Unknown)).
20. Place the plate into the droplet reader and press "Run." Select FAM probes at the start of the run (*see* **Note 14**).
21. During analysis, set thresholds for all samples based on the thresholds determined for the NTC (Fig. 2a) and then compute copies/μL for each gene assessed using QuantaSoft droplet reader software (Fig. 2b).

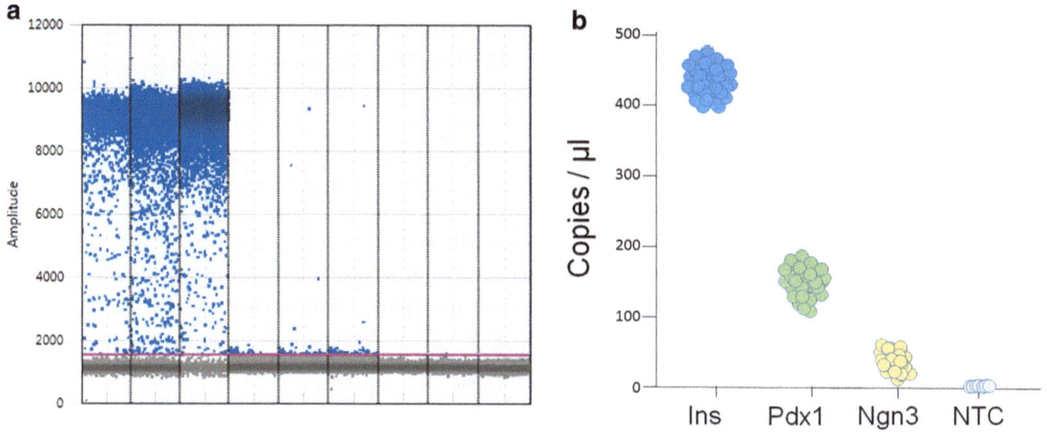

Fig. 2 Representative images of the results obtained from nine different wells after completion of digital droplet PCR (**a**). First three wells show majority of droplets with positive signal suggestive of high abundance gene. Middle three wells show low abundance gene with very few positive droplets, whereas last three wells contain water (NTC) with no signal. These NTC wells are used to set up threshold as shown here in purple line. Depiction of absolute number of copies/μL for different islet hormone/transcription factors in hIPCs along with no-template (negative) control after analysis of ddPCR data (**b**)

4 Notes

1. Add bovine serum albumin (BSA), if reconstituting EGF to lower concentrations. Store aliquots at −20 °C.
2. RNase Away helps remove RNA degrading enzymes during the procedure. This is highly recommended; however, it is not essential with appropriate handling.
3. One can use synthetic forward and reverse primers along with ddPCR EvaGreen supermix and the specific AutoDG™ oil for Evagreen. Our lab has optimised use of TaqMan primer-probe and ddPCR Supermix for Probes (no dUTP) as per the details provided in this chapter.
4. This protocol involves the use of Bio-Rad's droplet digital PCR and therefore all reagents and special plasticware are to be purchased as per the company's recommendations that suit the instrument for best results.
5. Cadaveric human islets are obtained from islet isolation centres after appropriate material transfer agreements and institutional ethics approval to use human islets for research experiments.
6. Storage medium is same as the medium in which islets are transported from islet isolation centres to the laboratory.
7. One can make as many flasks as needed for future experimental work with the seeding density of around 2500 IEQ/10 mL CMRL-SCM/flask.
8. Do not leave cells in trypsin for longer times (approximately >3–5 min) as trypsin–EDTA can be lethal/toxic to cells if exposed for too longer durations.
9. Due to the time involved in the RNA isolation of each sample, we recommend a maximum of eight samples to be processed at a time. Volume of TRIzol depends upon the number of cells. We recommend using 1 mL of TRIzol for 0.5–1 million cells.
10. If the starting cell number for RNA isolation is very less, use molecular biology grade glycogen at the first step of isolation for higher recovery during precipitation and better RNA yields.
11. It is recommended not to vortex tubes that contain TRIzol© due to its corrosive property.
12. Volume of the nuclease-free water to be added to the RNA pellet will vary based upon the size of the pellet. Use higher volumes or bigger pellets; however remember that it is always easy to dilute the RNA but not so to concentrate it.
13. Always store the plate support block at −20 °C and upside down so as to avoid ice crystals depositing in the wells.

14. Gene expression analysis can be performed for two genes, labeled with FAM and VIC fluorophores in a single reaction. The protocol here describes single fluorophore (FAM) reaction only. Further optimization would be necessary for dual-probe analysis.

Acknowledgments

The support provided to WKMW through the University of Sydney postgraduate awards, MVJ through the Australian Diabetes Society (ADS) Skip Martin fellowship and currently through the JDRF International postdoctoral fellowship, CM through NHMRC clinical Trials Centre, and AAH through the JDRF Australia Career Development Award as well as the visiting professorship through the Danish Diabetes Academy is highly acknowledged.

References

1. Sykes PJ et al (1992) Quantitation of targets for PCR by use of limiting dilution. BioTechniques 13(3):444–449
2. Saiki RK et al (1988) Primer-directed enzymatic amplification of DNA with a thermostable DNA polymerase. Science 239 (4839):487–491
3. Morley AA (2014) Digital PCR: a brief history. Biomol Detect Quantif 1(1):1–2
4. Vogelstein B, Kinzler KW (1999) Digital PCR. Proc Natl Acad Sci U S A 96(16):9236–9241
5. Manoj P (2016) Droplet digital PCR technology promises new applications and research areas. Mitochondrial DNA A DNA Mapp Seq Anal 27(1):742–746
6. Hindson BJ et al (2011) High-throughput droplet digital PCR system for absolute quantitation of DNA copy number. Anal Chem 83 (22):8604–8610
7. Pohl G, Shih Ie M (2004) Principle and applications of digital PCR. Expert Rev Mol Diagn 4(1):41–47
8. Huggett JF, Whale A (2013) Digital PCR as a novel technology and its potential implications for molecular diagnostics. Clin Chem 59 (12):1691–1693
9. Whale AS et al (2013) Methods for applying accurate digital PCR analysis on low copy DNA samples. PLoS One 8(3):e58177
10. Sanders R et al (2011) Evaluation of digital PCR for absolute DNA quantification. Anal Chem 83(17):6474–6484
11. Sanders R et al (2013) Evaluation of digital PCR for absolute RNA quantification. PLoS One 8(9):e75296
12. Dingle TC et al (2013) Tolerance of droplet-digital PCR vs real-time quantitative PCR to inhibitory substances. Clin Chem 59 (11):1670–1672
13. Gershengorn MC et al (2005) Are better islet cell precursors generated by epithelial-to-mesenchymal transition? Cell Cycle 4(3):380–382
14. Davani B et al (2009) Human islet-derived precursor cells can cycle between epithelial clusters and mesenchymal phenotypes. J Cell Mol Med 13(8B):2570–2581
15. Joglekar MV, Hardikar AA (2010) Epithelial-to-mesenchymal transition in pancreatic islet beta cells. Cell Cycle 9(20):4077–4079
16. Bar-Nur O et al (2011) Epigenetic memory and preferential lineage-specific differentiation in induced pluripotent stem cells derived from human pancreatic islet beta cells. Cell Stem Cell 9(1):17–23
17. Gershengorn MC et al (2004) Epithelial-to-mesenchymal transition generates proliferative human islet precursor cells. Science 306 (5705):2261–2264
18. Russ HA et al (2008) In vitro proliferation of cells derived from adult human beta-cells revealed by cell-lineage tracing. Diabetes 57 (6):1575–1583
19. Russ HA et al (2009) Epithelial-mesenchymal transition in cells expanded in vitro from lineage-traced adult human pancreatic beta cells. PLoS One 4(7):e6417

20. Joglekar MV et al (2009) Human fetal pancreatic insulin-producing cells proliferate in vitro. J Endocrinol 201(1):27–36
21. Wong WH, Hardikar AA, Joglekar MV (2016) Generation of human islet progenitor cells via epithelial-to-Mesenchymal transition. In: Hardikar AA (ed) Pancreatic islet biology. Springer International Publishing, Cham, pp 217–240
22. Hardikar AA et al (2003) Human pancreatic precursor cells secrete FGF2 to stimulate clustering into hormone-expressing islet-like cell aggregates. Proc Natl Acad Sci U S A 100 (12):7117–7122
23. Joglekar MV, Hardikar AA (2012) Isolation, expansion, and characterization of human islet-derived progenitor cells. Methods Mol Biol 879:351–366
24. Wong W et al (2014) Lineage-committed pancreatic progenitors and stem cells. In: Turksen K (ed) Adult Stem Cells. Springer, New York, pp 339–357

Chapter 5

Isolation and In Vitro Culture of Human Gut Progenitor Cells

Jessica Bruce, Gerard E. Kaiko, and Simon Keely

Abstract

The gastrointestinal epithelium is a highly regenerative organ, where each cell is replaced in a cycle of 4–6 days, depending on the mammalian species. This highly proliferative state is driven by gastrointestinal stem and progenitor cells, located at the base of crypts. These cells give rise to at least six types of differentiated epithelial cells, each with a distinct function in maintaining homeostasis between the intestinal interface and the luminal environment. The isolation and culture of these cells from mammalian gastrointestinal tissue is a novel technique, which allows for the generation and maintenance of an in vitro culture system for adult epithelial cells. There are two predominant methods for isolation and propagation of gastrointestinal epithelial cells, the first is the organoid system developed in 2009, and the second is a later version known as the L-WRN spheroid system. In this chapter, we describe the method to isolate and culture human gastrointestinal stem and progenitor cells and culture them as three-dimensional spheroids using L-WRN cell conditioned media.

Key words Spheroid, Organoid, Stem cell culture, Intestinal cell culture, Intestinal crypt, Epithelial isolation

1 Introduction

The gastrointestinal epithelium is a single layer of cells structurally organized into crypts and villi. The villi consist of at least six major types of highly specialized, post-mitotic cells (enterocytes, goblet cells, enteroendocrine cells, tuft cells, M cells, and Paneth cells), that work in concert to perform the secretory, immunoprotective, and absorptive functions of the gastrointestinal tract [1]. All specialized epithelial cell types originate from the same proliferative progenitor cells, intestinal stem cells (ISCs), which are located at the base of the intestinal crypts [2]. ISCs are multipotent stem cells, identifiable by their expression of leucine-rich repeat-containing G-protein coupled receptor 5 (LGR5), along with other makers including olfactomedin 4 (OLFM4) and achaete-scute family bHLH transcription factor 2 (ASCL2) [3, 4]. ISCs undergo asymmetric replication to form an identical replicate ISC and a daughter cell [5]. These daughter cells migrate out of the crypt as transient

Mugdha V. Joglekar and Anandwardhan A. Hardikar (eds.), *Progenitor Cells: Methods and Protocols*, Methods in Molecular Biology, vol. 2029, https://doi.org/10.1007/978-1-4939-9631-5_5,

amplifying cells which then terminally differentiate into the specialized epithelial cell types [1], before being shed from the villus tip in a process that takes approximately 5 days [6].

A highly specific microenvironment known as the stem cell niche supports the growth and renewal of ISCs. This niche consists of physical support from the extracellular matrix as well as a cocktail of cellular factors released by the surrounding cells and provides the signal for the self-renewal or adoption of a cell lineage [7].

In vitro gastrointestinal tissue culture systems have historically been heavily reliant on immortalized cancer-derived cell lines. These systems are based around culturing cells selected and developed based on their ability to mimic specific physiological functions of the epithelium. Generally, these cell lines are robust, easily transfected, and inexpensive to culture. However, drawbacks include their cancer lineage, tendency toward unknown genetic mutations, inability to recapitulate responses to primary cell growth/stem factors, and lack of physiological architecture. Thus, although these cell lines are derived from the gastrointestinal tract, they may not accurately represent any one section of the gut or account for the diversity of cell types that constitutes the in vivo epithelium [8]. The culture of primary epithelial cells, which more closely resemble the in vivo intestinal epithelium, is therefore preferable. Given the non-proliferative state of differentiated enterocytes, cultures of proliferative ISCs and progenitors are needed to establish primary epithelial cultures.

Until recently, the complexity of the stem cell niche and the factors required for ISC survival have been barriers to the development of a robust system for the culture of primary epithelial cells [9]. However, recent advances in our understanding of stem cell biology have identified at least three major factors, namely, Wnt, R-spondin, and noggin, that permit the isolation and long-term culture of ISCs [10, 11]. Wnt proteins provide the signal for the self-renewal and proliferation of stem cells [12, 13], R-spondins act as coactivators and modulators of the Wnt signalling pathway [14], and noggin inhibits the differentiation of ISCs into epithelial cell lineages by blocking bone morphogenetic protein (BMP) signalling, for retention of stem cell specificity [15]. Addition of Wnt, R-spondin, and noggin to cell culture medium, in combination with the use of the basement membrane substitute Matrigel, permits long-term, perpetual in vitro culture (and biobanking) of ISCs isolated from the gastrointestinal tract in three-dimensional (3D) structures [6, 10, 16].

The organoid system was originally developed by the Clevers lab in 2009 [11, 17]. This system utilizes a number of recombinant factors as additives to culture media, which enables the growth of heterogeneous intestinal structures often termed "mini-guts." In contrast, the gastrointestinal spheroid system developed in the Stappenbeck lab utilizes the commercially available L-WRN cell line developed by Miyoshi et al., to provide a cost-effective source of lipidated Wnt, R-spondin, and noggin in one media [10, 16]. This system facilitates the growth and differentiation of more homogenous cultures of gastrointestinal epithelial cells. The addition of media conditioned by this cell line onto isolated intestinal crypts in a mixed tissue digest provides the necessary levels of Wnt, R-spondin and noggin to facilitate the outgrowth and long-term culture of epithelial cells [6, 18].

To establish cultures, gastrointestinal crypts, are harvested from fresh or cryofrozen mammalian intestinal tissue via a combination of enzymatic and mechanical digestions. After enrichment, whole crypts are seeded in the basement membrane substitute Matrigel and cultured in 50% L-WRN conditioned media. Cells subsequently form three-dimensional spherical structures, with a luminal facing apical surface, of primarily non-differentiated ISCs and progenitor cells. Spheroids may be used in virtually any of the biological or molecular assays that have been developed for traditional cell lines [19], including transfection [10], and unlike traditional cell lines, have demonstrated genetic stability over 3 months [10]. Spheroids can be maintained as stem/progenitor cells for use in the study of stem cell dynamics and tissue regeneration [10, 16]. They can also be differentiated into polarized, non-dividing cells resembling regionally specific differentiated cells (e.g., colonocytes or enterocytes and goblet cells) [6, 18]. 3D spheroids can also be cultured into monolayers [18, 20] to facilitate studies of epithelial cell biology, microbe–epithelial or drug compound–epithelial interactions. Furthermore, this system can be used in bioengineered scaffolds to permit the construction of the crypt–villus unit to mimic the in vivo architecture of the intestinal tract [21].

In this chapter, we describe in detail, a method for the isolation and culture of gastrointestinal stem cells, which has been successful in establishing cultures from all portions of the colon, small intestine, and stomach.

2 Materials

2.1 General Materials (Required for All Steps)

1. Biosafety cabinet.
2. Water bath set to 37 °C.
3. Swinging bucket centrifuge.
4. Pipettes and pipette tips (able to dispense between 1 μL and 1000 μL).
5. Conical tubes (15 mL and 50 mL).
6. 1.5 mL Eppendorf tubes
7. Cell culture incubator (37 °C, 95% O_2, 5% CO_2).

2.2 Preparation of L-WRN Conditioned Media

1. L-WRN cells (ATCC: CRL-3276).
2. L Cell media: DMEM high glucose, containing 10% fetal bovine serum (FBS), 100 units/mL penicillin, 0.1 mg/mL streptomycin and 2 mM L-glutamine. Store at 4 °C for up to 1 month. Warm to 37 °C by incubating in a water bath prior to use.
3. Geneticin (G418).
4. Hygromycin.
5. PBS-EDTA: Add 500 μL of 0.5 M EDTA (pH 8.0) to a 500 mL bottle of PBS. Store at room temperature.
6. Trypsin–EDTA: Dilute 10× trypsin 1:10 in PBS–EDTA. Aliquot and store at −20 °C for up to 1 year. Once thawed, store at 4 °C.
7. Primary cell culture media: Advanced DMEM/F-12, containing 20% FBS, 100 units/mL penicillin, 0.1 mg/mL streptomycin, and 2 mM L-glutamine. Store at 4 °C for up to 1 month. Warm to 37 °C by incubating in a water bath prior to use.
8. 150 cm^2 cell culture flasks
9. 1 L autoclaved glass storage containers.

2.3 Isolation of Crypts from Biopsies

1. Washing Media: DMEM containing 10% FBS, 100 units/mL penicillin, streptomycin (0.1 mg/mL). Store at 4 °C for up to 1 month.
2. Collagenase Solution: Add 20 mg of collagenase type 1 to 10 mL washing media and supplement with 10 μL of gentamicin. This solution should be prepared fresh on the day of the isolation.
3. Surgical equipment: Fine forceps and tweezers. Surgical equipment should be autoclaved prior to use. Before contact with biopsies, equipment should be wiped down with an alcohol swab and allowed to dry. Following use, equipment should be washed in bleach.
4. 35 mm dishes
5. 24-well plates
6. Matrigel; containing 8–10 mg/mL protein and less than 1.5 endotoxin units/mL. Aliquots of Matrigel should be thawed in the fridge on ice the night before use to prevent the polymerization of the gel. Thawed aliquots can be stored at 4 °C for up to 1 month.
7. ROCK inhibitor Y27632: 10 mM stock in sterile water. Store aliquots at −20 °C.
8. TGFβIR inhibitor SB 431542: 10 mM stock in DMSO. Store aliquots at −20 °C.

2.4 Cryopreserving Spheroid Cultures

1. Freezing Media: 1 mL FBS, 1 mL of DMSO, 8 mL of washing media.
2. Cryovials.
3. Controlled rate freezing container.

3 Methods

3.1 Generation of L-WRN Conditioned Media

1. Thaw cryopreserved L-WRN cells in a 37 °C water bath. Once solution is thawed, transfer the cells immediately to a 50 mL conical flask containing 25 mL of prewarmed L cell media.
2. Transfer the cell suspension to a 150 cm^2 cell culture flask and incubate the flask in a cell culture incubator for approximately 1 day to allow cells to adhere.
3. Once the cells have adhered, aspirate the culture media from flask and replenish with fresh L cell culture media containing 500 μg/mL of both G418 and hygromycin.
4. Return the flask to the cell culture incubator and incubate the cells until they are confluent.
5. Once cells are confluent, aspirate and discard the culture media from the flask.
6. Add 20 mL PBS–EDTA to the flask. Cap the flask then swirl the liquid over the bottom of the flask to wash the cells.
7. Aspirate the PBS–EDTA then add 1 mL of trypsin–EDTA to the flask.
8. Again, cap the flask and swirl the trypsin over the bottom of the flask to coat the cells.
9. Place the flask in a cell culture incubator for three to 5 min to allow cells to detach (*see* **Note 1**).
10. Once cells are detached from the bottom of the flask, add 12 mL of fresh L cell culture media to inhibit the trypsin. Pipet the solution gently several times to mix.
11. Aspirate the cell suspension from the flask and place in a 50 mL conical tube.
12. Centrifuge the tube for 5 min at 233 rcf to pellet the cells. During this step, add 120 mL of L-cell culture media containing G418 and hygromycin at 500 μg/mL into five 150 cm^2 cell culture flasks.
13. Aspirate and discard the supernatant from the spun down cell pellet and resuspend in 5 mL of fresh L cell culture media.
14. Add 1 mL of L-WRN cell suspension to each of the five 150 cm^2 flasks. Swirl the flask several times to distribute the cells uniformly in the media.

15. Place flasks in the cell culture incubator.
16. Incubate the cells until they are over-confluent and some cell aggregates start to detach from the flask. This takes 3–4 days (*see* **Note 2**).
17. Once the flasks are over confluent, aspirate the cell culture media from each flask and wash the cells once with 10 mL of primary cell culture media per flask (*see* **Note 3**).
18. Once the cells have been washed, add 25 mL of fresh primary cell culture media to each of the flasks and return the flasks to the incubator.
19. After a 24-h incubation period, aspirate the media from each of the flasks into 50 mL centrifuge tubes. Replenish the flasks with 25 mL of fresh primary cell culture media and return the flasks to the cell culture incubator.
20. Centrifuge the 50 mL tubes containing media aspirated from cells at 1455 rcf for 5 min at room temperature.
21. Decant the spun down supernatant, into a 1 L storage bottle. This constitutes the first collection of L-WRN conditioned media. Store this at 4 °C.
22. Repeat **steps 18–20** at 24 h intervals, adding the L-WRN conditioned media to the same 1 L bottle, for 4 days. Following the fourth collection, the total volume of conditioned media in the bottle should be 500 mL (*see* **Note 4**).
23. Add fresh primary cell culture media in a 1:1 volume ratio to the conditioned media and mix well to produce 50% conditioned media.
24. Aliquot the media into 50 mL conical tubes. Store the media at −20 °C (*see* **Note 5**).

3.2 Isolation of Crypts from Human Gastrointestinal Biopsies

1. Collect biopsies (two to four biopsies with an area approximately 5 mm^2) in 5 mL of washing media and store on ice (*see* **Note 6**).
2. Transfer samples to a 35 mm dish on ice. Using the fine scissors, mince the biopsy pieces until they are small enough to pass through a P1000 pipette tip (*see* **Note 7**).
3. Add 1 mL of collagenase solution and pipet biopsies several times to distribute the biopsies within the solution.
4. Place the biopsies in a 37 °C incubator. At intervals of 5 min, vigorously pipette the solution containing the biopsies several times. This pipetting provides additional mechanical disruption to assist with crypt dissociation.
5. Examine the solution under the microscope to check for crypt dissociation. Sample images of crypt dissociation are provided in Fig. 1a.

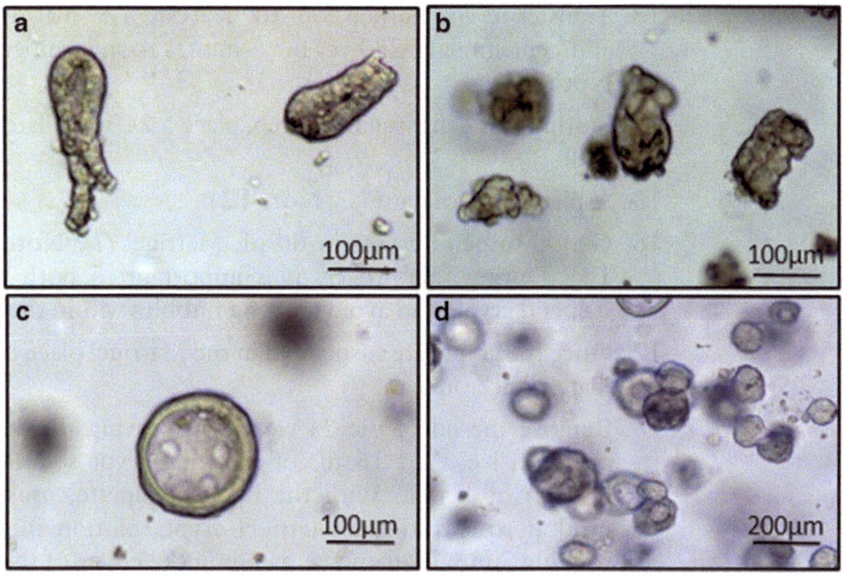

Fig. 1 Crypts immediately post-collagenase digestion (**a**). Crypts immediately after seeding in Matrigel (**b**). Single spheroid on day 2 post-isolation (**c**). Spheroid culture on Day 4, prior to splitting (**d**). Scale bars as indicated for each panel

6. If no dissociated crypts are visible, return dish to the incubator, and repeat **steps 4** and **5**. These steps can be repeated multiple times to maximize the number of crypts, however biopsies should not remain in the collagenase solution for a total period greater than 30 min.
7. Once crypts are dissociated, strain the solution through a 70 µm cell strainer into a 50 mL conical tube. Wash the strainer through with 9 mL of washing media to inhibit collagenase. Addition of washing media at a volume 10× greater than the volume of collagenase solution will inhibit collagenase. Crypts and small debris will now be in the 50 mL tube.
8. Transfer the solution containing the crypts into a fresh 15 mL conical tube and centrifuge the crypts at 58 rcf for 5 min at room temperature.
9. Check the bottom of the tube to see if a small pellet has formed (*see* **Note 8**).
10. Aspirate the supernatant, being careful not to disrupt the cell pellet (*see* **Note 9**).
11. Suspend the cell pellet in 10 mL of fresh washing media and centrifuge again at 58 rcf for 5 min.
12. Aspirate the supernatant (*see* **Notes 9** and **10**), and resuspend the pellet in 1 mL of fresh washing media.

13. Transfer crypt suspension to a fresh 1.5 mL Eppendorf and centrifuge at 233 rcf for 5 min at room temperature (*see* **Note 11**).
14. During this centrifugation step, place a 24-well plate on ice to chill.
15. Aspirate supernatant (*see* **Note 12**).
16. Gently suspend crypts in 60 μL Matrigel (*see* **Notes 13** and **14**). During this process it is important to both uniformly suspend crypts and avoid creating bubbles within the Matrigel.
17. Once the crypts are suspended in the Matrigel place this immediately back on ice.
18. Remove the lid of the 24-well plate, leaving the base of the place on ice. Add 15 μL of Matrigel–crypt solution to the bottom of 1 well. With the tip of a pipette, use a circular motion to spread the Matrigel–crypt solution into a circle roughly 7 mm in diameter, avoiding the edges of the well.
19. Repeat **step 18** for another three wells.
20. Place the lid of the plate back on. Hold the plate by the edges and in one swift movement invert the plate so that the bottom of the wells faces upward (*see* **Note 15**).
21. Incubate the plate at 37 °C for approximately 10 min to allow the Matrigel to polymerize.
22. Following incubation, confirm that the back of the plate is warm to touch (an indicator that the Matrigel has polymerized). Turn the plate back over and add 500 μL per well of 50% conditioned medium.
23. Return the plate to a cell culture incubator.

Monitor growth of crypts over the next 4 days. A visual timeline of crypt growth is provided in Fig. 1. On day 2 (day 0 is the day of isolation) aspirate culture media, avoiding the Matrigel dome and replace the culture media with 400 μL of fresh 50% conditioned media containing 10 μM ROCK inhibitor and 10 μM TGFβIR inhibitor.

3.3 Passaging Spheroids

Spheroids should be passaged to keep optimal growth conditions. Spheroids isolated from human tissue should be passaged on day 4 or when spheroids reach a density as shown in Fig. 1d, however this can be extended to day 5 if needed (e.g., over the weekend).

1. Aspirate culture media.
2. Add 500 μL of 0.5 mM PBS–EDTA to each well.
3. Scratch the Matrigel–spheroid dome from the bottom of the well and pipet a few times to disperse the Matrigel–spheroid within the PBS.

4. Wash the well down a few times, to collect all of the spheroids.
5. Transfer the PBS–spheroid suspension to a 15 mL conical tube and centrifuge at 233 rcf for 5 min at room temperature.
6. Remove the supernatant, being careful not to disrupt the cell pellet (*see* **Note 9**).
7. Add 200 μL of trypsin–EDTA per well to the tube and pipet the pellet a few times to suspend it in the trypsin.
8. Place the tube with cells in trypsin in a 37 °C water bath for approximately 90 s (*see* **Note 16**).
9. During this incubation set a P1000 pipette to 400 μL. Following the incubation, vigorously pipet the solution about 20 times. The aim of this is to disrupt the spheroid structure, reducing it to small clumps of cells. Visualize the cells under a microscope to confirm that this has occurred. If happy with the level of dissociation of the spheroids proceed to **step 10**, if the cells have not dissociated, vigorously pipet the mixture again.
10. Add washing media at a volume 10× that of the volume of trypsin used to quench the trypsin (e.g., if 800 μL of trypsin used, add 8 mL of washing media).
11. Centrifuge at 233 rcf for 5 min at room temperature. By now, the Matrigel will be digested and be contained within the supernatant.
12. Aspirate the supernatant (*see* **Note 9**) and resuspend the pellet in 1 mL of fresh washing media. Transfer this to a fresh 1.5 mL Eppendorf tube (*see* **Note 17**).
13. Centrifuge the cells at 233 rcf for 5 min at room temperature in a swinging bucket centrifuge to pellet the cells (*see* **Note 11**). During this step, place a 24-well plate on ice to chill.
14. Remove the supernatant, leaving a small volume so as not to disturb the cell pellet (*see* **Note 12**). Place the tube on ice.
15. Resuspend the pellet in Matrigel at a volume of 15 μL per well to be seeded. The volume of Matrigel used in this step should be 15 μL per well (*see* **Notes 13** and **14**).
16. Pipet droplets of the Matrigel–cell suspension into the desired number of wells of the prechilled 24-well plate.
17. Use a pipette tip to spread the drop into a circle about 7 mm in diameter, being careful to avoid getting the solution too close to the sides of the well (*see* **Note 15**).
18. Replace the lid onto the plate and in one, motion-flip the plate upside down.
19. Incubate the plate in a 37 °C incubator for approximately 10 min, or until the Matrigel has polymerized.

20. Once the Matrigel has solidified, add 400 μL of 50% conditioned media containing 10 μM ROCK inhibitor and 10 μM TGFβIR inhibitor to each well.
21. Incubate the plate in a cell culture incubator.

3.4 Cryopreserving Cells

1. Aspirate culture media from wells, avoiding the Matrigel dome.
2. Add 500 μL of fresh washing media to each well.
3. Scratch the Matrigel dome from the plate using a pipette tip. Angle the plate slightly so the media pools at the bottom, wash any spheroids still stuck to the plate with the media in the plate then transfer the suspension to a 15 mL tube.
4. Centrifuge the tube at 233 rcf for 5 min at room temperature.
5. Remove the supernatant, taking care not to disrupt the pellet (*see* **Note 9**).
6. Resuspend the pellet in 1 mL of freezing media per well.
7. Aliquot 1 mL of cell suspension into the desired number of cryovials (usually 1 well of a 24-well plate per cryovial), then place cryovials in a controlled rate freezing container.
8. Freeze cells at −80 °C over 48 h, then transfer to liquid nitrogen for long-term cryopreservation.

3.5 Reviving Cryopreserved Stocks of Cells

1. Retrieve frozen cell vial from stocks from liquid nitrogen or very cold freezer and place immediately on ice to keep cold.
2. Rapidly immerse the vial in a 37 °C water bath, to thaw the aliquot.
3. Once the cell suspension is thawed, transfer to a 15 mL tube containing 5 mL of prewarmed wash media.
4. Centrifuge the tube at 58 rcf for 5 min at room temperature.
5. Aspirate the supernatant (*see* **Note 9**).
6. Add 1 mL of fresh wash media and resuspend the cell pellet.
7. Transfer the cell suspension to a fresh 1.5 mL Eppendorf tube and centrifuge at 233 rcf for 5 min at room temperature using a 50 mL conical tube as an adaptor. Chill a 24-well plate on ice.
8. Aspirate the supernatant from the spun down cells, leaving a small volume to prevent disruption of the cell pellet (*see* **Note 12**).
9. Resuspend the pellet in Matrigel at a volume of 15 μL per well to be seeded (e.g., if seeding four wells, add 60 μL of Matrigel) (*see* **Notes 13** and **14**). Avoid creating bubbles and use a circular motion to stir the solution while pipetting. Place the tube immediately back on ice.

10. Add 15 μL of cells per well to the prechilled 24-well plate, then spread the Matrigel out to a 7 mm diameter circle, avoiding the edges of the well.
11. Place the lid back onto the plate then invert and incubate at 37 °C for approximately 10 min or until Matrigel has solidified.
12. Add 400 μL of 50% conditioned media containing 10 μM ROCK inhibitor and 10 μM TGFβIR inhibitor to each well and return the plate to the incubator.

4 Notes

1. Periodically, remove the flask from the incubator and tap the flask against the side of a bench to provide a mechanical force to assist in cell detachment. Observe and monitor the cells under a microscope; if all cells have detached proceed to **step 11**. If cells are not detached return the flask to the cell culture incubator and repeat **step 10**. Do not leave the cells in trypsin for more than 5 min.
2. To increase the volume of conditioned media collected, it is possible to passage the L-WRN cells seeded in **step 14**. To achieve this, once the L-WRN cells seeded in **step 14** have reached 80% confluency repeat **steps 5–13** for each flask. This will generate enough cells to seed up to twenty-five 150 cm^2 flasks which can then be used for **steps 14** through to **23**. Alternatively, EasyFill Cell Factory Nunclon trays (Thermo Scientific cat. no. 140400) can be used to upscale conditioned media production as described in [10].
3. G418 and hygromycin should be included in the L cell culture media but excluded from the primary culture media. To avoid trace amounts of these antibiotics in the conditioned media, it is important to wash the L-WRN cells in primary cell culture media prior to incubating the cells for collection of conditioned media.
4. Media can be collected from one passage of L-WRN cells for 12 days without a decrease in activity [10].
5. Conditioned media can be stored at −20 °C for at least 3 months without a decrease in activity. After thawing an aliquot, store at 4 °C and use within 1 week.
6. For samples that may be infected (e.g., samples from patients with known bacterial susceptibility or antibiotic resistance), gentamicin and primocin at a final concentration of 1:1000 and 1:500, respectively can be added to the collection media.
7. When mincing the biopsies, if the biopsies are sticky, a small amount of collagenase solution can be added to the dish for

lubrication. It is important to note the total amount of collagenase added to the biopsies so that washing media at a volume of 10× collagenase solution can be added to neutralize the enzyme and prevent damage to the cells.

8. Following the initial centrifugation of the crypts examine the bottom of the tube under a microscope to check for a pellet. If no pellet is visible after centrifugation, it may be necessary to centrifuge the tube again at 149–233 rcf.
9. When aspirating supernatant above crypt/spheroid pellets during wash steps leaving about 200 μL of media will prevent disruption of pellet and loss of crypts.
10. Following initial centrifugation of crypts, the supernatant can be saved and centrifuged at 233 rcf to ensure no crypts were lost.
11. When centrifuging crypts in a 1.5 mL Eppendorf, it is possible to use a 50 mL conical tube as an adaptor in a swinging-bucket centrifuge.
12. When aspirating the supernatant from the 1.5 mL tube prior to suspending the crypts in Matrigel, it is necessary to leave a small amount of media so as not to disrupt the cell pellet, but also to aspirate enough so that the Matrigel will still polymerize. During this step, the user should ensure that the ratio of Matrigel to media is no less than 2:1, that is, if 10 μL of media is left above the pellet, more than 20 μL of Matrigel should be added to the Eppendorf. Optimally the ratio should be 4:1 or higher.
13. Matrigel will polymerize to form a gel-like matrix when the temperature of the Matrigel starts to rise above 10 °C. For this reason, it is important to keep the tube containing Matrigel on ice when not in use. Holding the tube containing Matrigel at the top of the tube using your thumb and forefinger rather than at the base will also prevent the temperature of the Matrigel from rising unnecessarily.
14. Gently stirring the Matrigel with the pipette tip while aspirating and dispensing the cell pellet by pipetting can assist with suspending the cells in the Matrigel.
15. When plating out the crypt/Matrigel solution, inverting the plate keeps the crypts suspended in the Matrigel and prevents them descending through to contact the bottom of the plate.
16. During trypsinization of spheroids, the cells will die if left in trypsin too long. It is important to work quickly, keeping the total time that the cells are exposed to trypsin to no longer than 3 min. The optimal time of trypsin incubation will vary depending on the site of the gastrointestinal tract that the cells were collected from, the density of spheroids and the day of passage. The optimal length of incubation in trypsin should

be determined by the user, depending on how the spheroids respond.

17. At this point, it is possible to split the cells, so that once replated in Matrigel they are at a lower density. To achieve this, uniformly suspend the spheroids within the wash media then discard an appropriate volume of the cell suspension so that the cells will be reseeded at the required density. For example, if wishing to reseed the cells at one-third of their previous density, remove 600 μL of solution from the Eppendorf and discard. It is suggested to passage the cells at 1:3 or 1:4 of their preplating density.

Acknowledgments

This work was funded by NHMRC project grants (S.K. and G.E.K.) and a Cancer Institute New South Wales Career Development Fellowship (S.K.).

References

1. Rao JN, Wang JY (2010) Intestinal stem cells. In: Regulation of gastrointestinal mucosal growth. Morgan and Claypool Life Sciences, San Rafael
2. van Neerven SM, Vermeulen L (2017) Balancing signals in the intestinal niche. EMBO J 36 (4):389–391
3. Barker N (2013) Adult intestinal stem cells: critical drivers of epithelial homeostasis and regeneration. Nat Rev Mol Cell Biol 15:19
4. Barker N et al (2007) Identification of stem cells in small intestine and colon by marker gene Lgr5. Nature 449:1003
5. He S, Nakada D, Morrison SJ (2009) Mechanisms of stem cell self-renewal. Annu Rev Cell Dev Biol 25(1):377–406
6. Kaiko GE et al (2016) The colonic crypt protects stem cells from microbiota-derived metabolites. Cell 165(7):1708–1720
7. Ootani A et al (2009) Sustained in vitro intestinal epithelial culture within a Wnt-dependent stem cell niche. Nat Med 15:701
8. Kozuka K et al (2017) Development and characterization of a human and mouse intestinal epithelial cell monolayer platform. Stem Cell Reports 9:1976–1990
9. Bjerknes M, Cheng H (2006) [14] - intestinal epithelial stem cells and progenitors. In: Klimanskaya I, Lanza R (eds) Methods in enzymology. Academic Press, San Diego, pp 337–383
10. Miyoshi H, Stappenbeck TS (2013) In vitro expansion and genetic modification of gastrointestinal stem cells as organoids. Nat Protoc 8 (12):2471–2482
11. Sato T et al (2009) Single Lgr5 stem cells build crypt villus structures in vitro without a mesenchymal niche. Nature 459:262
12. Willert K et al (2003) Wnt proteins are lipid-modified and can act as stem cell growth factors. Nature 423:448
13. Pinto D et al (2003) Canonical Wnt signals are essential for homeostasis of the intestinal epithelium. Genes Dev 17(14):1709–1713
14. Kim K-A et al (2005) Mitogenic influence of human R-Spondin1 on the intestinal epithelium. Science 309(5738):1256–1259
15. Haramis A-PG et al (2004) De novo crypt formation and juvenile polyposis on BMP inhibition in mouse intestine. Science 303 (5664):1684–1686
16. Miyoshi H et al (2012) Wnt5a potentiates TGF-β signaling to promote colonic crypt regeneration after tissue injury. Science (New York, NY) 338(6103):108–113
17. Sato T, Clevers H (2013) Primary mouse small intestinal epithelial cell cultures. In: Randell SH, Fulcher ML (eds) Epithelial cell culture protocols, 2nd edn. Humana Press, Totowa, NJ, pp 319–328
18. VanDussen KL et al (2015) Development of an enhanced human gastrointestinal epithelial

culture system to facilitate patient-based assays. Gut 64(6):911–920

19. Clevers H (2016) Modeling development and disease with organoids. Cell 165(7):1586–1597
20. Moon C et al (2013) Development of a primary mouse intestinal epithelial cell monolayer culture system to evaluate factors that modulate IgA transcytosis. Mucosal Immunol 7:818
21. Wang Y et al (2017) A microengineered collagen scaffold for generating a polarized crypt-villus architecture of human small intestinal epithelium. Biomaterials 128:44–55

Chapter 6

Isolation and Characterization of Colony-Forming Progenitor Cells from Adult Pancreas

Janine C. Quijano, Jacob R. Tremblay, Jeffrey Rawson, and Hsun Teresa Ku

Abstract

Obtaining, growing, and analysis of pancreatic progenitor cells.

Adult stem and progenitor cells have been successfully used for cell-based therapies such as transplantation of hematopoietic stem cells for various diseases. Whether stem and progenitor cells in the adult pancreas can be identified and used for replacement therapy has been a highly controversial topic. To address this controversy, our laboratory has developed in vitro colony assays to detect and characterize individual pancreatic stem and progenitor-like cells. We found that a subpopulation of ductal cells in the adult murine pancreas has the abilities to self-renew and differentiate into multiple pancreatic lineages in three-dimensional space in methylcellulose-containing semisolid media. This protocol details the techniques used for culturing and characterizing these pancreatic stem and progenitor-like cells, which we have named pancreatic colony-forming units (PCFUs), as well as their progenies (colonies). The techniques presented here include dissociation of pancreases, sorting antibody-stained cells with a fluorescence-activated cell sorter, viral transduction of dissociated pancreatic cells, growth of PCFUs in semi-solid media, whole-mount immunostaining and Western blot analysis for proteins expressed in colonies, and kidney capsule transplantation of colonies for in vivo functional analysis.

Key words Pancreas, Progenitor, Stem cells, Organoid, Pancreatic colony-forming units, Differentiation, Self-renewal

1 Introduction

The pancreas contains three major cell lineages: acinar, ductal, and endocrine cells. Acinar cells secrete digestive enzymes, which are transported by the ducts into the duodenum to aid food digestion. The endocrine lineage is comprised of several hormone-producing cells, including the glucagon-secreting alpha cells and the insulin-secreting beta cells that are required to maintain glucose homeostasis. During pancreatic development, the early ductal cells are tripotent progenitor cells, capable of giving rise to the three major

Mugdha V. Joglekar and Anandwardhan A. Hardikar (eds.), *Progenitor Cells: Methods and Protocols*, Methods in Molecular Biology, vol. 2029, https://doi.org/10.1007/978-1-4939-9631-5_6,

lineages in adult pancreas [1]; however, whether the adult pancreas still retains tripotent progenitor cells in vivo is controversial. While some studies have shown that adult cells in the ductal compartment can function like progenitor cells and give rise to beta-like cells [2–4], other studies are inconclusive or show negative results [5–7].

To address the controversy of adult pancreatic stem and progenitor cells, our laboratory established a unique analytical tool: in vitro colony assay [8–12]. Depending on the extracellular matrix, growth factors, and small molecules included, our colony assays can be used to measure self-renewal and differentiation at a single cell level [13]. Self-renewal and differentiation are two criteria necessary to define a stem cell.

To perform our in vitro colony assays, adult pancreases are dissociated into mostly single cells and cultured in a semisolid medium. The semisolid medium contains methylcellulose, which is a viscous and biologically inert molecule that, together with other signaling molecules, enables a single progenitor cell to grow and form a colony of cells in a three-dimensional (3D) space. Importantly, semisolid medium restricts the movement and aggregation of cells plated, thereby allowing the lineage potentials of single progenitor cells to be accurately reflected in the resulting individual colonies. A single progenitor cell that is capable of giving rise to a colony is termed "pancreatic colony forming unit (PCFU)."

When assayed in a Matrigel-containing colony assay, PCFUs comprised approximately 1% of the total pancreatic cells in adult mice [8]. Addition to cell culture of a Wnt pathway agonist, R-Spondin1 [14], resulted in approximately 500,000 fold net expansion of PCFUs over 11 weeks, demonstrating self-renewal ability of PCFUs [8]. Most of the individual colonies expressed three major pancreatic lineage markers, suggesting that most of the originating PCFUs are tri-potent [8]. Using a fluorescence-activated cell sorter, we also demonstrated that PCFUs were enriched in the $CD133^{high}CD71^{low}$ cells, a subpopulation of ductal cells [11]. Together, these results demonstrate that PCFUs are a minor population cells in the adult pancreas and that adult PCFUs display in vitro activities of self-renewal and tri-lineage differentiation.

In this chapter, we will present procedures for (1) dissociation of adult murine pancreases, (2) sorting of $CD133^{high}CD71^{low}$ ductal cells, (3) viral transduction of freshly sorted cells, (4) growth of PCFUs in our colony assay, (5) protein expression analysis of colonies using Western blot and whole-mount immunostaining, and (6) transplantation of colonies under the kidney capsule of mice to test the in vivo functions of colonies.

2 Materials

2.1 Pancreas Dissociation into Single Cells

Note: all reagents, supplies, and instruments need to be sterile.

1. Wash Solution: Dulbecco's phosphate buffered saline (PBS), 0.1% bovine serum albumin (BSA), penicillin and streptomycin (Pen-Strep). Store stock solution at 4 °C.
2. 100-mm petri dishes, tweezers, scissors, and spring scissors.
3. DNaseI working solution: PBS, 0.1% BSA, Pen-Strep, 2000 U/mL DNaseI.
4. Prepare stock collagenase B (100 mg/mL) at least 1 day prior to experiment. Store stock solution in 1 mL aliquots at −20 °C.
5. 16G ½ needle, 10-mL syringe and Falcon cell strainers with 100-μm nylon mesh and 40-μm nylon mesh.

2.2 Live Cell Staining with Antibodies

Note: all reagents, supplies and instruments need to be sterile.

1. Antibody Blocking Solution: 1 mg/mL of LEAF™ Purified anti-mouse CD16/32, clone 93.
2. Antibodies to be used are as follows: 0.5 mg/mL anti-CD133-biotin clone 13A4, 0.2 mg/mL anti-CD71-PE-Cy7 clone R17217, 0.5 mg/mL rat IgG1 kappa:biotin clone eBRG1 (CD133 isotype control), 0.2 mg/mL rat IgG2a kappa:PE-Cy7 clone RTK2758 (CD71 isotype control), and 0.2 mg/mL APC-Streptavidin (STV).
3. DAPI solution: PBS, 0.1% BSA, Pen-Strep, 2000 U/mL DNaseI, 500 ng/mL 4′,6-diamidino-2-phenylindole (DAPI).
4. Collection media: Dulbecco's Modified Eagle Medium: Nutrient Mixture F-12 (DMEM:F-12), Pen-Strep, 10% fetal calf serum (FCS). Prepare collection media the day of sorting and aliquot 1.2 mL into 5-mL polystyrene tubes. Make sure to coat the inner surface of the polystyrene tubes with Collection media to prevent cell adhesion after sorting.

2.3 FACS Sorting

Note: all reagents, supplies, and instruments need to be sterile.

1. Our laboratory routinely use a fluorescence-activated cell sorter equipped with an 80 μm nozzle to sort cells.
2. Trypan blue (0.02% dissolved in PBS), hemocytometer and light microscope.

2.4 Lentiviral Transduction of PCFUs

Note: all reagents, supplies, and instruments need to be sterile.

1. Low-binding flat-bottom 96-well plate.
2. 500 μg/mL polybrene.
3. Appropriate lentivirus carrying the gene of interest.

4. Recovery Solution: DMEM:F12, Pen-Strep, 5% (vol/vol) FCS, 10 mM nicotinamide, 10 ng/mL activin-B, 0.1 nM exendin-4, 1 ng/mL vascular endothelial growth factor-A, and 750 ng/mL R-spondin 1 (RSPO-1) (*see* **Note 1**).
5. 10 mg/mL puromycin.

2.5 In Vitro *Culture of PCFUs*

Note: all reagents, supplies, and instruments need to be sterile.

1. Prepare conditioned media from mouse embryonic stem cell-derived pancreatic-like cells in advance [15]. Briefly, generate 6-day-old embryoid bodies (EBs) in suspension with 15% FCS and decreasing doses of monothioglycerol (6 mM for 2 days followed by 0.6 mM for the next 4 days). Transfer approximately 100 EBs per well into 6-well plates in DMEM/F-12, 15% knockout serum replacement, 2 mM L-glutamine, 50 U/mL penicillin, and 50 μg/mL streptomycin. On culture day 13, replace media with DMEM/F-12, 15% knockout serum replacement, 2 mM L-glutamine, 50 U/mL penicillin, 50 μg/mL streptomycin, 10 mM nicotinamide, 0.1 nM exendin-4, and 10 ng/mL activin B. Collect media off of cells on culture day 16 and filter through a 0.2 μm polyethersulfone membrane. Store conditioned media in 5 mL aliquots at −80 °C.
2. Prepare stock methylcellulose (3.3% in DMEM:F-12) a week prior to culture, with 100 mL aliquots stored at −20 °C and a working stock stored at 4 °C [15].
3. Serum-containing Liquid Culture Media: DMEM:F-12, 50% (vol/vol) conditioned media from mESC-derived pancreatic-like cells, 5% (vol/vol) FCS, 10 mM nicotinamide, 10 ng/mL activin-B, 0.1 nM exendin-4, 1 ng/mL vascular endothelial growth factor-A, and 750 ng/mL RSPO-1 (*see* **Note 1**).
4. Serum-free Liquid Culture Media: DMEM:F-12, 10% KnockOut Serum Replacement, 25 ng/mL EGF, 100 ng/mL Noggin, 10 mM nicotinamide, 10 ng/mL activin-B, 0.1 nM exendin-4, 10 ng/mL vascular endothelial growth factor-A, and 750 ng/mL RSPO-1.
5. Matrigel.
6. 24-well Costar plate, ultralow attachment.

2.6 Whole-Mount Fixation of PCFU-Derived Colonies

1. Cell Recovery Solution (Corning).
2. 4% paraformaldehyde (PFA) stored in aliquots at −20 °C. Thaw on the day of fixation.
3. PBS-T: PBS, 0.1% Triton X-100.

2.7 Whole-Mount Immunofluorescent Staining of PCFU-Derived Colonies

1. 96-well black-walled plate with clear bottom.
2. Blocking buffer: PBS-T, 5% donkey serum or PBS-T, 5% goat serum. Use the same species for the block as the secondary antibody was raised in.

3. Plate shaker, set to low speed.
4. Appropriate primary and secondary antibodies.
5. DAPI/PBS-T solution: PBS-T, 5 μg/mL DAPI.
6. Glass-bottom microwell dish.
7. Coverslip: 22 × 22-mm, 1-oz.

2.8 Western Blot Analysis of PCFU-Derived Colonies

1. Sample buffer: Dissolve 3.8 g Glycerol, 1 g SDS, and 1.2 mg Bromophenol Blue in 7 mL of 4× Tris, pH 6.8. Add 500 μL BME. Bring volume to 10 mL with distilled water. Store 0.5 mL aliquots at −20 °C.
2. DMEM:F-12 and Cell Recovery Solution (Corning).
3. Radioimmunoprecipitation assay (RIPA) lysis buffer: Dilute one tablet of SIGMAFAST Protease Inhibitor in 10 mL dH_2O to make a 10× protease inhibitor solution. Dilute one tablet of PhosSTOP phosphatase inhibitor in 1 mL RIPA buffer to make a 10× phosphatase inhibitor solution. Dilute 100 μL of 10× protease inhibitor solution with 100 μL 10× phosphatase inhibitor solution and 800 μL RIPA buffer.
4. Heat block.
5. Western blot apparatus including buffer tank, Glass Plates, Extra Thick Blot Paper—Filter paper, Power Pac Basic, and plastic wrap.
6. Polyvinylidene difluoride (PVDF) membrane.
7. TBS-T: 0.01% Tween 20 in Tris-buffered saline (TBS).
8. Dry milk: 5% (w/v) in TBS-T.
9. Appropriate primary and secondary antibodies conjugated with horseradish peroxidase (HRP), and Clarity Western ECL Substrate.
10. X-ray film.

2.9 Generating Hyperglycemia by Inducing Beta-cell Damage with Streptozotocin (STZ) in Mice

1. NOD.CB17-*Prkdc*^{scid} (NOD-scid) mice: 8–10 weeks old.
2. 50 mM sodium citrate buffer, pH 4.5: Dissolve 0.735 g sodium citrate dihydrate in 45 mL distilled water. Use 3 N hydrochloric acid to adjust pH to 4.5. Adjust volume to 50 mL with distilled water. Filter-sterilize the solution and store at 4 °C.
3. Streptozotocin (STZ): diluted to 5 mg/mL in 50 mM sodium citrate buffer, pH 4.5 (*see* **Note 2**).
4. Lo-Dose U-100 insulin syringe: ½ cc, 28G ½ in.
5. HemoCue Glucose 201 with microcuvettes.

2.10 Collection of PCFU-Derived Colonies for Transplantation into Mice

1. Wash Solution: Dulbecco's phosphate buffered saline (PBS), 0.1% bovine serum albumin (BSA), penicillin and streptomycin (Pen-Strep). Store stock solution at 4 °C.
2. 0.125% Trypsin.
3. 25 μL glass syringe with a fixed 22G 2-in. needle and PE50 tubing with 0.023-in. inner diameter.

2.11 Kidney Capsule Transplantation of PCFU-Derived Colonies into STZ-Treated Mice

1. Isoflurane and O_2 gas.
2. Sterile eye lubricant, such as Akwa Tears.
3. Disposable under pad.
4. Battery operated electric clippers.
5. 70% isopropanol, Betadine topical microbicide solution (a 10% povidone–iodine solution) and sterile gauze.
6. Sterile surgical scissors, tweezers, and razor blade.
7. Sterile 1-mL syringe with 30G 1-in. needle filled with saline solution (0.9% sodium chloride).
8. 25 μL glass syringe with a fixed 22G 2-in. needle.
9. Absorbable, sterile surgical sutures with attached needle and suture wound clips, which will be used to close the abdominal muscle and skin, respectively.
10. Buprenorphine (or equivalent analgesic).
11. HemoCue Glucose 201 with microcuvettes.

3 Methods

A flowchart of various methods described in this section is provided in Fig. 1.

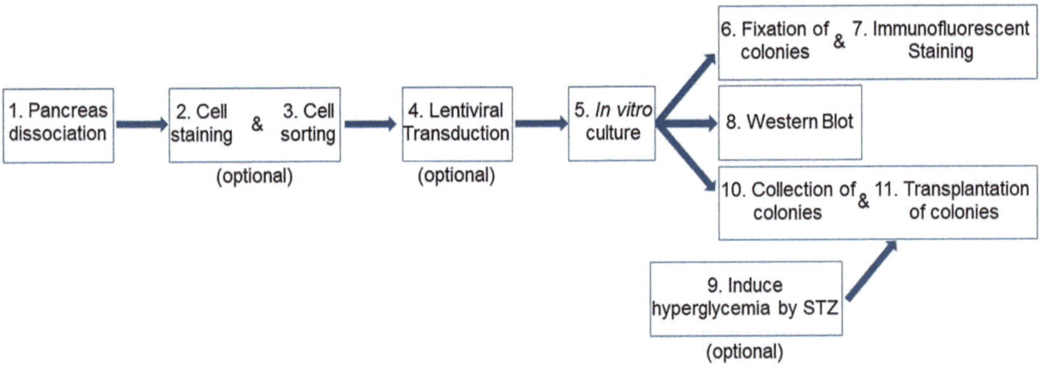

Fig. 1 Flow chart of the different procedures detailed within this chapter, with the numbers referring to the different subsections

3.1 Pancreas Dissociation into Single Cells

1. Euthanize five mice with CO_2 gas according to institutional approved protocol, such as an IACUC protocol (*see* **Note 3**).
2. Use a squeeze bottle to douse the abdomen of the mice with 70% isopropanol.
3. For each mouse, cut open the abdomen just below the rib cage. Remove the splenic and gastric lobes of the pancreas using scissors and forceps, and place the tissue into a 100-mm petri dish containing 10 mL cold Wash Solution.
4. Under a dissecting microscope, remove fat tissue from pancreases using a pair of fine-tipped forceps (*see* **Note 4**). Place pancreases into a new petri dish containing 10 mL cold Wash Solution.
5. In a laminar flow hood, wash pancreases twice with 10 mL cold Wash Solution. Using tweezers, move the tissue into the wash buffer in a petri dish and swirl to mix. After washes, remove excess liquid by dragging tissue along a dry sterile surface, for example the inside lid of petri dish. Place pancreases on a dry petri dish on ice and mince pancreases with sterile spring scissors for 3 min or until the tissue is evenly and finely minced. Transfer minced pancreatic tissue into a 50-mL conical tube containing 10 mL of DNaseI working solution.
6. Add collagenase B at 35 μL per mL of DNaseI working solution for a final concentration of 3.5 mg/mL. Place the 50-mL conical tube with tissues into a 37 °C water bath for 7 min. Invert conical tube every 3–4 min to mix cells.
7. Draw and expel cell solution seven times through a 16G ½ needle attached to a 10-mL syringe to hasten the dissociation of tissue. To do this, place the bevel of the needle against the wall of the conical tube, and expel the cells from the syringe with sufficient force. Place the 50-mL conical tube with cells into a 37 °C water bath for 7 min.
8. Draw and expel cell solution seven times through a 16G ½ needle attached to a 10-mL syringe to complete dissociation of tissue (*see* **Note 5**).
9. Add cold DNaseI working solution to bring the volume up to 25 mL. Spin down cells at 400 × *g* for 5 min at 4 °C. Remove supernatant and resuspend cells in 3–5 mL cold DNaseI working solution.
10. Filter solution though a 100-μm mesh followed by a 40-μm mesh to obtain mostly single cells.

3.2 Live Cell Staining with Antibodies

1. Block nonspecific antigens on cells with Antibody Blocking Solution by adding 10 μL per mL of cell suspension. Incubate cells on ice for 10 min, inverting the tube every 5 min.

2. Transfer 150 μL cell suspension into four different Eppendorf tubes for staining controls labeled "no staining," "1-color APC," "1-color PE-Cy7," and "Isotype," with the remainder of the cell suspension used for sorting (*see* **Note 6**).
3. Add primary antibody to appropriate tubes.
   (a) "1-color APC" control: add 0.75 μL anti-CD133-biotin.
   (b) "1-color PE-Cy7" control: add 1.5 μL anti-CD71-PE-Cy7.
   (c) "Isotype" control: add 0.9 μL CD133 isotype control and 1.5 μL CD71 isotype control.
   (d) "Sorting" sample: add anti-CD133-biotin at 5 μL per mL and anti-CD71-PE-Cy7 at 10 μL per mL.
4. Incubate cells for 20 min on ice, inverting the tube every 5 min.
5. Wash cells twice with DNaseI working solution. Resuspend "Isotype" control in 100 μL DNaseI working solution and "Sorting" sample in a 500 μL less volume of DNaseI working solution than the volume of cell suspension in **step 2**.
6. Add 6 μL/mL APC-STV to the "1-color APC control," "Isotype," and "Sorting" tubes. Incubate cells for 15 min on ice, inverting the tube every 5 min.
7. Wash cells twice with cold DNaseI working solution.
8. Resuspend cells in Eppendorf tubes of "no stain," "1-color APC," and "1-color PE-Cy7" with 500 μL of cold DNaseI working solution. Resuspend Eppendorf tube "Isotype" control in 500 μL of cold DAPI solution and "Sorting" sample in 6 mL of cold DAPI solution.
9. Filter cell solutions with a 40-μm mesh before sorting.

3.3 FACS Sorting

1. Acquire cell events from "no stain," "1-color APC," and "1-color PE-Cy7" controls. These data are used subsequently to set compensation of overlapping fluorescence signals.
2. Exclude cell debris by gating cells on forward and side scatter areas.
3. Exclude cell doublets by gating cells on forward and side scatter widths.
4. Exclude dead cells by gating out DAPI positive cells.
5. Use "Isotype" control to examine nonspecific fluorescence in PE-Cy7 and APC channels.
6. Set gates for sample collection, based on the isotype control.
7. Collect the sorted cell population of interest, that is, the $CD133^{high}CD71^{low}$ cells, into 5-mL polystyrene tubes containing collection buffer. Keep cells on ice.

8. Once sorting is completed, combine tubes that contain the same population of sorted cells and spin down at 400 × *g* for 5 min.
9. Remove supernatant and resuspend in a small volume (~100 μL) of Wash Solution. Take 10 μL of cell solution and count cells using trypan blue and a hemocytometer to determine the cell density [16].

3.4 Lentiviral Transduction of PCFUs

1. Aliquot 100,000 freshly sorted cells in a well of a low-binding flat-bottom 96-well plate using Recovery Solution or DMEM: F-12 containing 10% FCS. Incubate at 37 °C for 4 h (*see* **Note** 7).
2. Add 2 μg/mL polybrene to the cells. Add lentivirus at an MOI of 20. Bring total volume to 200 μL using Recovery Solution or DMEM:F-12 containing 10% FCS. Incubate at 37 °C overnight.
3. Add virally treated cells to 1 mL Recovery Solution or DMEM: F-12 containing 10% FCS. Spin cells at 400 × *g* for 5 min. Wash cells with Recovery Solution.
4. Plate cells at 2500 cells per 24-well in plating media (*see* Subheading 3.5). The next day add 1 μg/mL puromycin, or other appropriate selection molecule.

3.5 In Vitro *Culture of PCFUs*

1. Prepare Liquid Culture media for 2.5 mL total volume for each experimental group (*see* **Note 8**). Add cold Matrigel to Serum-containing or Serum-free Liquid Culture media at 5% (vol/vol) with a cold pipette tip. Add 1% (wt/vol) cold methylcellulose using a 1-mL syringe and a 16G ½ in. needle.
2. Add cells to culture media (*see* **Note 9**) and bring volume to 2.5 mL with DMEM:F-12, Pen-Strep. Mix vigorously by shaking tube for 1 min. Let tube sit in ice for approximately 5 min to allow air bubbles to clear.
3. Plate 500 μL per well of a 24-well Costar plate, using a 1-mL syringe and 16G ½-in. needle (*see* **Note 10**).
4. Incubate at 37 °C with 5% CO_2 for 2–3 weeks (*see* **Note 11**).
5. Colonies can be counted under a light microscope [15].

3.6 Whole-Mount Fixation of PCFU-Derived Colonies

1. Add 500 μL prewarmed Wash Solution on top of the semisolid medium to one well of colonies. Using a p1000, pipet up and down to disrupt the semisolid medium, and transfer liquid to a 15-mL conical tube. Pipet slowly to be careful and avoid creating air bubbles. Wash out the well with another 500 μL Wash Solution, and combine Wash Solution to the 15-mL conical tube. Combine 2 wells from the same group. Centrifuge colonies at 400 × *g* for 5 min.

2. Remove supernatant and resuspend colonies in 1 mL warm Wash Solution. Transfer colony solution to an Eppendorf tube. Centrifuge colonies at 400 × *g* for 5 min.
3. Remove supernatant and resuspend colonies in 500 μL cold Cell Recovery Solution. Incubate at 4 °C for 30 min, spinning tubes in the rack every 10 min (*see* **Note 12**). Allow colonies to settle to the bottom of the tube (*see* **Note 13**).
4. Remove supernatant and add 1 mL cold Wash Solution to colonies. Incubate for 5 min, without disturbing tube.
5. Remove supernatant and add 1 mL of 4% PFA. Incubate overnight at 4 °C (*see* **Note 14**).
6. Remove PFA and wash once with 1 mL PBS-T (*see* **Note 15**).
7. Fixed colonies can be stored at 4 °C for 1 week.

3.7 Whole-Mount Immunofluorescent Staining of PCFU-Derived Colonies

1. Transfer 100 μL of fixed colonies to one well of a 96-well black-walled plate with clear bottom. Incubate at room temperature for at least 5 min to allow for colonies to settle on the bottom of the well.
2. Remove PBS-T without disturbing the colonies (*see* **Note 16**). Add 200 μL of Blocking buffer to each well. On a plate shaker, incubate overnight at 4 °C or 2 h at room temperature.
3. Prepare primary antibody of interest in Blocking buffer.
4. Remove Blocking buffer from wells with colonies and add 200 μL of primary antibody of interest. Incubate overnight at 4 °C on a plate shaker.
5. Wash colonies three times with PBS-T, with a 10-min incubation between washes (*see* **Note 17**).
6. Prepare the secondary antibody in Blocking buffer, in a room with dim lighting (*see* **Note 18**).
7. Remove Wash buffer and add 200 μL of secondary antibody solution. On a plate shaker, incubate for 2 h at room temperature or overnight at 4 °C.
8. Wash colonies three times with PBS-T, with a 10-min incubation between washes (*see* **Note 17**).
9. Remove PBS-T from wells with colonies and add 200 μL DAPI/PBS-T solution (*see* **Note 19**).
10. Transfer 30 μL of colonies in DAPI/PBS-T solution onto a MatTek petri dish. Place a coverslip onto the MatTek petri dish, avoiding the creation of air bubbles under the coverslip (*see* **Note 20**).
11. Image colonies with a confocal laser scanning microscope.

3.8 Western Blot Analysis of PCFU-Derived Colonies

1. Use a p10 pipette tip to hand-pick every colony and place the colonies into an Eppendorf tube with 500 μL of DMEM:F-12 (*see* **Note 21**). Pellet colony solution at 400 × *g* for 5 min and wash twice with warm Wash Solution.
2. Using pippet tips rinsed with Wash Solution (*see* **Note 13**), resuspend colonies in 1 mL cold Cell Recovery Solution and incubate at 4 °C for 30 min, mixing the tube every 10 min (*see* **Notes 12** and **13**). Pellet colony solution at 400 × *g* for 5 min.
3. Aspirate supernatant and add RIPA Lysis buffer to pelleted colonies (*see* **Note 22**). Incubate on ice for 15 min, vortexing every 5 min.
4. Centrifuge lysate at 17,000 × *g* for 15 min at 4 °C. Save supernatant in a separate Eppendorf tube, and discard pellet.
5. Calculate protein concentration (*see* **Note 23**) and store protein samples at −80 °C until ready for processing.
6. Dilute Sample buffer into protein samples to a 1× concentration. Boil samples on a heat block at >95 °C for 15 min.
7. Add 10 μg protein per well on a SDS-PAGE gel and run gel at 70 V for 2 h (*see* **Note 24**).
8. Transfer protein onto PVDF membrane at 25 V for 30 min (*see* **Note 25**).
9. Block membrane in 5% milk for 2 h at room temperature on a rocker.
10. Dilute primary antibody in 5% milk and add to membrane. Incubate at 4 °C overnight on a rocker.
11. Wash three times in TBS-T for 5 min, on a rocker. Dilute secondary antibody in 5% milk and add to membrane. Incubate secondary antibody for 1 h at room temperature on a rocker.
12. Wash three times in TBS-T for 10 min, on a rocker.
13. Mix ECL components 1:1 and add to membrane. Incubate for 1 min. Drain ECL solution, and let extra liquid drip off (*see* **Note 26**).
14. Cover membrane with a clear plastic wrap and expose membrane in a dark room, using X-ray film.

3.9 Generating Hyperglycemia by Inducing Beta-Cell Damage with Streptozotocin (STZ) in Mice

1. Weigh STZ powder and calculate the volume of the sodium citrate buffer needed to make a 5 mg/mL solution. Keep the powder and buffer separately in two 5-mL polystyrene tubes.
2. Pour sodium citrate buffer into the tube containing STZ powder and shake vigorously to dissolve STZ (*see* **Note 2**).
3. With an insulin syringe, inject 50 mg/kg body weight STZ intraperitoneally into a NOD-SCID mouse.

4. Repeat **steps 1–3** once daily for 3 consecutive days for each mouse.
5. Two weeks after the last STZ injection, collect 4 μL blood from a tail incision to measure the concentration of blood glucose using HemoCue Glucose 201 system (*see* **Note 27**).
6. Only those mice that show consistent hyperglycemia (blood glucose above 250 mg/dL) for consecutive 3 days will be used for transplantation experiments (*see* **Note 28**).

3.10 Collection of PCFU-Derived Colonies for Transplantation into Mice

Note: Transplant adult PCFU-derived cells underneath the renal capsule at least 2 weeks after the last dose of STZ injection to ensure that hyperglycemia has been achieved.

1. Collect 3-week-old colonies for transplantation by adding 500 μL warm Wash Solution on top of semisolid medium to each well of a 24-well plate. Pipet up and down and pool cells from four wells into one 15-mL conical tube. Use 1 mL warm Wash Solution to wash out wells. Avoid generating bubbles during colony collection. Combine Wash Solution from four wells to the colony mixture in the 15-mL conical tube.
2. Spin colonies down at 400 × *g* for 5 min and aspirate supernatant without disturbing the pellet. Using a p200 pipette, remove the top half of the pellet (*see* **Note 29**).
3. Wash the cell colonies twice with 10 mL of warm Wash Solution.
4. After the last wash add 500 μL of warm Wash Solution and equal volume (500 μL) of prewarmed trypsin to reach a final concentration at 0.0625%. Incubate for 3 min at 37 °C.
5. Mix cell solution up and down twice with a p1000 pipette. Add 1 mL FCS to stop the trypsin reaction and spin cell solution at 400 × *g* for 5 min (*see* **Note 30**).
6. Wash twice with 10 mL warm Wash Solution. After the last wash, combine all cells into one tube and bring the volume of the cell solution to 10 mL with warm Wash Solution. Spin cells at 400 × *g* for 5 min. Resuspend cells in 2 mL of warm Wash Solution.
7. Determine cell concentration by adding 5 μL cell solution into 45 μL trypan blue. Add 10 μL of the above solution to a hemocytometer, count cell number (including all cells in a small cluster) and calculate cell density [16]. Aliquot 2×10^6 cells per Eppendorf tube and pellet cells at 400 × *g* for 5 min. Each aliquot of cells will be used for a graft in one mouse.
8. Remove as much supernatant as possible without disturbing the pellet (*see* **Note 31**).

3.11 Kidney Subcapsule Transplantation of PCFU-derived Colonies into STZ-treated Mice

1. Attach a 23G ½ in. needle to a 500 μL syringe and slide one end of an 8-in. piece of sterile PE50 tubing onto the needle. Draw the cell solution from one Eppendorf tube into the PE50 tubing. Bend the PE50 tubing in half and place in a 15-mL conical tube. Repeat for each aliquot of cells in an Eppendorf tube. Spin the 15-mL conical tube(s) at 400 × *g* for 5 min to "pellet" the colonies to the middle of the PE50 tubing (*see* **Note 32**).
2. Anesthetize diabetic NOD-SCID mice with constant 2% isoflurane gas mixed with 1.5% O_2.
3. Cover the eyes of the mice with an eye lubricant. Place mice on a clean disposable under pad. Clip hair ~3 cm in diameter on the left mid-dorsal skin with electric clippers.
4. After shaving the mouse, remove the loose hair by first dousing the mouse with 70% isopropanol, followed by wiping off the isopropanol with a sterile gauze. Repeat this one more time. Apply Betadine onto the shaved area and wipe off with a sterile gauze. Finally, clean the Betadine with 70% isopropanol and wipe off with a clean sterile gauze.
5. Move the mouse to a second clean area with a new clean disposable under pad. Change gloves to decrease the chances of contamination of the site. Clean the shaved area again with Betadine followed by 70% isopropanol.
6. Position the mouse in right lateral recumbency and cover the mouse with sterile gauze that frames the abdomen. Make an incision through the skin, perpendicular to the spine, just below the rib cage. Identify the location of the kidney and make a second vertical incision in the abdominal wall.
7. Using sterile tweezers, expose the left kidney so that it sits on the abdominal wall. With a 30G 1 in. needle, carefully open a small hole on renal capsule, 1–2 mm wide, on the anterior side of the kidney (*see* **Note 33**).
8. Attach one opening of the PE50 tubing, containing cells/colonies, to a Hamilton syringe and cut the other end of the tube, with a sterile razor blade, at the point where the cells are pelleted (*see* **Note 33**).
9. Insert the other opening of the PE50 tubing under the renal capsule, and slowly moving the tube around to create space while injecting the cells (*see* **Note 33**).
10. Use the side of the 30 G needle to carefully arrange the cells in as much of a single layer as possible, pushing the cells toward the posterior side of the kidney and away from the hole in the renal capsule (*see* **Note 34**).
11. Gently guide the kidney back inside the body by opening the abdominal wall with tweezers, taking care not to damage the transplanted graft.

12. Close the abdominal wall with absorbable surgical sutures, followed by closing the skin with suture wound clips.
13. Inject mice subcutaneously with 0.05 mg/kg body weight of buprenorphine or other required analgesic, and allow mice to recover under a warm light (*see* **Note 35**). Once the mouse is able to move around the cage, transfer it to the usual housing facility.
14. Repeat **steps 2–13** for each mouse that will be transplanted with a graft.
15. Monitor blood sugar levels every 3–4 days following transplantation for the first 4 weeks, and then once every week thereafter until the end of week 28. Perform tail bleeds and measure the blood glucose levels with the HemoCue 201 (*see* **Notes 27** and **28**).
16. Remove the suture wound clips 10–14 days posttransplantation.

4 Notes

1. Stock solutions of all growth factors are reconstituted to vendor specifications. Store aliquoted stock solutions at −80 °C. Thawed growth factors are used within a week when stored in 4 °C.
2. STZ in solution is unstable and degrades quickly. To maximize effectiveness, keep STZ powder and sodium citrate buffer separately and store on ice. Combine STZ and sodium citrate buffer, and use within 10 min to minimize STZ degradation.
3. Dissect the mice as soon as possible after euthanasia. This is important to avoid autodigestion of the acinar tissue due to postmortem changes.
4. Based on empirical experience, it is critical to remove excess fat tissue to ensure the viability of PCFUs in the subsequent steps.
5. View 10 μL of the solution under a microscope to determine if the cells are mainly single cells or large clusters of 10–50 cells. If there are still large clusters, repeat previous steps at 37 °C with periodic inversion and drawing up and then expelling the cells in a syringe. Maximum time in collagenase B solution should be 30 min to prevent over-digestion.
6. The single-positive color controls are necessary for compensation to correct for the overlapping of fluorescent emissions.
7. The 4 h of recovery is necessary for the cells to recover from the stress of dissociation and sorting prior to transduction. If necessary, a shorter interval of 3 h is possible, but suboptimal.

8. Either serum-containing liquid culture media or serum-free liquid culture media can be used with murine cells. While preparing culture media, keep all reagents on ice to prevent Matrigel from solidifying.
9. We usually plate unsorted cells at 10,000 cells per well of a 24-well plate to generate colonies that originate from single cells. Sorted $CD133^{high}CD71^{low}$ cells are plated at 1000 cells per well.
10. Avoid air bubbles if plating a fourth well in a group by angling the tube so that the liquid gathers in one corner at the bottom edge of the tube. Then, place the bevel of the needle against the tube and draw up the cell solution.
11. Colony density should be kept low, about 150–200 colonies per well. With such number of colonies, media changes are not necessary during the 3-week culture period. Also, we typically fill the far left and far right columns with sterile distilled water, to minimize evaporation from the cell culture due to the long incubation time.
12. Minimal movement is recommended to keep the colonies from sticking to plastic. Flicking the top of the tube with a finger while it sits in an Eppendorf rack will cause the tube to spin in place, creating a vortex and effectively mixing the solution.
13. The Cell Recovery Solution should remove excess Matrigel from the colonies. Once the Matrigel is removed, the colonies are fragile and will disintegrate into single cells easily. If the maintenance of the colony structure is important, do not centrifuge cells. Alternatively, allow gravity to pull colonies to the bottom of the tube. Typically 5 min is enough time for the colonies to settle to the bottom of the tube; if the colonies are small, then allow at least 10 min. Additionally, because colonies stick to plastics easily, always coat plastic surfaces (tubes and pipet tips) that come in contact with colonies with a buffer that contains proteins, such as the Wash Solution.
14. Colonies may appear to be clumped together. Do not try to separate them by pipetting or shaking because the colonies will stick to the tube and pipette tip.
15. The addition of Triton X-100 is to prevent colonies from sticking to plastic. Use a pipette to mix the colonies and break apart clumps. Be careful not to pipet too much or the colonies will disintegrate into single cells.
16. When removing liquid from a well, place the plate on a light microscope to visually confirm that colonies are not being disturbed. Create two bends in the pipette tip to create a lightly curved end to enable liquid manipulation while watching colonies simultaneously.

17. For each wash, remove as much liquid from the well and add 200 μL of PBS-T. This step is important to reduce background staining.
18. Most fluorescent molecules are sensitive to light and should be protected from light to avoid photobleaching.
19. If background from DAPI is too high when imaging with a fluorescent microscope, subsequent staining procedures should remove DAPI/PBS-T solution and resuspend in PBS-T before mounting.
20. Solution mounted on petri dishes may evaporate. Evaporation can be minimized by sealing the opening of petri dish with paraffin. Alternatively, 50 μL of colonies resuspended in 70% glycerol can be added onto a glass slide, covered by a 22 × 22-mm coverslip, and sealed with nail polish. However, the 3D architecture of the colony may be lost due to pressure on the colonies from the coverslip.
21. Colonies collected from a total of 16 wells of 24-well plate will typically yield enough proteins for multiple Western blot analyses.
22. To keep protein concentration from being too dilute, use the RIPA lysis buffer at 5×10^6 cells per mL of solution.
23. A BSA Protein Concentration Assay can be utilized for this procedure.
24. Using a 10% polyacrylamide gel and 70 V for 2 h is enough to separate large and small bands without the samples running off the gel.
25. Using a semidry transfer apparatus from Bio-Rad, the assembled blot sandwich will efficiently transfer in this amount of time.
26. Let liquid run off of the edge of the membrane onto Kimwipes paper. Once almost all of the liquid has gone, it is possible to dab the surface to further remove liquid using a Kimwipes paper. Insufficient liquid removal will result in high background and subpar image acquisition.
27. Keep the container of HemoCue microcuvettes on ice until use.
28. If blood glucose is above the detection limit of HemoCue 201, dilute blood before reading. For example, use 2 μL of blood with 5 μL of saline solution (Sigma) at a 3.5-fold dilution. Subsequently, multiply the HemoCue 201 number by 3.5 to obtain the amount of blood glucose (mg/dL).
29. Cell colonies will mostly be at the bottom of the Matrigel/methylcellulose pellet. If colonies are large with diameter >600 mm, then it is possible to visually inspect the pellet to

determine where the majority of colonies are. These washes are intended to remove as much of the matrix as possible, with minimal loss of colonies.

30. The majority of Matrigel and methylcellulose should not pellet after trypsinization. The goal here is to obtain smaller cell clusters and single cells that are free of Matrigel and methylcellulose. The next few washes should remove any remaining Matrigel and methylcellulose. If Matrigel and methylcellulose persist, repeat steps from trypsin digestion to wash.
31. The total volume of the cell solution should not exceed 30 μL.
32. Keep the cell solution in PE50 tubing at room temperature until they are transplanted into the kidney capsule.
33. Use either a syringe or sterile swabs to periodically keep the kidney capsule wet with saline solution when the kidney is outside the abdominal cavity.
34. Because the hole in the renal capsule is not sealed, pushing the cells away from the opening will minimize cell loss.
35. A heating pad placed under the cage is another way to warm mice during recovery.

Acknowledgments

J.C.Q. is supported by a Juvenile Diabetes Research Foundation (JDRF) Postdoctoral Fellowship 3-PDF-2016-174-A-N. This work is funded in part by a National Institutes of Health (NIH) Grant R01DK099734 to H.T.K.

References

1. Shih HP, Wang A, Sander M (2013) Pancreas organogenesis: from lineage determination to morphogenesis. Annu Rev Cell Dev Biol 29:81–105. https://doi.org/10.1146/annurev-cellbio-101512-122405
2. Xu X, D'Hoker J, Stange G, Bonne S, De Leu N, Xiao X, Van de Casteele M, Mellitzer G, Ling Z, Pipeleers D, Bouwens L, Scharfmann R, Gradwohl G, Heimberg H (2008) Beta cells can be generated from endogenous progenitors in injured adult mouse pancreas. Cell 132(2):197–207. https://doi.org/10.1016/j.cell.2007.12.015
3. Inada A, Nienaber C, Katsuta H, Fujitani Y, Levine J, Morita R, Sharma A, Bonner-Weir S (2008) Carbonic anhydrase II-positive pancreatic cells are progenitors for both endocrine and exocrine pancreas after birth. Proc Natl Acad Sci U S A 105(50):19915–19919. https://doi.org/10.1073/pnas.0805803105
4. Zhang M, Lin Q, Qi T, Wang T, Chen CC, Riggs AD, Zeng D (2016) Growth factors and medium hyperglycemia induce Sox9+ ductal cell differentiation into beta cells in mice with reversal of diabetes. Proc Natl Acad Sci U S A 113(3):650–655. https://doi.org/10.1073/pnas.1524200113
5. Solar M, Cardalda C, Houbracken I, Martin M, Maestro MA, De Medts N, Xu X, Grau V, Heimberg H, Bouwens L, Ferrer J (2009) Pancreatic exocrine duct cells give rise to insulin-producing beta cells during embryogenesis but not after birth. Dev Cell 17(6):849–860. https://doi.org/10.1016/j.devcel.2009.11.003
6. Kopp JL, Dubois CL, Schaffer AE, Hao E, Shih HP, Seymour PA, Ma J, Sander M (2011) Sox9+

ductal cells are multipotent progenitors throughout development but do not produce new endocrine cells in the normal or injured adult pancreas. Development 138(4):653–665. https://doi.org/10.1242/dev.056499

7. Furuyama K, Kawaguchi Y, Akiyama H, Horiguchi M, Kodama S, Kuhara T, Hosokawa S, Elbahrawy A, Soeda T, Koizumi M, Masui T, Kawaguchi M, Takaori K, Doi R, Nishi E, Kakinoki R, Deng JM, Behringer RR, Nakamura T, Uemoto S (2011) Continuous cell supply from a Sox9-expressing progenitor zone in adult liver, exocrine pancreas and intestine. Nat Genet 43 (1):34–41. https://doi.org/10.1038/ng.722
8. Jin L, Feng T, Shih HP, Zerda R, Luo A, Hsu J, Mahdavi A, Sander M, Tirrell DA, Riggs AD, Ku HT (2013) Colony-forming cells in the adult mouse pancreas are expandable in Matrigel and form endocrine/acinar colonies in laminin hydrogel. Proc Natl Acad Sci U S A 110 (10):3907–3912. https://doi.org/10.1073/pnas.1301889110
9. Jin L, Feng T, Zerda R, Chen CC, Riggs AD, Ku HT (2014) In vitro multilineage differentiation and self-renewal of single pancreatic colony-forming cells from adult C57BL/6 mice. Stem Cells Dev 23(8):899–909. https://doi.org/10.1089/scd.2013.0466
10. Ghazalli N, Mahdavi A, Feng T, Jin L, Kozlowski MT, Hsu J, Riggs AD, Tirrell DA, Ku HT (2015) Postnatal pancreas of mice contains tripotent progenitors capable of giving rise to duct, acinar, and endocrine cells in vitro. Stem Cells Dev. https://doi.org/10.1089/scd.2015.0007
11. Jin L, Gao D, Feng T, Tremblay JR, Ghazalli N, Luo A, Rawson J, Quijano JC, Chai J, Wedeken L, Hsu J, LeBon J, Walker S, Shih HP, Mahdavi A, Tirrell DA, Riggs AD, Ku HT (2016) Cells with surface expression of CD133highCD71low are enriched for tripotent colony-forming progenitor cells in the adult murine pancreas. Stem Cell Res 16 (1):40–53. https://doi.org/10.1016/j.scr.2015.11.015
12. Wedeken L, Luo A, Tremblay JR, Rawson J, Jin L, Gao D, Quijano J, Ku HT (2017) Adult murine pancreatic progenitors require epidermal growth factor and nicotinamide for self-renewal and differentiation in a serum- and conditioned medium-free culture. Stem Cells Dev 26(8):599–607. https://doi.org/10.1089/scd.2016.0328
13. Tremblay JR, LeBon JM, Luo A, Quijano JC, Wedeken L, Jou K, Riggs AD, Tirrell DA, Ku HT (2016) In vitro colony assays for characterizing tri-potent progenitor cells isolated from the adult murine pancreas. J Vis Exp. https://doi.org/10.3791/54016
14. Binnerts ME, Kim KA, Bright JM, Patel SM, Tran K, Zhou M, Leung JM, Liu Y, Lomas WE 3rd, Dixon M, Hazell SA, Wagle M, Nie WS, Tomasevic N, Williams J, Zhan X, Levy MD, Funk WD, Abo A (2007) R-Spondin1 regulates Wnt signaling by inhibiting internalization of LRP6. Proc Natl Acad Sci U S A 104 (37):14700–14705. https://doi.org/10.1073/pnas.0702305104
15. Winkler M, Trieu N, Feng T, Jin L, Walker S, Singh L, Ku HT (2011) A quantitative assay for insulin-expressing colony-forming progenitors. J Vis Exp 57:e3148. https://doi.org/10.3791/3148
16. Marchenko S, Flanagan L (2007) Counting human neural stem cells. J Vis Exp 7:262. https://doi.org/10.3791/262

Chapter 7

A Novel Gene Delivery Approach Using Metal Organic Frameworks in Human Islet-Derived Progenitor Cells

Arpita Poddar, Mugdha V. Joglekar, Anandwardhan A. Hardikar, and Ravi Shukla

Abstract

The ability to regenerate insulin-producing β cells is the ultimate goal for treatment of type 1 diabetes. Several sources of stem cells have been investigated by studying their differential potential to form insulin-producing β cells that can be used for replacement therapy. Progenitor cells derived from human islets that are lineage committed have been shown to be better alternatives with regard to their differentiation capabilities for the generation of insulin-producing β-like cells. Controlling the differentiation of progenitor cells is a vital approach in exploiting cellular expansion, mesenchymal transition and β-cell generation. One of the most powerful and useful methods involve the intracellular delivery of biomolecules like genes, miRNAs, siRNAs, proteins, and peptides. However, the delivery vehicle used for such approaches is the most significant factor that determines the in vivo efficacy. Current delivery systems, although promising, are deterred by issues like toxicity, sustained release, loading capacity, and cost-effectiveness. In this chapter, we show an alternative nanomaterial called metal organic frameworks (MOFs) as gene delivery systems in human islet-derived progenitor cells (hIPCs). Based on our results, we believe that nanoscale MOFs can function as controlled cellular delivery agents that deliver, protect, and maintain functional activity of genes or other bioactive molecules into the cytoplasm or nucleus of progenitor cells. Here, we describe the details for the synthesis, characterization, and transfection of selected, biocompatible MOFs in hIPCs.

Key word hIPCs, Gene delivery, Metal organic frameworks

1 Introduction

The burgeoning field of nanomedicine hold promise in stem cell biology and has the potential to be an invaluable tool in exploiting unique approaches to stem cell expansion, differentiation, and transplantation [1]. Several powerful methods for the controlled differentiation of progenitor cells involve the intracellular delivery of bioentities like genes, miRNAs, siRNAs, proteins, peptides, or other small molecules [2–4]. Nanoparticles, quantum dots, carbon nanotubes, nanofibers, and nanosubstrates are some of the most commonly explored nanotechnological delivery agents in stem cell

Mugdha V. Joglekar and Anandwardhan A. Hardikar (eds.), *Progenitor Cells: Methods and Protocols*, Methods in Molecular Biology, vol. 2029, https://doi.org/10.1007/978-1-4939-9631-5_7, © Springer Science+Business Media, LLC, part of Springer Nature 2019

biology, yet sustained release, loading capacity, cost-effectiveness, and cytotoxicity continue to be large obstacles hindering proper advancement [5–8]. Moreover, most nanoparticle methods generally require the use of coprecipitating agents or utilize postsynthesis modification to generate the desired composites of particles or bioentities for transfection [9]. Here, we explore the advent of an emerging and highly promising nanoscale system called metal organic frameworks (MOFs) and our results demonstrate their potential as delivery agents for transfecting human islet derived progenitor cells (hIPCs).

In the last few decades, porous materials have garnered increasing attention due to considerable features which endow them with exceptional properties. Prominent among them are MOFs which are porous hybrid solid composites built from metal ions anchored to organic bridging ligands resulting in a well-defined coordination geometry [10]. This coordination structure gives rise to a multidimensional open framework with inherent stability, porosity, high surface area, pore volume, and multiple adsorption sites [11]. These features have proven them to be promising platforms for biological and biomedical applications, giving rise to a biomaterial system with high loading capacity and versatile functionality coupled with biocompatibility [12, 13]. The facile synthesis of MOFs with different compositions of metals and linkers allows for immense diversity in structure, shapes, sizes, and chemical properties. They are highly tunable in nature, can be synthesized under mild synthetic conditions and functionalise their pore structure to enable specific targeting. MOFs can store simulaneously hydrophilic, hydrophobic and amphiphilic entities and can be adapted to the physicochemical properties of each biomolecule or drug and its medical application [13]. Furthermore, they are biodegradable due to the presence of relatively labile metal ligand bonds; this feature makes it possible to rapidly degrade the composite material and release the cargo.

MOFs have been widely used for drug and siRNA delivery as well as for encapsulation of proteins, nucleic acids, and prokaryotic and eukaryotic cells [14–18]. They have been effectively used for delivering cytotoxic drugs and siRNAs to cancer cells [17]. While MOFs have been used as carriers for therapeutic moieties, the MOF structure itself can act as imaging agent [19]. This has huge potential in a new stage of molecular medicine which requires the association of therapeutics and diagnostics in a single theranostic agent.

Zeoliticimidazolate frameworks (ZIFs) are a class of MOFs that have attracted much attention as ideal drug delivery nanocarriers in recent years. ZIFs are one of the most promising MOFs and when synthesized at the nanoscale level serve as excellent vehicles with ultrahigh loading capacity and controlled release profile of the cargo material [20]. ZIF is built with zinc as the metal ion and imidazole as the organic linker [21]. The presence of zinc makes it one of the more promising MOFs to investigate for the treatment of diabetes. Zinc containing compounds have shown promotion of insulin signal transduction and decreased cytokine production, resulting in an insulin-like effect by inducing hypoglycemic properties [22]. This antidiabetic effect has recently been explored by Briones et al. where a novel zinc based MOF was synthesized which displayed potential in vivo antidiabetic activity as well as low in vitro cell toxicity [23]. In our work, we utilize a recent synthesis technique called biomimetic mineralization where ZIF-8 is synthesized to encapsulate and protect biomacromolecules in a process that mimics the natural self-assembly process of biomineralization [15]. For the first time, we show that biomimetically mineralised ZIF-8 can be used to encapsulate entire gene sets and cause transfection in hIPCs. This simple entrapment method is effective for intracellular delivery where progressive release over sustained time periods is essential.

We herein demonstrate a MOF-based gene delivery to hIPCs using a green fluorescent protein (GFP) plasmid representing a complete gene set. To synthesize the GFP@ZIF-8 MOF system, a biomimetically mineralized process is carried out in aqueos media that takes around 1 h. Following characterisation using small angle X-ray scattering (SAXS) and Fourier transformed infrared spectroscopy (FTIR) to confirm structure, scanning electron microscopy (SEM) to confirm morphology, and agarose gel elctrophoresis to confirm the presence of DNA, GFP@ZIF-8s are transfected into hIPCs. This method of transfection involves gradual release as opposed to initial bursts as commonly seen in cases of lipid-based transfection reagents. Sustained release is higly beneficial where therapeutic efficay of the cargo is maintained over prolonged time points. Further harnessing of these promising results and the potential of MOFs by developing a controlled cellular delivery agent that delivers, protects, and maintains functional activity of bioactive molecules into the cytoplasm or nucleus of progenitor cells is essential.

2 Materials

2.1 Cell Culture

1. L-15 (Leibovitz) medium.
2. Dulbecco's Modified Eagle Medium (DMEM).
3. CMRL-1066.
4. Fetal calf serum (FCS).
5. GlutaMAX™-1at final concentration of 2 mM.
6. Antibiotic solution: penicillin–streptomycin (pen-strep). 50 U/mL penicillin and 50 μg/mL streptomycin as the final working concentration in medium.

7. Trypsin–EDTA: 500 mL of trypsin–EDTA contains 4 g NaCl, 0.2 g KCl, 0.5 g D-glucose, 0.18 g $NaHCO_3$, 0.03 of KH_2PO_4, 0.05 g of Na_2HPO_4, 0.19 g EDTA, and 1.25 g bovine pancreatic trypsin. Volume is made up with Milli-Q water, and the solution is sterile filtered through 0.2-μm filters.
8. Human epithelial growth factor (hEGF): 100 ng/μL.
9. CMRL serum-containing medium for hIPCs culture: 450 mL CMRL-1066, 50 mL of FCS, 5 mL of GlutaMAX, 5 mL of pen-strep, and 50 μL of hEGF.

2.2 GFP@ZIF-8

1. Milli-Q water, type 1, ultrapure water 18.2 MΩ•cm at 25 °C.
2. UltraPure™ DNase/RNase-free distilled water.
3. Ethanol, ACS grade (>99.5%).
4. Zinc nitrate hexahydrate, ACS grade ≥98% ((Zn(NO3) 2.6H2O), 2-Methylimidazole (MeIM).
5. pCDNA5frt-EGFP-N1 6549bp plasmid.
6. 40 mM of $Zn(NO_3)_2.6H_2O$: Dissolve 11.9 mg in 1 mL ultrapure DNase/RNase-free water. Prepare fresh before use.
7. 160 mM of MeIM: Dissolve 13.15 mg in 1 mL ultrapure DNase/RNase-free water. This is stable at room temperature and can be stored for at least 3 months.

2.3 Characterization

1. 1×TAE (Tris–Acetic acid–EDTA buffer): Stock of 50×TAE is made by dissolving 242 g Tris base, 57.1 mL glacial acetic acid, and 100 mL 0.5 M EDTA (pH 8.0) solution, and bringing the final volume up to 1 L (*see* **Note 1**).
2. 1% (w/v) agarose gel: Add 1 g agarose in 100 mL 1×TAE buffer and heating till solution is completely clear.

2.4 Transfection

1. Opti-MEM I (1×) Reduced Serum Medium.
2. Lipofectamine™ 3000 Transfection Reagent.
3. Antibiotic-free media: 450 mL CMRL-1066, 50 mL of FCS, 5 mL of GlutaMAX, and 50 μL of hEGF.
4. Tissue culture-treated multiple 24-well plates.

3 Methods

3.1 Synthesis of GFP@ZIF-8

3.1.1 DNA Selection

1. Select a standard GFP plasmid vector such as pCDNA5frt-EGFP-N1 (CAT) with a reporter gene (chloramphenicol acetyltransferase; CAT) (*see* **Note 2**) (Fig. 1).
2. Dissolve plasmid DNA in ultrapure DNase/RNase-free water (*see* **Note 3**). It is recommended to prepare 500 ng/μL–1 μg/μL concentration of plasmid DNA in ultrapure DNase/RNase-free water.

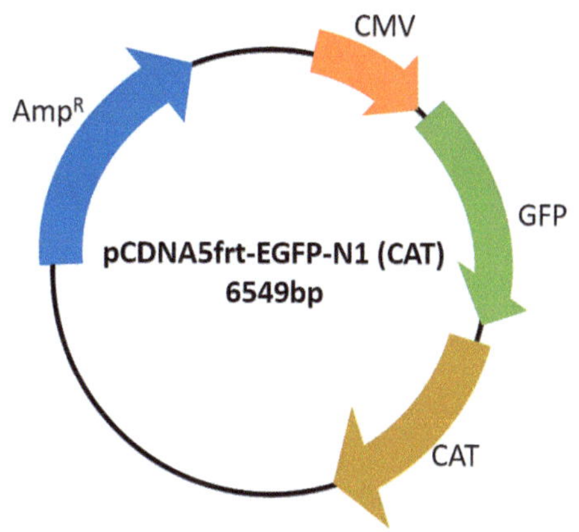

Fig. 1 A map of pCDNA plasmid with enhanced GFP (EGFP) and CMV promoter

3.1.2 GFP@ZIF-8 Synthesis (See Fig. 2)

1. In a sterile 1.5 mL Eppendorf tube, add 1 μL of plasmid followed by 10 μL MeIM solution. Resuspend 5–6 times to ensure proper mixing then quickly add 10 μL $Zn(NO_3)_2.6H_2O$ solution. Resuspend gently till the clear solution turns turbid (*see* **Note 4**). This step is carried out at room temperature (*see* **Note 5**).
2. Allow the tube to incubate at room temperature for 10 min (*see* **Note 6**).
3. Centrifuge the tube at 10,000 × *g* for 10 min (*see* **Note 7**).
4. Resuspend the pellet in ethanol and wash (×2) by centrifuging at 10,000 rcf for 10 min (*see* **Note 8**).
5. Resuspend the pellet which is the synthesized GFP@ZIF-8 in 20 μL DNase/RNase-free water for characterization or Opti-MEM I (1×) Reduced Serum Medium (Gibco) for transfection.

3.2 Characterization of DNA@ZIF-8

3.2.1 Agarose Gel Electrophoresis to Confirm DNA Presence in GFP@ZIF-8

1. Add appropriate DNA stain like ethidium bromide to the 1% agarose gel when it reaches a temperature of roughly 50 °C.
2. Pour the gel into the casting tray, place the comb. After gel polymerizes (it usually takes about an hour), carefully remove the comb to expose sample wells and place the gel into the tank of the electrophoresis apparatus. Pour 1×TAE to cover the gel and to act as running buffer.
3. Run the gel with the following samples: a DNA ladder (size marker), DNA amount used for DNA@ZIF-8 synthesis (naked DNA), supernatant following initial incubation during synthesis and final pellet resuspended in water (*see* **Note 9**).
4. Add appropriate amount of loading dye to the samples depending on the DNA concentration used and load samples onto wells.

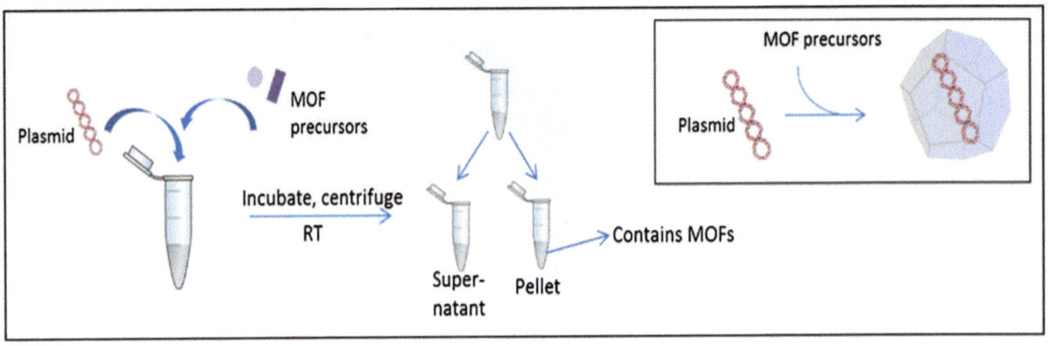

Fig. 2 Schematic representation of biomimetic mineralization of GFP@ZIF-8. Inset: ZIF-8 encapsulation of GFP

5. Connect the apparatus to the power supply and set voltage at ~100 V. Allow the gel to run for about an hour or until the loading dye is roughly half way through (*see* **Note 10**).
6. Turn off the apparatus and carefully remove the gel and place it either on a UV transilluminator or a Gel Doc system.
7. DNA bands will be visible for naked DNA and the pellet. Supernatant will either show absent or faint band (Fig. 3a). This will confirm the presence of DNA in ZIF-8.

3.2.2 Small-Angle X-ray Scattering (SAXS) to Confirm GFP@ZIF-8 Structure

1. After synthesis, resuspend the GFP@ZIF-8 in ~5 μL Milli-Q water.
2. Carefully, load the 5 μL into a glass capillary. Use the narrow end of a glass Pasteur pipette to ensure the sample is at an appropriate position for detection inside the capillary.
3. Place the capillary in the SAXS sample holder.
4. Take background measurement of an empty capillary and four measurements (different positions) for each sample capillary. The measurements should be averaged and subtracted from background.
5. SAXS/WAXS beamline with 9.3 keV, 2675 mm camera length, a Pilatus 1 M detector and transmission mode can be used.
6. Appropriate software like Scatterbrain can be used to obtain the data.
7. The plot obtained shows crystalline phases identical to conventionally synthesized pure ZIF-8 (*see* **Note 11**). This confirms that the synthesized particle is ZIF-8 (Fig. 3b).

3.2.3 Fourier Transformed Infrared (FTIR) Spectroscopy to Confirm GFP@ZIF-8 Structure

1. Sample quantity of DNA@ZIF-8 synthesized is generally low for FTIR, so best results are obtained using the diffuse reflectance (DRIFT) mode which is a specific accessory type present with all FTIR instruments.

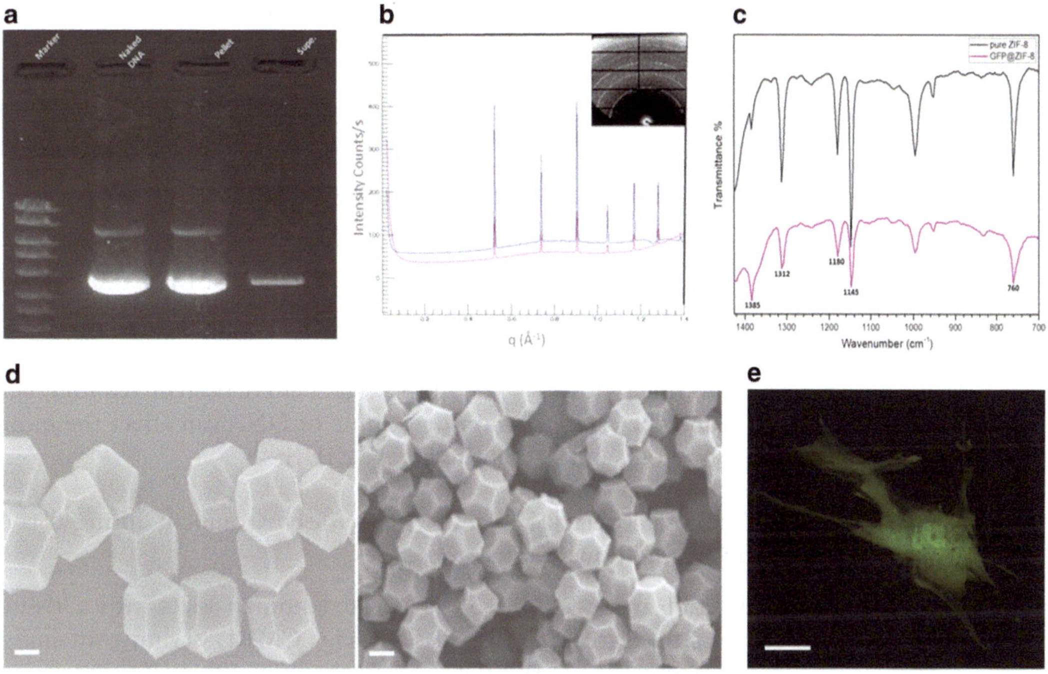

Fig. 3 (**a**) Representative gel image where 500 ng GFP has been used. Loaded wells contain DNA ladder, naked plasmid DNA, DNA in the pellet, and DNA in supernatant, from left to right. Maximum GFP is present in the pellet (GFP@ZIF-8). (**b**) SAXS patterns of pure ZIF 8 (black) and biomimetically mineralized GFP@ZIF-8 (purple). Inset: 2D representation of SAXS patterns of the materials. (**c**) FTIR patterns of pure ZIF 8 (black) and biomimetically mineralized GFP@ZIF-8 (purple). (**d**) SEM images of pure ZIF 8 (left) and biomimetically mineralized GFP@ZIF-8 (right). GFP@ZIF-8 synthesized with 1 μL of 1.5 μg/μL GFP incubated for 10 min during synthesis. Scale bar 200 nm. (**e**) hIPC image transfected with GFP@ZIF-8 taken at 48 h posttransfection. Scale bar 20 μm

2. After synthesis of GFP@ZIF-8, there is no need to resuspend the pellet. Potassium bromide (KBr) is directly added to the pellet. This mixture is kept at 60–70 °C for 1–2 h to get rid of moisture.
3. The KBr and pellet are mixed thoroughly (*see* **Note 12**). This can be done either with the help of a pipette tip inside the 1.5 mL tube or by transferring the tube contents onto a clear piece of paper and mixing the contents by folding the paper.
4. Transfer the KBr mixture to the instrument sample holder. Make sure to run pure KBr as background first. Data can be collected at 128 scans with 4 cm^{-1} resolution.
5. The FTIR data plot should demonstrate coordination between the zinc ions and the MeIM and be identical to conventionally prepared pure ZIF-8 (*see* **Note 11**). These MOF-specific peaks should be apparent for methyl bending at 1385 cm^{-1}, ring expansion and N–H wagging at 1312 cm^{-1}, C–N stretching and N–H wagging at 1180 cm^{-1}, C–N stretching of the C5

carbon at 1146 cm^{-1}, and C=N out-of-plane bending and N–H bending at 760 cm^{-1} (Fig. 3c).

3.2.4 Scanning Electron Microscopy (SEM) to Confirm GFP@ZIF-8 Morphology

1. Resuspend the GFP@ZIF-8 pellet in Milli-Q water.
2. Drop cast ~2 μL onto a silicon wafer and allow to air-dry.
3. Place wafer on a carbon-coated SEM stub.
4. Sputter coat with 5–6 nm of iridium.
5. Image under high resolution visualization at EHT 5.0 kV.
6. The particles should be visible as rhombic dodecahedral structures characteristic of ZIF-8 (Fig. 3d). The time of incubation during synthesis when the precursors have been mixed together and the concentration or volume of DNA can affect the morphology and size of the DNA@ZIF-8.

3.3 Human Islet-Derived Progenitor Cell (hIPC) Transfection and Imaging

Use biosafety class II cabinet with sterile environment during all cell culture and transfection procedures.

1. Add around 2000–3000 isolated human islets into a tissue culture treated T75 flask using CMRL serum-containing medium. Incubate the flasks in the humidified CO_2 incubator with medium changes after 3–4 days. Islets attach to the flask and islet cells start migrating out in first 6–7 days. These cells undergo epithelial-to-mesenchymal transition to generate highly proliferative islet-derived progenitor cells (*see* **Note 13**). Once the cells reach confluency, detach them with trypsin–EDTA and continue to passage them at a ratio of 1:2.
2. Seed hIPCs at such a concentration so as to be 40–50% confluent at transfection in culture well plates. It is recommended to use 24-well plates as this volume and surface area is best suited for GFP@ZIF-8 transfection. However, 96-well and 6-well plates can also be used (*see* **Note 14**).
3. On the day of transfection, replace the cell media with 100/500/2000 μL of antibiotic-free media depending on 96/24/6-well plates respectively (*see* **Note 15**).
4. Prepare GFP@ZIF-8 with desired concentration of DNA (0.5–5 μg/μL). Replace the ethanol with water washing in **step 5** of GFP@ZIF-8 synthesis (Subheading 3.1.2).
5. Resuspend the pellet in 10/50/200 μL of Opti-MEM I (1×) Reduced Serum Medium depending on 96/24/6-well plates respectively. Resuspend thoroughly and vortex briefly for 2–3 s.
6. Add into the wells in a dropwise manner spirally. Very gently swirl the plate 1–2 times.
7. Incubate cells at 37 °C for 3–3.5 h to allow for cellular uptake.
8. Replace the media containing GFP@ZIF-8 with fresh CMRL serum-containing media.

9. Incubate cells at 37 °C for 2–4 days.
10. Analyze transfected cells by checking for green fluorescence (Fig. 3e) (*see* **Note 16**).

4 Notes

1. Preparing the EDTA solution can take some time so it is best to keep it ready before hand. The 50× TAE stock solution can be diluted 49:1 with Milli-Q water to make a 1× TAE working solution.
2. When plasmids are used for transfection the promoter type (CMV vs. EF1α), type of selection (Enhanced GFP vs. Emerald GFP) and size of expression vector (5 kb vs. 8 kb) can have significant effects on transgene expression.
3. Nucleic acids like DNA are frequently dissolved in Tris–EDTA (TE) buffers. Although these help in long term storage of DNA, the EDTA present can chelate the zinc ions in ZIF-8 and degrade the crystals. Hence it is recommended to dissolve the nucleic acid in ultrapure DNase/RNase-free water.
4. Make sure to add MeIM to DNA before adding the zinc solution. Free zinc ions (in the absence of MeIM) can cause breakage in the DNA backbone. In the absence of DNA in this step, no turbidity will result. Presence of turbidity is an indication of ZIF-8 formation.
5. Take special note of the DNA concentration in the final reaction mixture as the amount of DNA effects the crystallinity of ZIF-8. Increasing concentration increases size and shifts morphology toward a more cubic phase.
6. Increasing time to more than 1 h of incubation can also affect size and morphology similarly. 10–15 min is the recommended time.
7. Save the supernatant if using a fluorescent biomolecule to help with calculating the encapsulation efficiency or for characterization.
8. When used for transfection, it is recommended to wash with water. MOFs are like sponges with their high porosity and if ethanol remains adsorbed on the surface, it can interfere with cellular transfection. Washing with water also causes the particles to remain dispersed uniformly. Take special care to not lose any particles when the supernatant is being discarded.
9. In case of plasmids, although a DNA ladder can be used, it cannot determine size. Plasmid DNAs are supercoiled and migrate through the gel irrespective of size. If DNA size

needs to be determined, it is recommended to digest the plasmid with appropriate restriction enzymes prior to running on gels.

10. Make sure the electrodes are connected in the correct orientation (negatively charged DNA moves toward the positive electrode).
11. Conventional ZIF-8 (pure; without any biomolecule) can be synthesized by preparing MeIM (160 mM, 20 mL) in methanol at room temperature and a separate solution of zinc acetate dihydrate (40 mM, 20 mL) in methanol. Combine these two solutions, vortex for 10 s, and incubate at room temperature for 24 h to grow ZIF-8 crystals. Centrifuge and wash pellet with fresh methanol. This can be used as a standard for comparison with biomimetically mineralized ZIF-8.
12. There is no need to grind the pellet with KBr as diffuse reflectance is used.
13. More detailed methodologies for generating hIPCs are available in our previously published protocols [24].
14. It is best to transfect cells at early passages.
15. The presence of antibiotics during DNA@ZIF-8 transfection can sometimes adversely affect hIPCs. A possible explanation could be that antibiotics get associated or adsorbed with DNA@ZIF-8 crystals and be taken up by the cells.
16. As with all successful cellular transfections, conditions should be optimized for each hIPC source and expression levels of gene of interest. It is critical to achieve optimum transfection with minimal toxicity, which depends on the amount of DNA and ZIF-8 precursors used. The above detailed protocol is a general method to achieve transfection in hIPCs, and it is highly recommended to carry out initial optimizations with seeding density, DNA volume, and transfection durations.

Acknowledgments

A.P. and R.S. acknowledge the joint scholarship support to A.P. from RMIT University and the Commonwealth Scientific and Industrial Research Organisation. Fellowship support from JDRF international to M.V.J. and from JDRF Australia CRN to A.A.H. is acknowledged.

References

1. Yi DK et al (2017) Recent progress in nanotechnology for stem cell differentiation, labeling, tracking and therapy. J Mater Chem B 5 (48):9429–9451
2. Moghimi SM, Hunter AC, Murray JC (2005) Nanomedicine: current status and future prospects. FASEB J 19(3):311–330
3. Thiagarajan L, Abu-Awwad Hosam Al-Deen M, Dixon James E (2017) Osteogenic programming of human mesenchymal stem cells with highly efficient intracellular delivery of RUNX2. Stem Cells Transl Med 6 (12):2146–2159
4. Delehanty JB et al (2010) Peptides for specific intracellular delivery and targeting of nanoparticles: implications for developing nanoparticle-mediated drug delivery. Ther Deliv 1(3):411–433
5. Kenry, Lim CT (2017) Nanofiber technology: current status and emerging developments. Prog Polym Sci 70:1–17
6. Gao X et al (2005) In vivo molecular and cellular imaging with quantum dots. Curr Opin Biotechnol 16(1):63–72
7. Ramon-Azcon J et al (2014) Applications of carbon nanotubes in stem cell research. J Biomed Nanotechnol 10(10):2539–2561
8. Mashinchian O et al (2015) Regulation of stem cell fate by nanomaterial substrates. Nanomedicine 10(5):829–847
9. Vijayaraghavan R et al (2010) Long-term structural and chemical stability of DNA in hydrated ionic liquids. Angew Chem Int Ed 49(9):1631–1633
10. Falcaro P et al (2016) Application of metal and metal oxide nanoparticles@MOFs. Coord Chem Rev 307(Part 2):237–254
11. Doherty CM et al (2014) Using functional nano- and microparticles for the preparation of metal–organic framework composites with novel properties. Acc Chem Res 47 (2):396–405
12. Horcajada P et al (2010) Porous metal-organic-framework nanoscale carriers as a potential platform for drug delivery and imaging. Nat Mater 9(2):172–178
13. Horcajada P et al (2012) Metal–organic frameworks in biomedicine. Chem Rev 112 (2):1232–1268
14. Liang K et al (2017) Biomimetic mineralization of metal-organic frameworks around polysaccharides. Chem Commun 53 (7):1249–1252
15. Liang K et al (2015) Biomimetic mineralization of metal-organic frameworks as protective coatings for biomacromolecules. Nat Commun 6:7240
16. Liang K et al (2016) Metal–organic framework coatings as cytoprotective exoskeletons for living cells. Adv Mater 28(36):7910–7914
17. He C et al (2014) Nanoscale metal-organic frameworks for the co-delivery of cisplatin and pooled siRNAs to enhance therapeutic efficacy in drug-resistant ovarian cancer cells. J Am Chem Soc 136(14):5181–5184
18. Liang K et al (2016) Amino acids as biomimetic crystallization agents for the synthesis of ZIF-8 particles. CrystEngComm 18 (23):4264–4267
19. Della Rocca J, Liu D, Lin W (2011) Nanoscale metal–organic frameworks for biomedical imaging and drug delivery. Acc Chem Res 44 (10):957–968
20. Li S et al (2016) Novel biological functions of ZIF-NP as a delivery vehicle: high pulmonary accumulation, favorable biocompatibility, and improved therapeutic outcome. Adv Funct Mater 26(16):2715–2727
21. Park KS et al (2006) Exceptional chemical and thermal stability of zeolitic imidazolate frameworks. Proc Natl Acad Sci U S A 103 (27):10186–10191
22. Maret W (2017) Zinc in pancreatic islet biology, insulin sensitivity, and diabetes. Prev Nutr Food Sci 22(1):1–8
23. Briones D et al (2016) Highly active antidiabetic metal–organic framework. Cryst Growth Des 16(2):537–540
24. Joglekar MV, Hardikar AA (2012) Isolation, expansion, and characterization of human islet-derived progenitor cells. Methods Mol Biol 879:351–366

Chapter 8

Encapsulation and Transplantation of Pancreatic Progenitor Cells

Luke Carroll, Auvro R. Mridha, and Bernard E. Tuch

Abstract

Type 1 diabetes, characterized by autoimmune destruction of pancreatic beta cells, affects 41 million people worldwide. Beta cell replacement therapies have immense potential as a treatment option because pancreatic progenitors derived from human pluripotent stem cells can provide a near limitless supply of transplantable tissue. The key limitation of this approach is the need for lifelong use of immunosuppressive drugs that have undesirable side effects. Microencapsulation is an option for providing protection for transplanted cells from mechanical stress and immune attack. Traditionally, pluripotent cells are differentiated on a 2D matrix before being transferred into an immunoisolation device. Here, we describe a method of differentiating pluripotent stem cells into pancreatic progenitors while the cells are encapsulated in alginate microspheres. This method provides several advantages including the need for fewer steps compared to the traditional approach, protection against mechanical/physical damage during differentiation in bioreactors, and immune-protection of cells once transplanted into the host.

Key words Differentiation, Pancreatic progenitors, Alginate, Encapsulation

1 Introduction

The usefulness of human islets for transplanting into people with Type 1 Diabetes is limited by two major factors, availability of donor tissue and immunosuppression posttransplantation. The shortage of donor tissue can be addressed by the differentiation of pluripotent stem cells from a variety of sources into pancreatic cells. Current protocols differentiate stem cells on a 2D matrix in culture dishes and then encapsulate postdifferentiation [1].

To differentiate from pluripotent stem cells, pancreatic cells must transition through a number of progenitor states. At each stage change is initiated by signaling molecules in the culture media. It is important for the differentiated cells to express PDX1 as this transcription factor is crucial for pancreatic development, including beta cell maturation and activation of the insulin gene [2]. To reach a *PDX1*-positive pancreatic progenitor stage the cells

Mugdha V. Joglekar and Anandwardhan A. Hardikar (eds.), *Progenitor Cells: Methods and Protocols*, Methods in Molecular Biology, vol. 2029, https://doi.org/10.1007/978-1-4939-9631-5_8,

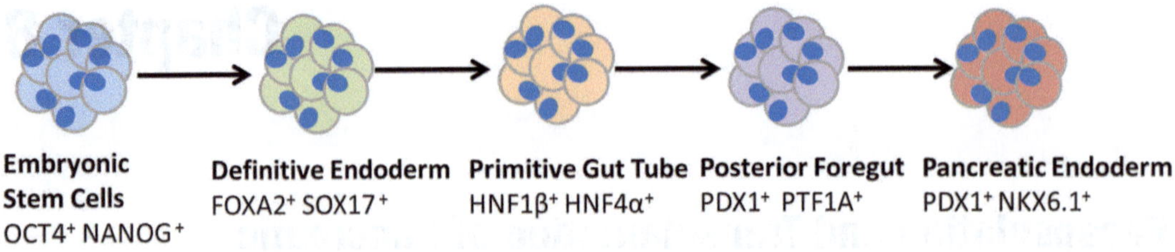

Fig. 1 Developmental stages during stem cell differentiation with relevant gene markers. As the cells move from pluripotent stem cells to differentiated pancreatic precursors, transcription factors can be used to define cell stages. Markers shown are not an exhaustive list however are key examples. Recent reviews have constructed expression patterns for all pancreatic lineages [4]

move through four distinct stages. First, the undifferentiated pluripotent stem cells are differentiated into the definitive endoderm primary germ layer. From this stage the cells differentiate into primitive gut tube, then to posterior foregut and then finally pancreatic endoderm which is representative of the pancreatic bud [3] (Fig. 1).

Generating transplantable tissue is only half of the constraint. Traditionally recipients of engrafted tissues are required to take immunosuppressive drugs as a means of preventing rejection. For all recipients, the benefit of the tissue therapy given must be greater than the risks of the immunosuppression administered [5]. Placing cells inside an immunoisolation device, such as microcapsules, prior to transplantation is one means of preventing immune destruction of transplanted cells.

One material which is used as an immunoisolation barrier is alginate. Reconstituted from a powder, this hydrogel is negatively charged and can be mixed with cells. From this suspension microcapsules can be formed with the use of an air jet or when an electrostatic charge is applied. These microcapsules are subsequently solidified in a bath containing positive ions, most commonly Ca^{2+} or Ba^{2+}. The formed alginate microcapsules are porous, enabling the passage of small molecules like oxygen, nutrients, and signaling molecules into the capsules, while allowing metabolites and wastes to leave the capsules [6]. Large molecules including immune cells and IgM are too large to pass through the pores of the microcapsules.

To test the efficacy of the differentiated cells in vivo, it is important to have a consistent experimental diabetes animal model associated with low morbidity. The induction of diabetes in mice through multiple low doses of streptozotocin (STZ) produces such a model through beta cell destruction [7].

The current standard practice for islet transplantation (Edmonton Protocol) [8] uses the portal vein to infuse islets in. There are a number of alternative transplant sites that are currently being investigated, including the mesentery. Due to capsule rigidity and size, they are not suitable for insertion into the circulatory system, and currently the optimal site for transplantation is the peritoneal cavity. This ensures excellent exposure to interstitial fluid as well as blood supply for oxygen and nutrients.

2 Materials

2.1 Encapsulation

1. Alginate: Ultrapure alginate manufactured in GMP conditions was sourced from NovaMatrix®. Specifically the alginate used in this protocol was PRONOVA UP MVG, which is a medium-viscosity sodium alginate containing a minimum of 60% of its monomer units as guluronate. It is sterilized during preparation in the laboratory however minimum specifications include ≤100 endotoxin units per gram.
2. Alginate solution: Rehydrate alginate by adding 2.2% (w/v) ultrapure alginate powder and Milli-Q water in 50 mL conical tube and leaving overnight on a roller at medium speed. The following day add 0.2 mL of 0.9% sterile NaCl solution per 1.8 mL of alginate solution. The mixture is then briefly vortexed before being centrifuged at 500 × *g* for 5 min to precipitate undissolved alginate. Sterilize the solution by passing through a 0.22 μm filter (*see* **Note 1**) and store at 4 °C.
3. Barium chloride solution: Dissolve 4.16 g of barium chloride, 2.09 g of 3-(*N*-morpholino)propanesulfonic acid (MOPS), and 6.95 g NaCl in 1 L of Milli-Q water (20 mM $BaCl_2$, 10 mM MOPS, 119 mM NaCl). Sterilize by passing through a 0.22 μm filter and store at 4 °C.
4. 14G plastic catheter: Latex free, medical grade PVC catheter, 15 cm in length appropriate for gentle mixing in a 15 mL conical tube.
5. 3 mL syringe: Latex free, clear syringe barrel with bold scale markings. Additionally an integrated plunger stop helps prevent the plunger from being accidently pulled out when drawing solutions and capsule suspensions into the syringe.
6. Encapsulation device: Air-driven droplet generator for sterile bead production. The cell–alginate suspension is forced through the device to the jet nozzle unit for air-jet driven droplet formation.
7. Phosphate buffer saline (PBS): Dissolve 8 g sodium chloride, 0.2 g potassium chloride, 1.42 g disodium phosphate, and 0.24 g monopotassium phosphate in 1 L of Milli-Q water (137 mM NaCl, 2.7 mM KCl, 10 mM Na_2HPO_4, 1.8 mM KH_2PO_4). Sterilize by passing through a 0.22 μm filter and store at 4 °C.

2.2 Cell Culture and Differentiation

1. Knockout serum replacement (KOSR) medium.
2. Knockout Dulbecco's modified Eagle medium (DMEM) supplemented with 20% knockout serum replacement, 1% insulin–transferrin–selenium (ITS) (v/v), 1% nonessential amino acids (NEAA) (v/v), and 4 ng/mL basic fibroblast growth factor (bFGF) (*see* **Note 2**).
3. RPMI medium standard formulation with no additives.
4. Fetal bovine serum (FBS).
5. High-glucose DMEM.
6. Activin A (ActA).
7. Wnt3A (WNT).
8. Keratinocyte growth factor (KGF).
9. Retinoic acid (RA).
10. Cyclopamine-KAAD (Cyc).
11. Noggin (Nog).
12. B27 supplement (B27).
13. TPB ((2S,5S)-(E,E)-8-(5-(4-(trifluoromethyl)phenyl)-2,4-pentadienoylamino)benzolactam).
14. Differentiation medium A contains 100 ng/mL activin A and 25 ng/mL Wnt3A in RPMI medium.
15. Differentiation medium B contains 100 ng/mL activin A and 0.2% FBS (v/v) added to RPMI medium.
16. Differentiation medium C contains 50 ng/mL KGF and 2% FBS added to RPMI medium.
17. Differentiation medium D contains 2 nM retinoic acid, 0.25 nM cyclopamine-KAAD, 50 ng/mL noggin, and 1% B27 (v/v) added to high-glucose DMEM.
18. Medium E contains 0.05 μM TPB, 50 ng/mL noggin, and 1% B27 (v/v) added to high-glucose DMEM.
19. Rho-associated, coiled-coil containing protein kinase (ROCK) inhibitor.
20. 10 cm culture plates.

2.3 Diabetes Induction

1. Streptozotocin (powder).
2. Glacial acetic acid.
3. Milli-Q water.
4. Anhydrous sodium acetate.
5. Sodium chloride.
6. Sodium acetate buffer: To prepare the acetate buffer, two solutions are used. Solution A is prepared by mixing 0.8 mL glacial acetic acid with 49.2 mL of Milli-Q water (0.3 M).

Solution B is prepared by adding 0.82 g of anhydrous sodium acetate or 1.36 g of sodium acetate trihydrate in a 50 mL volume of Milli-Q water (0.2 M). To prepare the final acetate buffer add 15.25 mL of solution A and 9.75 mL of solution B together and bring volume to 40 mL with Milli-Q water. Add 0.45 g of NaCl and mix into solution in an appropriately sized beaker. Adjust pH to 4.5, make final volume to 50 mL with Milli-Q water, filter-sterilize with a 0.22 μm filter, and store at 4 °C wrapped in aluminum foil.

7. Portable glucometer.
8. 25–27G needle.
9. Long-acting human insulin.
10. Scales accurate to 0.1 g.

2.4 Transplantation

1. Sterile drapes, surgical gloves, and instruments.
2. Analgesics: buprenorphine and meloxicam.
3. Saline.
4. Isoflurane for anesthesia.
5. Oxygen for isoflurane delivery.
6. Anesthetic gas evaporator.
7. Chlorhexidine.
8. Dissolvable sutures.
9. Wound clips.
10. Antiseptic cream.
11. Razor blades and/or hair removal cream.

3 Methods

Maintenance culture conditions for pluripotent stem cells are a continually evolving. The authors of this protocol cultured both human embryonic and induced pluripotent stem cells on cell cycle arrested human fetal fibroblast feeder layer. Passaging occurred weekly by either mechanical dissociation using a fine needle tip or enzymatically [9].This feeder layer was discarded prior to suspending the stem cells in alginate [10]. This protocol begins with the stem cells in a single cell pellet isolated from maintenance conditions. An overview of the protocol can be seen in flow diagram below (Fig. 2).

3.1 Encapsulation

1. Isolate stem cell pellet in 15 mL conical tube, add 1 mL of alginate per 2×10^6 cells.
2. Mix the stem cells using a 14G plastic catheter attached to a syringe. Draw the mixture into a 3 mL syringe and place the nozzle of the syringe into the encapsulation device (Fig. 3).

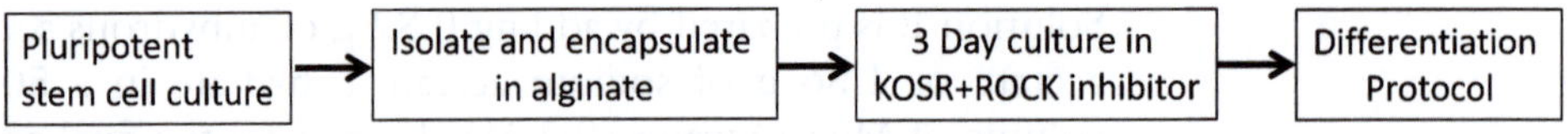

Fig. 2 Flow diagram of the steps involved in this protocol. Breaking down the flow of the protocol into distinct steps helps understanding and improvement on timings and expectations. The four distinct steps move the cells from continuous culture to encapsulated pluripotent cells ready for differentiation

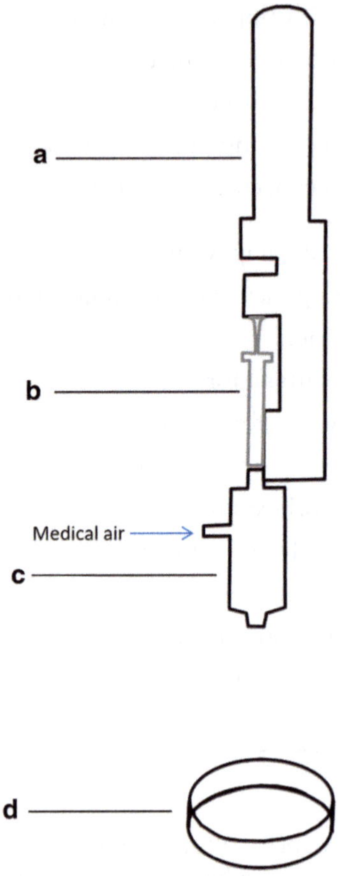

Fig. 3 Encapsulation device setup with syringe driver and barium chloride bath. (**a**) Constant speed syringe driver applying force to plunger on (**b**) syringe filled with alginate–cell suspension. This is forced through (**c**) the encapsulation device which mixes with medical air to form droplets which fall into (**d**) cell culture dish filled with barium chloride

3. Attach a constant-speed automated plunger to the syringe and force the alginate solution out of the syringe through the encapsulation device at a flow rate of 1 mL/min. Simultaneously, pass sterile air through the device at a rate of 6 L/min and a pressure of 80 kPa. These pressure and flow rates were optimized to produce droplet spheres of 400 ± 150 μm. These droplet spheres are then collected into a 30 mL bath of

sterile 20 mM barium chloride, placed 15 cm below the encapsulation device. Time in barium chloride bath should be minimized, recommend no more than 3 mL of cell–alginate suspension per generation cycle (*see* **Note 3**).

4. The microspheres are then washed in PBS three times (*see* **Note 4**) and cultured at 37 °C, 5% CO_2 in 10 cm dishes containing KOSR medium + 10 μM Rho-associated, coiled-coil containing protein kinase (ROCK) inhibitor for 3 days with media changed daily (*see* **Note 5**).

3.2 Generation of Pancreatic Progenitors

1. Following the 3-day incubation postencapsulation, change culture medium to medium A; the first differentiation medium. Change differentiation medium daily, wash with PBS between different medium stages and follow the protocol summarised in the Table 1.

3.3 Induction of Diabetes

1. STZ when in solution has a short half-life. To ensure a reliable dosage, the STZ solution is freshly prepared at 5 mg/mL in sodium acetate buffer prior to injection (*see* **Note 6**). NOD/SCID mice weighing 20 g or more are injected intraperitoneally at a dose of 50 mg/kg for 4 consecutive days.
2. Measure blood glucose levels (BGL) using a portable glucometer (e.g., OneTouch Verio IQ) by pricking the tail vein with a 25–27G needle. BGL are checked 3 days after final STZ dose and every second day thereafter. If mice are not diabetic 7 days after final injection an additional dose of 50 mg/kg is administered. Mice are considered diabetic when three separate (*see* **Note 7**) BGL readings greater than 15 mM are recorded.

3.4 Intraperitoneal Transplantation

All surgical procedures are performed within a class II biosafety cabinet that is UV-sterilized prior to use. Sterile drapes, surgical gloves, and instruments are used during all steps of surgical procedures.

1. Cells in the alginate microcapsules, which have reached the end of the differentiation protocol (Subheading 3.2) are washed once in PBS and then loaded into a sterile syringe using sterile saline. After differentiation, live cells should number ~80% of initial encapsulated cells (Unpublished data).
2. Give mice an injection of buprenorphine 0.1 mg/kg and meloxicam 0.5 mg/kg in sterile saline solution prior to surgery to ensure pain relief before mice regain consciousness. The mice are then anesthetized with 5% isoflurane at 2.5 mL/min which is reduced to 2% once the mouse has become unresponsive (*see* **Note 8**). The lower abdomen below the rib cage is shaved (*see* **Note 9**) and then swabbed with chlorhexidine.

Table 1
An overview of the differentiation protocol to generate pancreatic progenitors from pluripotent stem cells

	Stage 1		Stage 2	Stage 3	Stage 4
Differentiation stage	Definitive endoderm		Primitive gut tube	Posterior foregut	Pancreatic endoderm
Medium	A	B	C	D	E
Reagents	ActA+WNT RPMI	ActA RPMI 0.2% FBS	KGF RPMI 2% FBS	RA + Cyc + Nog DMEM 1% B27	Nog + TPB DMEM 1% B27
Duration	1 day	2 days	3 days	3 days	4 days

Each stage of the differentiation consists of distinct stages with specific media requirements to move the cells to the next stage. An overview of the media is provided with detailed breakdown in Subheading 2.2. Time is given in days (D), with beginning of the first day as D0. Abbreviations of additives can be found in Subheading 2.2

3. Make a midline skin incision, 1–2 cm long, in the lower abdomen below the rib cage.
4. Carefully tent up the musculoperitoneal layer to avoid damage to the bowel, and insert a 14G needle and catheter into the peritoneal cavity directly beneath the cutaneous incision.
5. Remove the needle and attach the syringe with the encapsulated cells to the catheter. Inject the encapsulated cells into the peritoneal cavity through the catheter (*see* **Note 10**).
6. Remove the catheter and close the peritoneum with synthetic dissolvable sutures in an interrupted or continuous pattern, taking extreme care to avoid perforation of the underlying bowel and ensuring saline and microcapsules not to escape from the peritoneal cavity (*see* **Note 11**).
7. Close the skin incision with 2–4 wound clips or interrupted sutures, and apply antiseptic cream.
8. Place the mice in a clean cage and keep the cage warm until the mice have fully recovered from anesthesia.
9. Remove the clips or sutures 7–10 days postsurgery, when the wound is healed.
10. Postoperative care should be in line with institutional ethics committee approval. Briefly, weight and BGL should be checked three times a week. 0.5 U of human exogenous long acting insulin can be administered if the weight of the animal drops below 10% of preinduction of diabetes weight. C-peptide of differentiated cell origin has been found in blood samples taken from the superficial temporal vein at 9 months posttransplantation (*see* **Note 12**).

4 Notes

1. Alginate is a highly viscous solution, so it is recommended to use a syringe filter attached to a 10 mL syringe. Be careful not to apply too much pressure to break the filter. Additionally take necessary precautions to prevent a stress injury if filtering a large volume of alginate.
2. Supplement media with bFGF immediately before applying to cells. Ensure that the bFGF is aliquoted to avoid freeze–thaw cycles.
3. Depending on the size and depth of the collection dish it may be necessary to gently swirl the dish to prevent alginate capsules from gelling to one another once in the barium chloride solution.
4. The alginate capsules have a greater density than PBS or barium chloride and will settle in a 50 mL conical tube after each wash cycle. Alternatively a cell strainer can be used to separate the capsules.
5. Resuspend with 10× media to capsule volume. For example, use 10 mL of medium for 1 mL of alginate capsules.
6. STZ is extremely toxic and should be handled with care; use a face mask and refer to the relevant MSDS guidelines while handling.
7. Separate BGL readings must be at least 16 h apart.
8. A 5 mL syringe makes an excellent nose cone to deliver oxygen–isoflurane mix.
9. Hair removal cream can be applied to ensure the area is free from hair; take care to limit the time of exposure as it may sensitize the skin.
10. Additional PBS may be required to completely flush syringe and catheter of capsules.
11. Capsules settle to under gravity and excess solution may escape through the wound. Ensure the fluid remaining inside the intraperitoneal cavity does not exceed 1% of body weight as per NHMRC guidelines on animal surgery.
12. Circulating human C-peptide was detected using an ultrasensitive ELISA. To have detectable levels roughly 2×10^6 total cells were transplanted, this number did not correct BGL (unpublished data).

Acknowledgments

Thank you Dr. Michael Morris for assisting in the preparation of the manuscript.

References

1. Vegas AJ, Veiseh O, Gürtler M, Millman JR, Pagliuca FW, Bader AR, Doloff JC, Li J, Chen M, Olejnik K (2016) Long-term glycemic control using polymer-encapsulated human stem cell-derived beta cells in immune-competent mice. Nat Med 22 (3):306–311
2. Stoffel M, Stein R, Wright CV, Espinosa R, Le Beau MM, Bell GI (1995) Localization of human homeodomain transcription factor insulin promoter factor I (IPF1) to chromosome band 13q12. 1. Genomics 28 (1):125–126
3. D'Amour KA, Bang AG, Eliazer S, Kelly OG, Agulnick AD, Smart NG, Moorman MA, Kroon E, Carpenter MK, Baetge EE (2006) Production of pancreatic hormone–expressing endocrine cells from human embryonic stem cells. Nat Biotechnol 24(11):1392–1401
4. Jennings RE, Berry AA, Strutt JP, Gerrard DT, Hanley NA (2015) Human pancreas development. Development 142(18):3126–3137
5. Hirshberg B, Rother KI, Digon BJ, Lee J, Gaglia JL, Hines K, Read EJ, Chang R, Wood BJ, Harlan DM (2003) Benefits and risks of solitary islet transplantation for type 1 diabetes using steroid-sparing immunosuppression. Diabetes Care 26(12):3288–3295
6. Goosen MF, O'Shea GM, Gharapetian HM, Chou S, Sun AM (1985) Optimization of microencapsulation parameters: semipermeable microcapsules as a bioartificial pancreas. Biotechnol Bioeng 27(2):146–150. https://doi.org/10.1002/bit.260270207
7. Rossini AA, Like AA, Chick WL, Appel MC, Cahill GF (1977) Studies of streptozotocin-induced insulitis and diabetes. Proc Natl Acad Sci 74(6):2485–2489
8. Shapiro AM, Ricordi C, Hering BJ, Auchincloss H, Lindblad R, Robertson RP, Secchi A, Brendel MD, Berney T, Brennan DC, Cagliero E, Alejandro R, Ryan EA, DiMercurio B, Morel P, Polonsky KS, Reems JA, Bretzel RG, Bertuzzi F, Froud T, Kandaswamy R, Sutherland DE, Eisenbarth G, Segal M, Preiksaitis J, Korbutt GS, Barton FB, Viviano L, Seyfert-Margolis V, Bluestone J, Lakey JR (2006) International trial of the Edmonton protocol for islet transplantation. N Engl J Med 355 (13):1318–1330. https://doi.org/10.1056/NEJMoa061267
9. Costa M, Sourris K, Hatzistavrou T, Elefanty AG, Stanley EG (2008) Expansion of human embryonic stem cells in vitro. Curr Protoc Stem Cell Biol 5(1):1C. 1.1–1C. 1.7
10. Tomishima M (2012) Splitting hPSCs with dispase. Harvard Stem Cell Institute, Cambridge (MA)

Chapter 9

Differentiation of Urothelium from Mouse Embryonic Stem Cells in Chemically Defined Conditions

Badwi B. Boumelhem, Stuart T. Fraser, and Stephen J. Assinder

Abstract

The urothelium of the bladder and urethra are derived from the definitive endoderm during development. Cellular signaling molecules important to the developmental specification of the urothelium are also implicated in the dysregulation of the tissue repair mechanism characteristic of bladder disease. Hence, a complete understanding of the regulation of urothelium development is central to understanding the processes of bladder disease, and in development of simple chemically defined methods for use in regenerative medicine. Key to this is a suitable in vitro model that readily allows for the prosecution of biologically pertinent questions. Here a method for differentiating urothelium from mouse embryonic stem cells in chemically defined conditions is described. The method includes a description of flow cytometry and RT-PCR analysis of definitive endoderm markers Cxcr4, c-Kit, and *FoxA2*, and of terminally differentiated urothelial cell markers *Upk1b* and *Upk2*.

Key words Urothelium, Uroepithelium, Bladder, Uroplakin, Interstitial cystitis/painful bladder syndrome, Cancer

1 Introduction

The urothelium (uroepithelium) is a specialized epithelium that lines the urinary tract including the urinary bladder and proximal urethra. The urothelium of mammals provides an impermeable protective barrier against the futile reabsorption of water and against urine toxicity, by separating plasma from hypertonic urine [1, 2]. This "urine–blood" barrier formed by the urothelium is provided by both the cellular arrangement, as a thick multilayered pseudostratified transitional epithelium [1, 3], and by asymmetric unit membranes of the fully differentiated superficial "umbrella" cells. These asymmetric units are formed by uroplakin protein complexes [4, 5]. Damage to this superficial cell layer induces a proliferative response of the urothelium to rapidly reestablish the urine–blood barrier [6, 7]. This response is typically seen following

Mugdha V. Joglekar and Anandwardhan A. Hardikar (eds.), *Progenitor Cells: Methods and Protocols*, Methods in Molecular Biology, vol. 2029, https://doi.org/10.1007/978-1-4939-9631-5_9,

urinary tract infections (UTIs) caused by uropathogenic Escherichia coli [6, 8].

Dysregulation of repair in response to insults has been shown to result in chronic bladder disease including interstitial cystitis/painful bladder syndrome (IC/PBS) following UTIs [9]; bladder cancers where known risk factors include IC/PBS; and associated inflammatory responses [10–12]. Dysregulation of cellular signaling molecules key to developmental processes are common to both diseases. Of particular note are bone morphogenetic protein 4 (BMP4), sonic hedgehog (SHH), and wingless type (WNT) family members [7]. Hence, an understanding of the development process, and an ability to recapitulate it in vitro, provides a powerful tool for the study of disease processes.

The isolation of embryonic stem (ES) cell lines and the Nobel Prize-winning development of nuclear reprogramming of mature, patient-derived cells into induced pluripotent stem (iPS) cells, have led to a revolution in the field of regenerative medicine. Individuals lacking urothelial organs such as the urinary bladder could, in the future, have three-dimensional urinary bladder-like structures derived in vitro from their own cells, avoiding immune rejection issues. Such advances in regenerative medicine however demand an understanding of the underlying developmental biological processes required to generate urothelial progenitors in vivo. Only then, can these conditions be replicated in vitro to generate large numbers of urothelial progenitors for experimental or clinical use. Ideally, differentiation conditions are chemically defined, do not contain products of animal origin (to prevent zoonoses) and are cost-effective for scaling up.

Mammalian development requires the development of the ectoderm, mesoderm, and endoderm germ layers. The bipotential mesendoderm progenitor forms prior to specification of either the mesoderm or the definitive endoderm [13]. The urothelium of the lower urinary tract that lines the bladder and proximal urethra is derived from the definitive endoderm. The endoderm can be divided into three regions: the foregut, midgut, and hindgut endoderm. The hindgut endoderm ends as a caudal expansion called the cloaca. The cloaca is a transitory endoderm-derived cavity that subdivides into the urogenital sinus (UGS) and the anal canal [14]. The UGS arises between E12.5-E13.5 of development in the mouse and approximately during the seventh week of gestation in humans [15, 16]. It is from the UGS that the urothelium of the bladder and proximal urethra develops [17] (Fig. 1).

Urogenital tissues, such as those of the bladder and the prostate, rely on paracrine signaling between the urogenital mesenchyme and epithelium for normal growth and development [18–22]. As stated above, the urothelium of the bladder forms the urine/blood barrier primarily via the asymmetric unit membranes on the apical surface of urothelium. The integrity of this

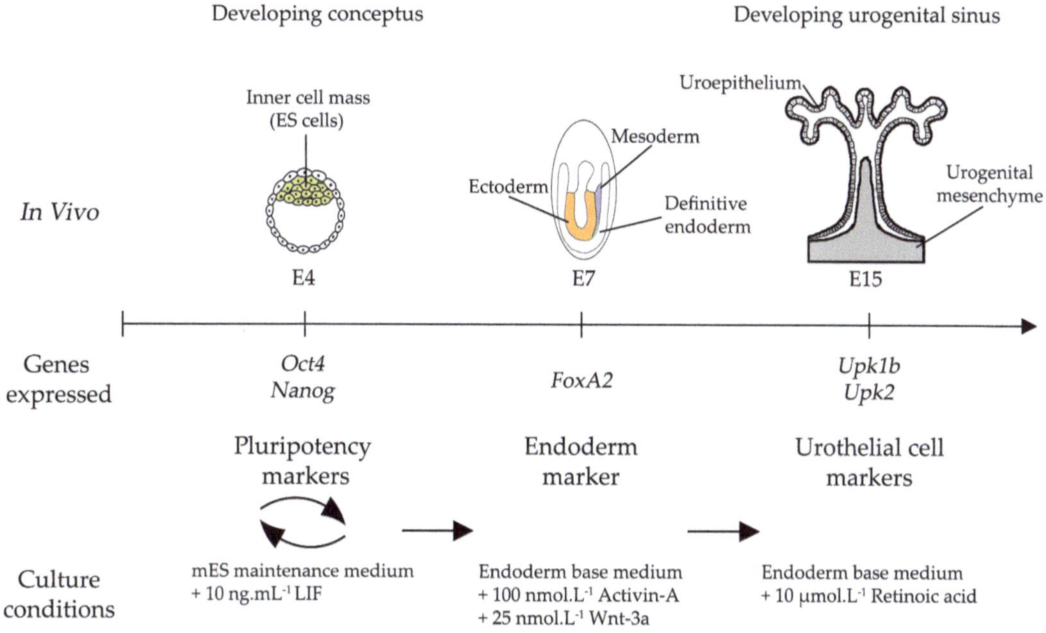

Fig. 1 Timeline of urothelium differentiation from mouse ES cells in vitro. Pluripotency markers *Oct4* and *Nanog* are expressed on undifferentiated mouse ES cells. The presence of leukemia inhibitory factor (LIF) in mouse ES cell cultures prevents spontaneous differentiation of mouse ES cells into cells from the three germ layers: the endoderm (green), the ectoderm (orange) and the mesoderm (purple). Endoderm (*FoxA2* expression) induction from mouse ES cells is achieved by removal of LIF and addition of Activin-A and Wnt-3a in cell cultures. Urothelial cell differentiation is driven by retinoic acid and confirmed with expression of uroplakins *Upk1b* and *Upk2*. Diagrammatic representation of corresponding stages in vivo is provided

barrier is dependent upon the uroplakin proteins (*Upk1a*, *Upk1b*, *Upk2*, *Upk3*) [22, 23]. Uroplakins are the first identifiable markers of the terminally differentiated urothelial cells in vivo [23]. Retinoic acid appears to be central to this process. This well-known factor of embryogenesis and cell differentiation is known to induce urothelial cell differentiation of mouse ES cell-derived definitive endoderm [24–27].

Whilst methodologies for the in vitro differentiation of urothelium from human ES cells and iPS cells, and mouse ES cells have been reported, studies of the molecular factors that are important to this process are limited [24–26, 28]. As the bladder urothelium is of endodermal origin, the first step in the generation of urothelium is the temporal differentiation of ES and iPS cells into definitive endoderm with Activin A [29]. Following induction of definitive endoderm, published protocols [24–26, 28] supplemented cultures with retinoic acid and urothelial-specific medium for differentiation of urothelium in mouse and human ES and iPS cells respectively [24–26]. Additionally, published protocols have used mouse fibroblast feeder layers and Matrigel to promote urothelium differentiation [24–26], or have employed xenografting of reconstituted tissue under the kidney capsule of immune-deficient recipient mice [28].

Here, we describe a simple, cost-effective method of in vitro differentiation of the urothelium of the lower urinary tract from mouse ES cells under chemically defined conditions in the absence of feeder cells or costly Matrigel.

2 Materials

All cell culture must be performed under sterile conditions within a laminar flow, biosafety cabinet. Cell cultures must be maintained in humidified incubators set to 37 °C with 5% CO_2. The media should be sterile-filtered and warmed to 37 °C prior use (but should be stored at 4 °C when not in use). All the reagents and materials required for cell culture must be handled aseptically.

2.1 Maintenance of Mouse Embryonic Stem (ES) Cells

1. 15 mL conical tubes.
2. 6-well cell culture plates.
3. Ultrapure water with 0.1% gelatin.
4. TrypLE® Express.
5. ES maintenance media—For 500 mL: 430 mL of Dulbecco's Modified Eagle Medium (DMEM), 50 mL of heat-inactivated fetal bovine serum (FBS), 10 mL of 100× GlutaMAX™, 5 mL of 10 mM sodium pyruvate, 5 mL of 100 × penicillin–streptomycin (100× Pen/Strep contains 10,000 units penicillin and 10 mg streptomycin/mL), and 6.3 μL monothioglycerol (15.75 mg/L final concentration). Sterilize by filtering through a 0.22 μm filter before use (*see* **Note 1**).

2.2 Differentiation of Mouse ES Cells into Definitive Endoderm

1. 15 mL conical tubes.
2. 35 mm petri dishes.
3. Sterile transfer pipettes.
4. Endoderm differentiation media—For 500 mL: 387.5 mL of Iscove's Modified Dulbecco's Medium, 100 mL of Knockout™ Serum Replacement, 5 mL of 100× penicillin–streptomycin, 5 mL of 100× GlutaMAX™, 2.5 mL of ascorbic acid (5 mg/mL), 19.5 μL of monothioglycerol (47.25 mg/L final concentration).
5. Endoderm differentiation media supplemented with 100 ng.mL^{-1} of Activin-A and 25 ng.mL^{-1} Wnt-3A (*see* **Note 2**).

2.3 Flow-Cytometric Analyses of Definitive Endoderm Differentiation from Mouse ES Cells

1. 15 mL conical tubes.
2. Phosphate Buffered Saline (PBS), pH 7.4.
3. Fluorescence-activated cell sorting (FACS) buffer. For 500 mL: 0.05 g of bovine serum albumin (BSA) in 500 mL of PBS. Sterilize by filtering through a 0.22 μm filter before use.
4. Cell Dissociation Buffer.
5. Fluorescence-conjugated antibodies recognizing mouse CXCR4 and c-Kit surface antigens.

2.4 Differentiation of Urothelium from Mouse ES-Derived Definitive Endoderm

1. 15 mL conical tubes.
2. 6-well cell culture plates.
3. Ultrapure water with 0.1% gelatin.
4. Endoderm differentiation media supplemented with 100 μm. L^{-1} of all-trans retinoic acid.

2.5 Reverse-Transcriptase Polymerase Chain Reaction (RT-PCR) of Definitive Endoderm and Urothelium Derived from Mouse ES Cells

1. 1.7 mL Eppendorf tubes
2. TRIzol®.
3. Pestle and mortar.
4. RNA extraction kit.
5. RNA-to-cDNA reverse transcriptase kit.
6. Amplitaq Gold® 360 Master Mix.
7. Gene specific forward and reverse primers (*see* Table 1).
8. Agarose.
9. Tris–acetate–EDTA buffer (pH 8.0)—For 1 L of 10× stock: 48.4 g of Tris base, 11.42 mL of glacial acetic acid, 20 mL of EDTA (0.5 M, pH 8.0), make up to 1 L with MilliQ H_2O.
10. Ethidium bromide.
11. Hyperladder™.

2.6 Equipment

1. Hemocytometer.
2. Benchtop centrifuge.
3. Phase contrast light microscope such as Zeiss Axiovert35.
4. Four-color flow cytometer such as FACSCalibur.
5. UV transilluminator.

3 Methods

A timeline depicting urothelium differentiation from mouse ES cells in a stage-dependent manner is outlined below in Fig. 1.

Table 1
Sequences of primers used for RT-PCR analyses for the detection of undifferentiated mouse ES cells, definitive endoderm and urothelium

Gene	Forward (5′–3′)	Reverse (5′–3′)	Predicted amplicon size (bp)	NCBI reference
ActB	CCTCTATGCCAACACAG TGC	CCTGCTTGCTGA TCCACATC	120	NM_007393.3
Nanog	AAGTACCTCAGCC TCCAGCA	GTGCTGAGCCCTTC TGAATC	163	NM_028016.2
FoxA2	CTACACACACGCCAAACC TC	GGCACC TTGAGAAAGCAGTC	201	NM_010446
Upk1b	TCCGTCAGAC TGGCAGAAAT	GTCCAGGTTGAGAGGC TCTT	118	NM_178924.4
Upk2	AGCCTGTTAATTGCC TTGCC	TGTCACCTGATA TGCGCTGA	195	NM_009476.2
Oct4	CACGAGTGGAAAGCAAC TCA	AGATGGTGGTCTGGC TGAAC	246	NM_013633.3

Sequences were obtained from the Nucleotide database provided by the National Centre for Biotechnology Information (NCBI reference). Primer sequences were generated using Primer3 software (http://primer3.ut.ee/). The melting temperature of the primers was 70 °C

3.1 Maintenance of Mouse Embryonic Stem (ES) Cells

1. Reagents used for cell culture should be warmed to 37 °C before use.
2. Cell culture is performed under sterile conditions in a laminar flow tissue culture cabinet.
3. The mouse embryonic stem (ES) cell line used in this protocol is 129/svj in the absence of a feeder layer.
4. Prior to passaging of mouse ES cells, add 1 mL of gelatin to 3 wells of a 6-well plate and let it sit for 10 min at room temperature before discarding (*see* **Note 3**).
5. Aspirate the maintenance media and wash the mouse ES cells with 1 mL of PBS.
6. Discard the PBS and add 0.5 mL of TrypLE® Express to each well.
7. Incubate the mouse ES cells for 3–5 min at 37 °C to initiate dissociation.
8. Quench the dissociation by adding 1 mL of maintenance media.
9. Transfer the cells to a sterile 15 mL conical tube and centrifuge for 5 min at 250 × *g*.

10. Discard the supernatant and add 100 units.mL^{-1} LIF to the mouse ES cells. (Note: the presence of 100 units.mL^{-1} leukemia inhibitory factor (LIF) inhibits the differentiation of stem cells and maintains undifferentiated state.)
11. Resuspend mouse ES cells in 6 mL of mouse ES cell maintenance media and plate onto three wells of a gelatinized 6-well plate.

3.2 Differentiation of Mouse ES Cells into Endoderm

1. Dissociate the mouse ES cells as outlined in **steps 5–9** in Subheading 3.1.
2. Resuspend mouse ES cells in 1 mL of maintenance media.
3. To differentiate into definitive endoderm in the absence of LIF, plate 1.5×10^4 cells.mL^{-1} into a 35 mm petri dishes and incubate at 37 °C and 5% CO_2.
4. The mouse ES cells will form aggregates termed embryoid bodies (EBs).
5. To induce endoderm differentiation, add 100 ng.mL^{-1} of Activin-A and 25 ng.mL^{-1} of Wnt-3A to EB cultures for 5 days following the removal of LIF.
6. Change media every 2 days (*see* **Note 4**).
7. Confirm definitive endoderm differentiation by the detection of *FoxA2* gene expression and surface protein expression of CXCR4 and c-Kit by RT-PCR and flow cytometry respectively (Fig. 2).

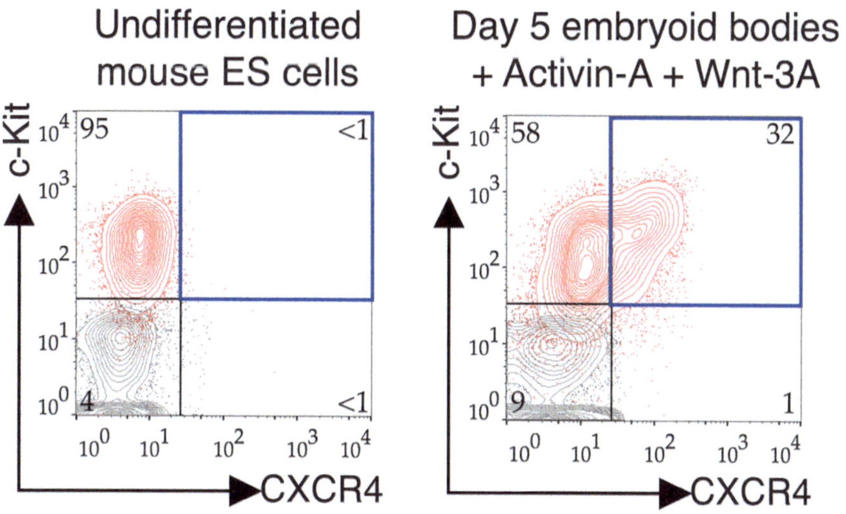

Fig. 2 Activin-A and Wnt-3A promote the differentiation of mouse ES cells into endoderm. Representative flow-cytometric plots of surface protein expression of endoderm markers c-Kit and CXCR4 between undifferentiated mouse ES cells and day 5 EBs with or without Activin-A and Wnt-3A treatment. The black contour plots represent the no-stain negative control while the red contour plots represent the c-Kit, CXCR4 stained samples. Flow cytometry was performed on a four color flow cytometer and data collected using CellQuest software

3.2.1 Analysis of Endoderm Markers c-Kit and CXCR4 by Flow Cytometry

Differentiation of pluripotent ES cells to definitive endoderm is marked by surface expression of c-Kit and CXCR4, and flow-cytometric analysis of c-Kit and CXCR4 is used to determine this developmental progression in vitro [29, 30].

1. Transfer Day 5 EBs into a 15 mL conical tube and let the EBs settle by gravity (10–15 min).
2. Discard the supernatant.
3. Add 1 mL of Cell Dissociation Buffer and incubate the EBs for 10 min at 37 °C.
4. Gently shake the 15 mL tubes every 2–3 min.
5. Add 2 mL of FACS buffer to quench the dissociation.
6. Centrifuge EBs for 5 min at 250 × *g*.
7. Discard the supernatant and resuspend in 1 mL of FACS buffer.
8. Make APC-conjugated c-Kit (BioLegend) and PE-conjugated CXCR4 (eBioscience) definitive endoderm surface markers to 1 mL at a final concentration of 1.3 μg.mL^{-1} and 3.25 μg. mL^{-1} respectively These concentrations may differ based on the source of antibodies and must be optimized/titrated using appropriate controls (*see* **Note 5**).
9. Stain 100 μL of dissociated EBs with 100 μL of the antibody mix and incubate on ice for 45 min in the dark.
10. Prepare the following controls: no stain controls (no antibody), isotype controls and single stain (single fluorophore) controls.
11. Following incubation, wash the EBs with 1 mL of FACS buffer and centrifuge for 5 min at 250 × *g*.
12. Discard the supernatant and the resuspend the cells in 400 μL of FACS buffer containing 0.01% propidium iodide.
13. Transfer the cells to round-bottom tubes for flow-cytometric analyses.

3.3 Differentiation of Endoderm to Urothelium

1. Following definitive endoderm differentiation, plate day 5 EBs on gelatin coated 6-well plates to form a monolayer.
2. Treat Activin-A and Wnt-3A EBs with 10 μmol.L^{-1} retinoic acid for 5 days to induce urothelial cell differentiation (Fig. 3) (*see* **Note 6**).
3. Observe treated cultures by light microscopy for the presence of cobblestone-like morphology (Fig. 3).
4. Confirm urothelial cell generation by the expression of *Upk1b* and *Upk2* by RT-PCR (Fig. 4).

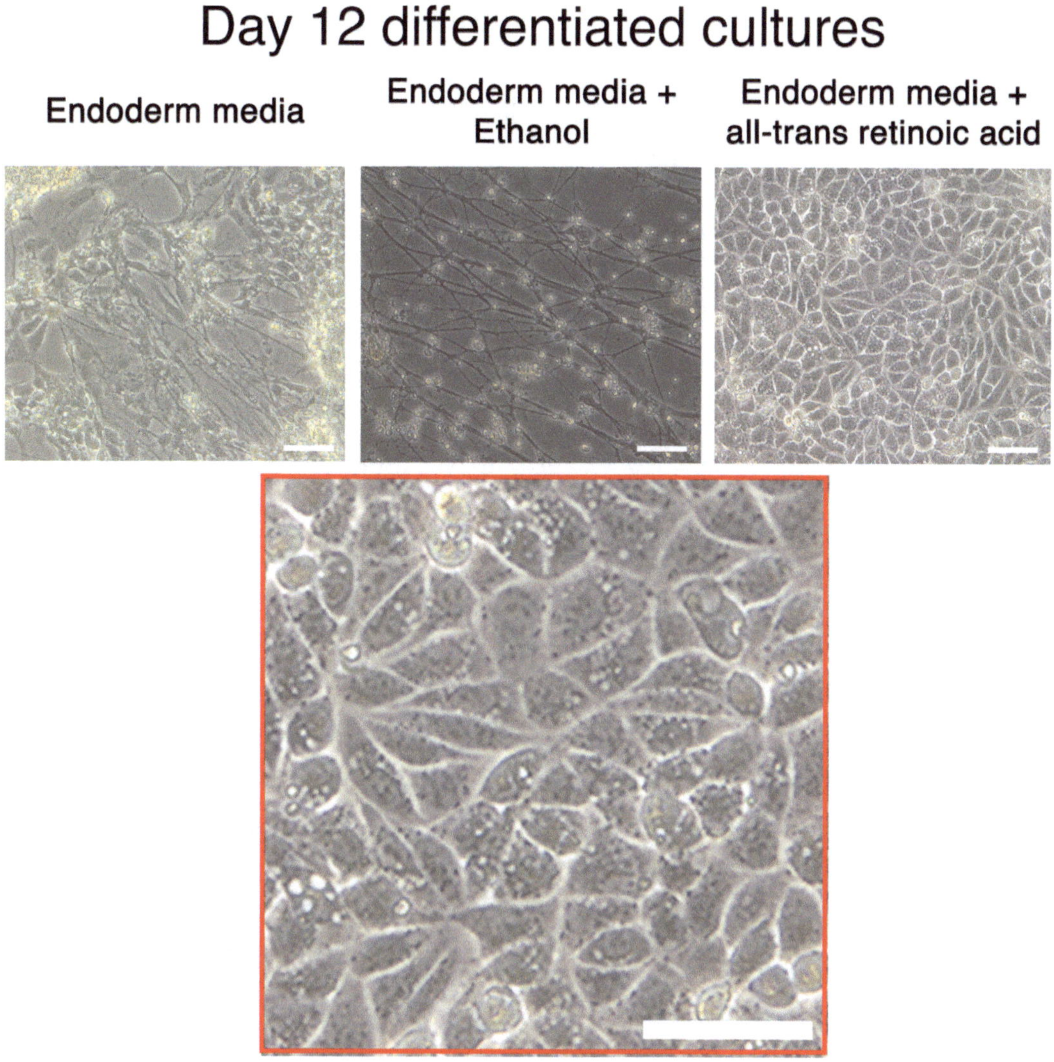

Fig. 3 Retinoic acid induces a cobblestone-like phenotype in mouse ES cell-derived endoderm cultures. Representative phase contrast micrographs of day 12 differentiated mouse ES cell cultures treated with endoderm media alone, endoderm media and ethanol (vehicle control) or endoderm media and all-trans retinoic acid. The red-boxed region (digitally magnified) highlights the cobblestone-like cells indicative of urothelial cell differentiation. Scale bars represent 20 μm

3.4 Reverse-Transcriptase Polymerase Chain Reaction (RT-PCR) Analysis

1. Dissociate undifferentiated mES, day 5 EBs and urothelial cells with 1 mL of TRIzol and transfer into a 1.7 mL Eppendorf tube.
2. For positive controls, homogenize bladder tissue in 1 mL of TRIzol reagent using a pestle and mortar and transfer to a 1.7 mL Eppendorf tube.
3. Isolate RNA according to manufacturer's instructions.

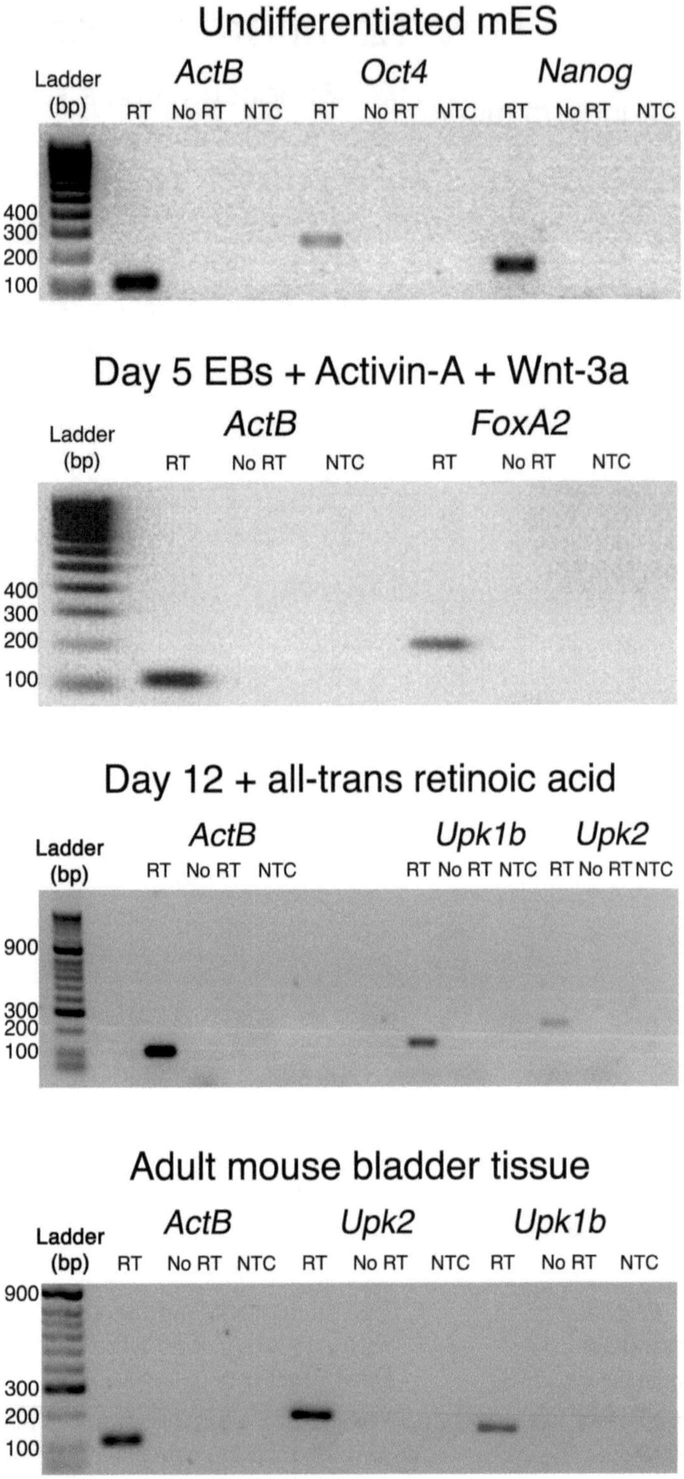

Fig. 4 Expression of uroplakins 1b and 2 are detected in day 12 mouse ES cell cultures treated with all-trans retinoic acid. Total RNA was harvested from undifferentiated mouse ES cells, day 5 embryoid bodies treated with Activin-A and Wnt-3a, day 12 cultures treated with 10 μmol.L^{-1} retinoic acid and adult

4. Perform reverse transcription of RNA samples using the RNA-to-cDNA Kit according to manufacturer's instructions.
5. Mix 1 μg of cDNA with universal components of Amplitaq Gold® 360 Master Mix as per manufacturer's instructions.
6. To each mix, add gene specific forward and reverse primers (Table 1).
7. Briefly centrifuge the reaction mixes (including no-RT and no-template control) prior to a reaction protocol of 1 cycle at 95 °C for 10 min; 45 cycles of 30 s at 95 °C, 30 s at 65 °C, and 1 min at 72 °C; followed by 1 cycle at 72 °C for 7 min.
8. Analyze the amplicons generated from RT-PCR by agarose gel submarine electrophoresis.
9. Mix 10 μL of samples with 2 μL of DNA loading buffer (containing 30% (v/v) glycerol, 0.25% (w/v) bromophenol blue, and 0.25% (w/v) xylene cyanol FF) and load onto a 2% (w/v) agarose-cum-Tris–acetate–EDTA (pH 8.0) (TAE) buffer gel containing 0.1% (w/v) ethidium bromide.
10. Add 5 μL of the HyperLadder™ 50 bp on each gel.
11. Separate loaded samples at 100 V for 1 h.
12. Visualize products were under an ultraviolet transilluminator.
13. Size with reference to HyperLadder™.

4 Notes

1. ES maintenance and endoderm differentiation media can be stored up to 1 month at 4 °C without affecting mouse ES cell differentiation or yield of definitive endoderm differentiation.
2. To maximize the shelf life of Activin-A and Wnt-3A, store in aliquots at −18 °C and thaw out once needed.
3. Gelatinization of culture plates are required to efficiently form a monolayer of undifferentiated (and differentiating) ES cells. It is crucial then to use gelatin that has been screened and optimized for ES cell culture to ensure no possible contamination. In this protocol, we used Ultrapure Water with 0.1% Gelatin (cat# ES-006-B, Merck Millipore).

Fig. 4 (continued) mouse bladder tissue (positive control) and gene expression for pluripotency markers (*Nanog and Oct4*), definitive endoderm (*FoxA2*) and urothelial cell markers (*Upk1b* and *Upk2*) examined. Predicted amplicon sizes for *Nanog*, Oct 4, *FoxA2*, *Upk1b*, and *Upk2* are 163 bp, 246 bp, 201 bp, 118 bp, and 195 bp respectively. Beta-actin (*ActB*; predicted amplicon size of 120 bp) is run alongside as a RT-PCR positive control in addition to negative no-reverse transcription (No-RT) and no-template (NTC) controls

4. To avoid disruption of embryoid bodies when changing media, gently swirl the dishes to pool them toward the center. Using a sterile transfer pipette, aspirate 1–1.5 mL of media from the edge of the plate and replenish with fresh, warm media. This will minimize mechanical disruption of EBs.
5. The yield of c-Kit$^+$, CXCR$^+$ definitive endoderm progenitors will vary according to the cell line used. For the mouse ES cell line used here, a yield between 30 and 45% with Activin-A and Wnt-3A treatment is typically achieved. Yields of definitive endoderm differentiation from a range of mouse ES cells lines has been published by Christodolou et al. [30].
6. Ensure media is changed every day after plating EBs to form a monolayer. This will minimize cell death caused by all-trans retinoic acid.

References

1. Hicks RM (1975) The mammalian urinary bladder: an accommodating organ. Biol Rev 50:215–246
2. Hicks RM, Ketterer B, Warren RC (1974) The ultrastructure and chemistry of the luminal plasma membrane of the mammalian urinary bladder: a structure with low permeability to water and ions. Philos Trans R Soc Lond Ser B Biol Sci 268:23–38
3. Visnjar T, Kocbeck P, Kreft ME (2012) Hyperplasia as a mechanism for rapid resealing urothelial injuries and maintaining high transepithelial resistance. Histochem Cell Biol 137:177–186
4. Sun TT, Zhao H, Provet J, Aebi U, Wu XR (1996) Formation of asymmetric unit membrane during urothelial differentiation. Mol Biol Rep 23:3–11
5. Wu XR, Kong XP, Pellicer A, Kreibech G, Sun TT (2009) Uroplakins in urothelial biology, function and disease. Kidney Int 75:1153–1165
6. Colopy SA, Bjorling DE, Mulligan WA, Bushman W (2014) A population of progenitor cells in the basal and intermediate layers of the murine bladder urothelium contributes to urothelial development and regeneration. Dev Dyn 243:988–998
7. Wang C, Ross WT, Myosorekar IU (2017) Urothelial generation and regeneration in development, injury and cancer. Dev Dyn 246:336–343
8. Mysorekar IU, Mulvey MA, Hultgren SJ, Gordon JI (2002) Molecular regulation of urothelial renewal and host defenses during infection with uropathogenic Escherichia coli. J Biol Chem 277:7412–7419
9. Dasgupta J, Tincello DG (2009) Interstitial cystitis/bladder pain syndrome: an update. Maturitas 64:212–217
10. Kaufman DS, Shipley WU, Feldman AS (2009) Bladder cancer. Lancet 374:239–249
11. Keller J, Chiou H-Y, Lin H-C (2013) Increased risk of bladder cancer following diagnosis with bladder pain syndrome/interstitial cystitis. Neurourol Urodyn 32:58–62
12. Khandelwal P, Abraham SN, Apodaca G (2009) Cell biology and physiology of the uroepithelium. Am J Physiol Renal Physiol 297:F1477–F1501
13. Boumelhem BB, Assinder SJ, Hammans C, Tanudiastro MP, Le DTM, Brigden KWL, Fraser ST (2017) The mesendoderm: a wellspring of cell lineages for regenerative medicine. In: Frontiers in stem cell and regenerative medicine research, vol 4. Bentham Science Publishers, Sharjah, pp 3–67
14. Gupta A, Bischoff A, Peña A et al (2014) The great divide: septation and malformation of the cloaca, and its implications for surgeons. Pediatr Surg Int 30:1089–1095
15. Marker PC, Donjacour AA, Dahiya R et al (2003a) Hormonal, cellular, and molecular control of prostatic development. Dev Biol 253:165–174
16. Cao M, Liu B, Cunha G et al (2008) Urothelium patterns bladder smooth muscle location. Pediatr Res 64:352–357
17. Haraguchi R, Motoyama J, Sasaki H, Satoh Y, Miyagawa S, Nakagata N, Moon A, Yamada G (2007) Molecular analysis of coordinated bladder and urogenital organ formation by hedgehog signalling. Development 134:525–533

18. Cunha GR, Fujii H, Neubauer BL et al (1983) Epithelial-mesenchymal interaction in prostatic development. I. Morphological observations of prostatic induction by urogenital sinus mesenchyme in epithelium of the adult rodent urinary bladder. J Cell Biol 96:1662–1670
19. Chung LW, Cunha GR (1983) Stromal-epithelial interactions: II. Regulation of prostatic growth by embryonic urogenital sinus mesenchyme. Prostate 4:503–511
20. Tanaka ST, Ishii K, Demarco RT et al (2010) Endodermal origin of bladder trigone inferred from mesenchymal-epithelial interaction. J Urol 183:386–391
21. Li W, Cavasotto CN, Cardozo T et al (2005) Androgen receptor mutations identified in prostate cancer and androgen insensitivity syndrome display aberrant ART-27 coactivator function. Mol Endocrinol 19:2273–2282
22. Wu X-R, Kong X-P, Pellicer A et al (2009) Uroplakins in urothelial biology, function, and disease. Kidney Int 75:1153–1165
23. Moll R, Wu XR, Lin JH et al (1995) Uroplakins, specific membrane proteins of urothelial umbrella cells, as histological markers of metastatic transitional cell carcinomas. Am J Pathol 147:1383–1397
24. Mauney JR, Ramachandran A, Yu RN et al (2010) All-trans retinoic acid directs urothelial specification of murine embryonic stem cells via GATA4/6 signaling mechanisms. PLoS One 5:e11513
25. Kang M, Kim H, Han Y-M (2014) Generation of bladder urothelium from human pluripotent stem cells under chemically defined serum- and feeder-free system. Int J Mol Sci 15:7139–7157
26. Osborn SL, Thangappan R, Luria A et al (2014) Induction of human embryonic and induced pluripotent stem cells into urothelium. Stem Cells Transl Med 3:610–619
27. Soprano DR, Teets BW, Soprano KJ (2007) Role of retinoic acid in the differentiation of embryonal carcinoma and embryonic stem cells. Vitam Horm 75:69–95
28. Oottamasathien SS, Wang YY, Franco WKK et al (2007) Directed differentiation of embryonic stem cells into bladder tissue. Dev Biol 304:11–11
29. D'Amour KA, Agulnick AD, Eliazer S (2005) Efficient differentiation of human embryonic stem cells to definitive endoderm. Nat Biotechnol 23:1534–1541
30. Christodoulou C, Longmire TA, Shen SS (2011) Mouse ES and iPS cells can form similar definitive endoderm despite differences in imprinted genes. J Clin Invest 121:2313–2325

Chapter 10

Isolation and Characterization of Progenitor Cells from Human Adipose Tissue

Nitya Shree and Ramesh Bhonde

Abstract

Adipose progenitor cells have gained a lot of importance recently due to their ability to repair and regenerate injured/diseased tissues especially in the case of metabolically dysregulated conditions. Here, we describe a method to isolate and characterize adipose tissue-derived progenitor cells for their possible therapeutic use.

Key words Adipose tissue, Progenitor cells, Stromal vascular fraction, Mesenchymal stem cells, Regenerative medicine

1 Introduction

A progenitor cell is similar to a stem cell and has the capacity to differentiate into a specific cell type. Progenitor cells have a limited capacity to divide, whereas stem cells can divide indefinitely [1]. Lately, progenitor cells from various sources have been identified, isolated, and characterized and are being studied extensively for their clinical applications. There are multiple research articles that demonstrate the use of mesenchymal stem cells derived from bone marrow, umbilical cord, dental pulp and many others for the treatment of metabolic syndrome [2–4]. Adipose tissue is known to be a heterogeneous organ which is distributed throughout the body and is responsible for endocrine and metabolic functions. During the course of development, the three types of adipose tissues, that is, Brown adipocytes, White adipocytes and Beige adipocytes are derived from separate lineages and progenitor cells [5]. In recent years, progenitor cells from adipose tissue have been identified to be more favorable due to their abundance and easy accessibility [6]. The vascular resident adipose progenitor cells residing within white adipose tissue are capable of proliferating and differentiating either into white or beige adipocytes. These

Mugdha V. Joglekar and Anandwardhan A. Hardikar (eds.), *Progenitor Cells: Methods and Protocols*, Methods in Molecular Biology, vol. 2029, https://doi.org/10.1007/978-1-4939-9631-5_10, © Springer Science+Business Media, LLC, part of Springer Nature 2019

newly differentiated adipocytes have the tendency to control diabetes and related conditions (viz., obesity) [7]. Adipose tissue is an abundant source of stromal vascular fraction (SVF) which is a component of the lipoaspirate obtained from liposuction of adipose tissue. Although lipoaspirate is the waste product of liposuction, it contains different subsets of cells, which have regenerative potential. It also contains a heterogeneous population of cells derived from enzymatically digested adipose tissue [8]. Here, we demonstrate that stromal vascular fraction can be isolated from lipoaspirate of the adipose tissue by collagenase digestion method and plated to obtain adherent cells (adipose progenitor cells). Furthermore, these cells are confirmed to be bona fide adipose progenitor cells by characterizing them using flow cytometry with specific cell surface markers. The isolated progenitor cells are also confirmed based on their ability for trilineage differentiation (viz., adipocytes, chondrocytes, and osteocytes). These cells can also be differentiated into hepatocytes, islet beta cells, and neurons using a mixture of specific cocktail to direct them to specific cell type [9, 10]. These cells can be used for a number of therapeutic applications (viz., diabetes, ischemia, and skin regeneration) [11–15]. The progenitors isolated from adipose tissue have been shown to have a better differentiation ability [16].

2 Materials

Prepare all the reagents fresh at room temperature (unless mentioned otherwise). Handle all reagents and culture media aseptically and in the biosafety cabinet while using cells.

1. Growth media: 25 mM DMEM high glucose, 10% fetal bovine serum, 2 mM L-glutamine, and 1% –10,000 units/mL penicillin and 10,000 μg/mL streptomycin (Pen-Strep).
2. Phosphate buffered saline (PBS) containing Ca^{2+} and Mg^{2+} ions.
3. Digestion Buffer: 0.075% collagenase Type II solution is prepared in PBS.
4. Oil Red O Stain: Prepare main stock of the stain by dissolving 350 mg of Oil Red O in 100 mL of 100% isopropanol. Filter through a 0.22 μm filter and store at 4 °C. Working stock of the staining solution is prepared by mixing main stain stock and distilled water (6:4). Incubate at room temperature for 20 min and filter through a 0.22 μm filter and use (*see* **Note 1**).
5. Alcian Blue Stain: Prepare 1% Alcian Blue solution in 0.1 N HCl for final use (*see* **Note 1**).
6. Alizarin Red Stain: Prepare 2% Alizarin Red solution in distilled water, mix well, and adjust the pH to 4.1–4.3 with HCl or

NH_4OH. Filter the dark brown solution using Whatman filter paper No.1 and store in dark until further use (*see* **Note 1**).

7. Fixing reagent: 4% Paraformaldehyde (PFA).
8. Plasticware: 35 mm tissue culture dish, 50 mL and 15 mL conical tubes, T-25 cm^2 culture flask.
9. Specialized differentiation kits: All the specialized differentiation kits for adipogenic differentiation, chondrogenic differentiation, and osteogenic differentiation are sourced commercially from companies such as Gibco.
10. Flow cytometry reagents: Fluorochrome-tagged antibodies for CD90, CD105, HLA-DR, CD34, and respective isotype controls.

3 Methods

3.1 Progenitor Cell Isolation

1. The lipoaspirate is processed to obtain SVF as established earlier [8].
2. For the isolation of SVF, the lipoaspirate is washed thoroughly with an equal volume of PBS solution (1:1).
3. The extracellular matrix is digested using the digestion buffer containing 0.075% collagenase type II in PBS containing calcium ions for 30 min with continuous gentle shaking in water bath set at 37 °C (*see* **Note 2**).
4. Digestion is neutralized using equal volume of Dulbecco's modified Eagle's medium (DMEM), containing 10% FBS and the suspension is centrifuged at 1200 × *g* for 10 min.
5. The pellet containing SVF is washed with PBS and resuspended in growth media.
6. The resulting solution is then plated in a 25 cm^2 culture flask. Cells are incubated in a humidified chamber maintained at 37 °C with 5% CO_2. After 24 h of incubation cells are observed under the microscope. Diagrammatic representation of the procedure is shown in Fig. 1.

3.2 Flow Cytometry Analysis

1. Once the cultured cells have attained 90% confluency, they are trypsinized using trypsin–EDTA solution (0.5%) followed by incubation at 37 °C for ~2 min (*see* **Note 3**).
2. The effect of trypsin is neutralized using the culture media and the detached cell suspension is centrifuged at 200 × *g* for 5 min.
3. The supernatant is discarded and the cell pellet is then washed with cold PBS.

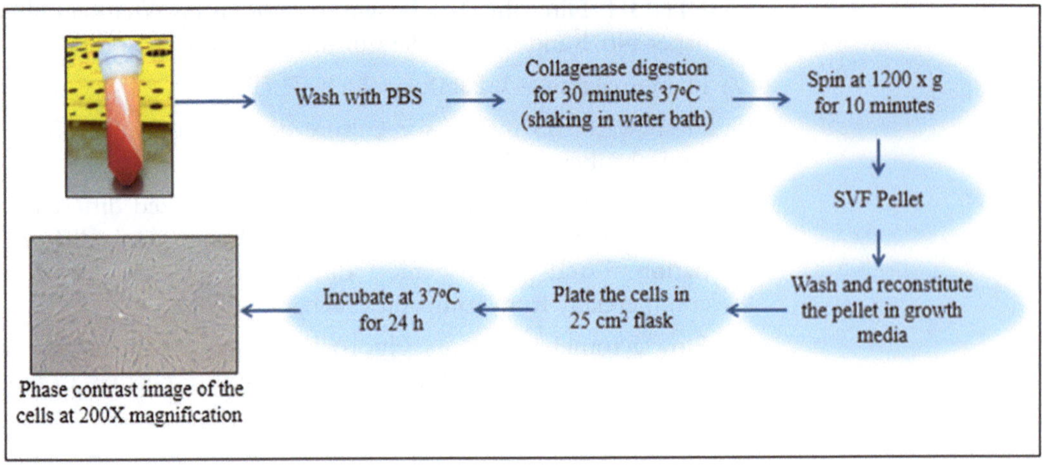

Fig. 1 Isolation of progenitor cells from adipose tissue. Lipoaspirate washed with PBS and digested with collagenase type II for 30 min at 37 °C in water bath with gentle shaking. The solution is spun at 1200 × *g* for 10 min, the resulting pellet is the SVF which is further washed with PBS and reconstituted in growth media. The solution is then plated in a 25 cm^2 flask

4. The pellet is then dislodged and further incubated with the primary antibody (viz., CD105, CD90, CD34, and HLA-DR) for 1 h on ice and then fixed with 4% paraformaldehyde. Cells are incubated on ice.
5. The cell suspension is spun again and washed with cold PBS. Percentage of positive and negative cells is analyzed by flow cytometry using FACS CALIBUR (BD Biosciences, San Jose, CA, USA) and data is analyzed using FACS Diva software (BD Biosciences, San Jose, CA, USA). 10,000 cells are acquired for each run.
6. The cells are positive for CD105 and CD90. The cells are negative for HLA-DR and CD34 markers, confirming the cells as bona fide progenitor cells.

3.3 Trilineage Differentiation

1. The cells are seeded in 35 mm tissue culture dishes at a density of 2×10^4 cells per dish for differentiation into adipocytes, chondrocytes, and osteocytes to confirm the trilineage differentiation potential of the cells.
2. The cells are differentiated using adipogenic, osteogenic, and chondrogenic medium as per the manufacturer's protocol to induce the formation of adipocytes, osteocytes, and chondrocytes, respectively (Fig. 2).
3. Oil Red O staining for adipocytes: After 8 days of differentiation, growth media is removed from the plate. The cells are quickly washed with PBS. The cells are then fixed by adding 2 mL of fixing reagent for 1 h at room temperature. The fixing reagent is then removed and the cells are quickly washed with

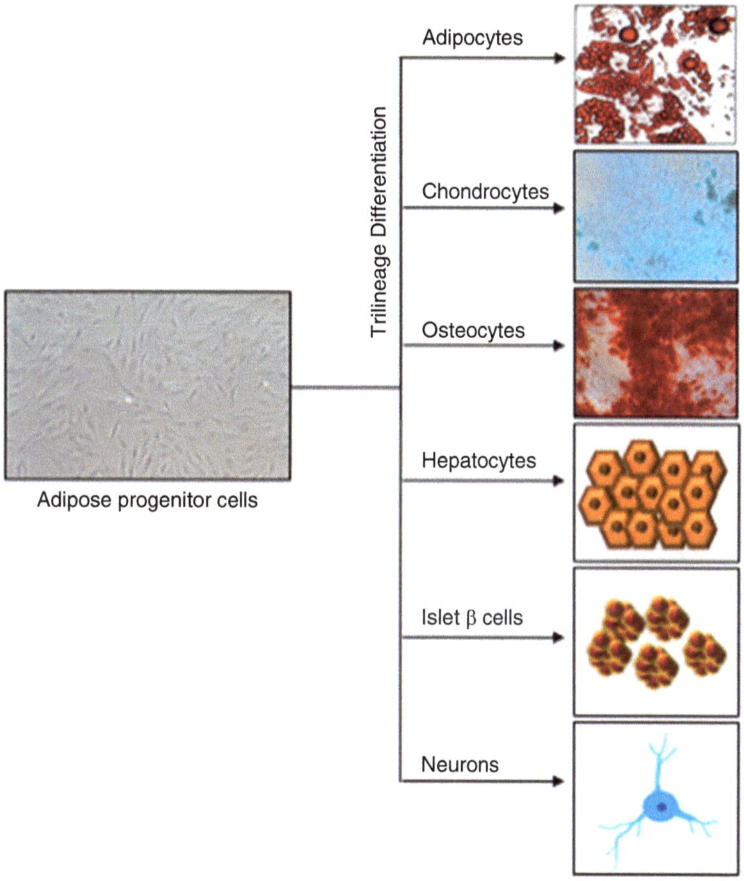

Fig. 2 Differentiation potential of adipose progenitor cells. Progenitor cells can differentiate mainly into three lineages, namely, adipocytes, chondrocytes, and osteocytes (as described in the method and shown in the upper panel as trilineage differentiation). These cells can also be directed to other cell types like hepatocytes, islet β cells, and neurons

60% isopropanol. The cells are allowed to dry for 2–3 min. Two milliliters of Oil Red O stain is added to the cells slowly and incubated at room temperature for 15 min. The stain is removed and the cells are washed five times with distilled water and observed under the microscope. Differentiated cells show matured adipocytes with oil droplets stained red (*see* **Note 4**).

4. Alcian Blue staining for chondrocytes: After 14 days of differentiation, the cells are washed with PBS once. The cells are then fixed with the fixing reagent for 30 min at room temperature. After incubation, the fixing reagent is removed; the cells are then quickly washed with PBS and stained with Alcian Blue for 30 min at room temperature. After incubation, the cells are washed with 0.1 N HCl and distilled water is added to

neutralize the acidity. The cells are then visualized under the microscope. Blue staining indicates the synthesis of proteoglycans by chondrocytes (*see* **Note 4**).

5. Alizarin Red staining for osteocytes: After differentiation, growth media is removed. The cells are washed with PBS (without Ca^{2+} or Mg^{2+}) quickly. The cells are then fixed using the fixing reagent for 30 min at room temperature. The cells are then washed with distilled water. Alizarin Red stain is added to the cells and incubated in dark for 45 min at room temperature (*see* **Note 5**). After incubation, the stain is removed and the cells are washed five times with distilled water. The cells are visualized under the microscope. Osteoblasts derived from the cells show extracellular calcium deposits stained bright orange red (*see* **Note 4**).

4 Notes

1. All staining reagents can be stored in dark, protected from direct light for ~ 2 months.
2. Collagenase digestion step requires digestion buffer containing calcium ions to enhance the dissociation efficiency. Incubation at 37 °C for 30 min is to complete the disaggregation process. Do not overdigest with collagenase, as that might result in reduced viability of the cells.
3. Trypsinization of the cells is a crucial step where trypsinizing the cells for longer than 3–4 min results in the lysis of the cells due to which the cells open up, DNA is released, and they clump.
4. Staining procedure requires gentle addition and aspiration of the liquid to avoid disruption of the cell monolayer.
5. For Alizarin Red stain, the correct pH of the solution is critical. Check pH if the solution is older than 1 month.

References

1. Seaberg RM, Van Der Kooy D (2003) Stem and progenitor cells: The premature desertion of rigorous definitions. Trends Neurosci 26 (3):125–131
2. Hao P, Liang Z, Piao H, Ji X, Wang Y, Liu Y, Liu R, Liu J (2014) Conditioned medium of human adipose-derived mesenchymal stem cells mediates protection in neurons following glutamate excitotoxicity by regulating energy metabolism and GAP-43 expression. Metab Brain Dis 29:193–205
3. Chen P, Huang Q, Xu XJ, Shao ZL, Huang LH, Yang XZ, Guo W, Li CM, Chen C (2016) The effect of liraglutide in combination with human umbilical cord mesenchymal stem cells treatment on glucose metabolism and β cell function in type 2 diabetes mellitus. Zhonghua Nei Ke Za Zhi 55:349–354
4. Xie Z, Hao H, Tong C, Cheng Y, Liu J, Pang Y, Si Y, Guo Y, Zang L, Mu Y (2016) Human umbilical cord-derived mesenchymal stem cells elicit macrophages into an anti-inflammatory phenotype to alleviate insulin resistance in type 2 diabetic rats. Stem Cells 34:627–639
5. Berry R, Rodeheffer MS, Rosen CJ, Horowitz MC (2015) Adipose tissue residing progenitors

adipocyte lineage progenitors and adipose derived stem cells ADSC. Curr Mol Biol Rep 1(3):101–109

6. Zuk PA, Zhu M, Ashjian P, De Ugarte DA, Huang JI, Mizuno H (2002) Human adipose tissue is a source of multipotent stem cells. Mol Biol Cell 13:4279
7. Berry DC, Jiang Y, Graff JM (2016) Emerging roles of adipose progenitor cells in tissue development, homeostasis, expansion and thermogenesis. Trends Endocrinol Metab 27 (8):574–585
8. Bourin P, Bunnell BA, Casteilla L, Dominici M, Katz AJ, March KL, Red H, Rubin JP, Yoshimura K, Gimble JM (2013) Stromal cells from the adipose tissue-derived stromal vascular fraction and culture expanded adipose tissue-derived stromal/stem cells: a joint statement of the International Federation for Adipose Therapeutics and Science (IFATS) and the International Society for Cellular Therapy (ISCT). Cytotherapy 15:641–648
9. Zuk PA, Zhu M, Mizuno H, Huang J, Futrell JW, Katz AJ, Benhaim P, Lorenz HP, Hedrick MH (2001) Multilineage cells from human adipose tissue: implications for cell-based therapies. Tissue Eng 7(2):211–228
10. Chandra V, Swetha G, Muthyala S, Jaiswal AK, Bellare JR, Nair PD, Bhonde RR (2011) Islet-like cell aggregates generated from human adipose tissue derived stem cells ameliorate experimental diabetes in mice. PLoS One 6(6): e20615
11. Cheng K, Kuo T, Kuo K, Hsiao C (2011) Human adipose-derived stem cells: Isolation, characterization and current application in regeneration medicine. Genom Med Biomarkers Health Sci 3:53e62
12. Thadani JM, Marathe A, Vyas BK, Vyas RB, Ansarullah A, Kshatriya P, Modiv D, Vakodikar S (2016) Therapeutic application of autologous ADMSCS (adipose derived mesenchymal stem cells) and islet like cell aggregates (ICAS) derived from them in type 1 and 2 diabetes mellitus–a proof of concept study. J Stem Cell Res Dev 3:008
13. Zhou L, Song Q, Shen J, Xu L, Xu Z, Wu R, Ge Y, Zhu J, Wu J, Dou Q, Jia R (2017) Comparison of human adipose stromal vascular fraction and adipose-derived mesenchymal stem cells for the attenuation of acute renal ischemia/reperfusion injury. Sci Rep 7:44058
14. Gaur M, Dobke M, Lunyak VV (2017) Mesenchymal stem cells from adipose tissue in clinical applications for dermatological indications and skin aging. Int J Mol Sci 18(1):208
15. Mao F, Tu Q, Wang L, Chu F, Li X, Li HS, Xu W (2017) Mesenchymal stem cells and their therapeutic applications in inflammatory bowel disease. Oncotarget 8(23):38008–38021
16. Manini I, Gulino L, Gava B, Pierantozzi E, Curina C, Rossi D, Brafa A, D'Aniello C, Sorrentino V (2011) Multi-potent progenitors in freshly isolated and cultured human mesenchymal stem cells: a comparison between adipose and dermal tissue. Cell Tissue Res 344(1):85–95

Chapter 11

Identification and Analysis of Mouse Erythroid Progenitor Cells

Chanukya K. Colonne, Jia Hao Yeo, Campbell V. McKenzie, and Stuart T. Fraser

Abstract

The most common cell type in the human body, the red blood cell or erythrocyte, has a life span of approximately 3 months. To compensate for this massive cellular requirement and short life span, the major blood producing tissues contain vast numbers of erythroid progenitor cells. Erythroid progenitors differentiate progressively from hematopoietic stem cells to committed erythroid progenitors to reticulocytes lacking a nucleus and finally to functionally mature erythrocytes in the circulation. Different erythroid progenitor activity, representative of distinct stages of erythropoiesis, can be observed using semisolid colony assays. Distinct stages of erythroid maturation can also be monitored by flow cytometry. Here, we discuss the range of different technical approaches that are used to identify and quantify erythroid progenitors, with particular focus on the mouse as a model system.

Key words Red blood cells, Erythroid, Progenitor, Bone marrow, Spleen, Anemia

1 Introduction

1.1 Crafting Red Blood Cells

Any researcher who spends time investigating the remarkable production of blood cells will notice the presence of the suffix *poiesis* in numerous scientific terms. What precisely does this suffix indicate? Terms ending in *poiesis* such as hematopoiesis, erythropoiesis and myelopoiesis refer to the "making or crafting" of those particular cell types. This suffix is of ancient Greek origin: *poiesis* referred to any situation in which something (a person, a house, a city, heroic fame) was crafted or made. In biological terms, *poiesis* is used to indicate the ongoing constant production of a particular cell type and to distinguish this process from the initial development of particular cell types. This latter process is given the suffix *genesis*—also from the ancient Greek—"to give rise to or to create from nothing." For example, in the developing embryo, the appearance of the first blood cells is termed "hemogenesis" or "hematogenesis" (e.g., hemogenic endothelium) whereas the constant

Mugdha V. Joglekar and Anandwardhan A. Hardikar (eds.), *Progenitor Cells: Methods and Protocols*, Methods in Molecular Biology, vol. 2029, https://doi.org/10.1007/978-1-4939-9631-5_11, © Springer Science+Business Media, LLC, part of Springer Nature 2019

production of blood from an existing stem cell pool, such as seen in the adult bone marrow, is termed "hematopoiesis." The adjective forms are therefore hematopoietic (e.g., hematopoietic stem cells—HSC), erythropoietic and myelopoietic or thrombopoietic ("crafting thrombocytes" or platelets).

The term "stem cell" was developed by Alexander Maximow, who hypothesized that there was a cell type capable of giving rise to many other different cell types by examining the yolk sacs of mammalian embryos and recognizing that there was an origin to the rapidly developing blood that he observed in this early extraembryonic tissue [1]. The term progenitor can be broken down into the Greek *pro* or first, and *gen* to give rise to. The term *blast* can often be seen in publications discussing blood production. Again from the ancient Greek, *blast* means a bud about to open, or a seed that is sprouting. In particular, the addition of blast indicates that a particular cell type has the potential to give rise to more mature cells but has not fully transformed into its final product.

The isolation of soluble growth factors that could expand the blood cell-producing progenitors in culture was a major advance on our understanding of the hematopoietic system. Carnot and Defrande in 1906, proposed the existence of a soluble factor that could enhance the production of blood cells. This factor was first termed "hemopoietin." The prefix *hemo* refers to the Greek word for blood; *poie* refers to making. The suffix *-tin* or *-in* comes from the Latin *inus*; pertaining or referring to a specific item. Hemopoietin was eventually found to support the production of red blood cells alone and hence became known as erythropoietin: *erythros* being Greek for red color; *poiet* referring to "crafting"; and *-in*, "pertaining to." Hence, erythropoietin is the "molecule pertaining to making red cells" and is commonly abbreviated to Epo [2]. We will leave it to the reader to break down the origins of the word "thrombopoietin." Interleukins comes from *inter* (between) and *leukos* (white) and again the suffix *-in* (pertaining to) [3].

1.2 Hematopoiesis

Hematopoiesis is the continuous, ongoing process of making blood cells [4]. We will be referring to this process in mammals, with a specific focus on the mouse, as this is the best described system however various forms of hematopoiesis has been described from insects to humans. This process is initiated in utero in the early mammalian embryo and continues through the entire life span of the mammal in order to maintain life [5]. This self-renewing biological system results in the production of erythrocytes, which transport oxygen and carbon dioxide; leukocytes, a broad classification of cells of the immune system involving granulocytes, monocytes, lymphoid cells; as well as megakaryocytes, involved in platelet production which aids with blood clotting.

The hematopoietic system is based on a hierarchy with rare hematopoietic stem cells (HSC) with self-renewal ability sitting at the apex, and differentiating into the various levels of progenitor

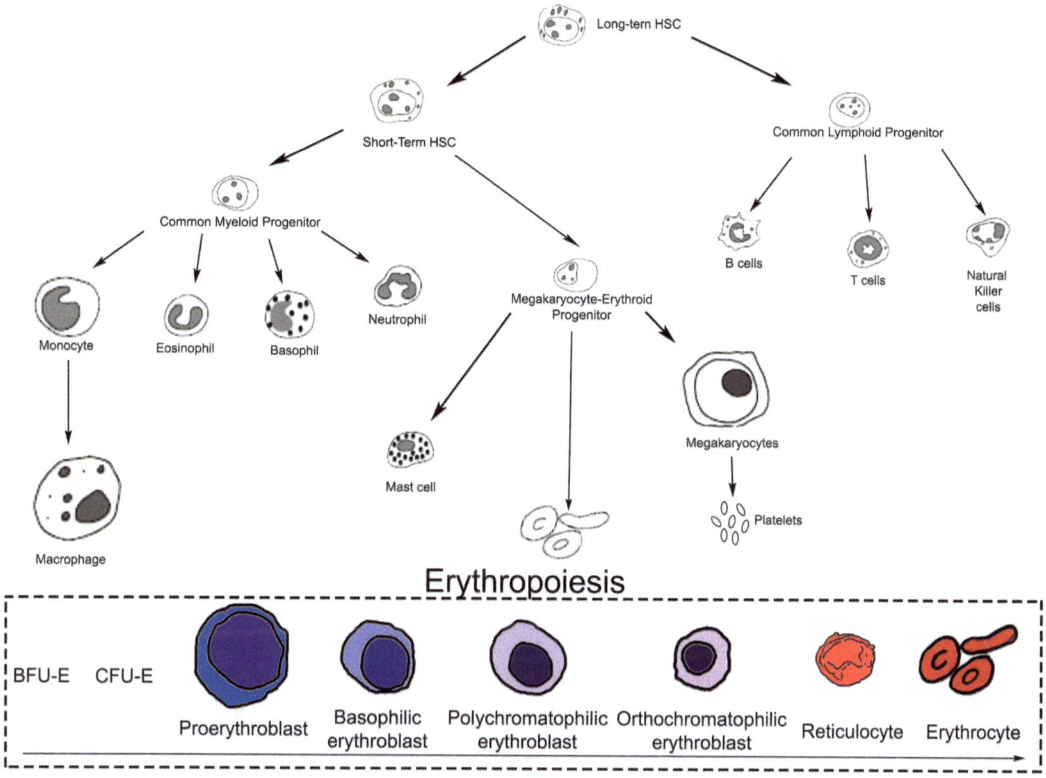

Fig. 1 The erythropoietic hierarchy found in the bone marrow of the adult mouse. Top panel shows the developmental hierarchy of the blood system from rare hematopoietic stem cells through to each major lineage. Boxed area shows the progression of the erythroid program from BFU-E, to erythroblast to erythrocyte

cells, before producing mature terminally differentiated functional cells [4, 5] (Fig. 1). There are three distinct arms of the hematopoietic system; the erythroid lineage which produces erythrocytes and platelets; the myeloid lineage, which culminates in granulocytes and macrophages; and the lymphoid system, which creates B, T, natural killer, as well as innate lymphoid cells. Each lineage requires a specific palate of individual yet at times overlapping genetic pathways that are initiated via a myriad of growth factors, cytokines, chemokines, extracellular matrix, and adhesion molecules that control each developmental process. This chapter will be focusing on the erythroid lineage and the methods and techniques of isolating and characterizing mouse erythroid progenitors.

1.3 Erythropoiesis

Erythropoiesis is the production of erythrocytes (red blood cells) [6]. Erythrocytes, the most frequent blood cell type in the human body, primarily function to facilitate transport of oxygen via the respiratory system to end organ tissues and carbon dioxide, the waste product of cellular metabolism. Mammalian adult (definitive) erythrocytes circulate without a nucleus or organelles. This allows the erythrocyte to form a biconcave discoidal shape, allowing passage through narrow capillaries. Erythrocytes contain vast amounts of hemoglobin, the iron-complexed tetrameric protein which binds oxygen and carbon dioxide. Adult erythrocytes and platelets share a common progenitor, the megakaryocyte/erythroid progenitor (MEP) [4]. The MEP arises from the common myeloid progenitors (CMP), which in turn arise from the multipotent hematopoietic stem cells (HSC) in the bone marrow and other erythropoietic niches. MEPs differentiate into a number of distinct erythroid progenitors. Terminal maturation of erythroid cells requires the condensation and expulsion of the nucleus resulting in reticulocytes which take their final form as erythrocytes in circulation (Fig. 1).

1.4 The Erythroid Lineage

MEPs give rise to two distinct erythroid progenitors; the slowly cycling BFU-E (burst-forming unit—erythroid); and the more mature, rapidly cycling CFU-E (colony-forming unit—erythroid) [7]. The proerythroblast is the first identifiable erythroid precursor [8]. From the proerythroblast stage, erythroblasts divide and mature, progressing from the proerythroblasts to the basophilic stage, then the polychromatic and orthochromatic erythroblast stage. Cellular changes in developing erythroid cells include progressive reduction in cell size; loss of RNA and nucleic material; hemoglobinization and reorganization of the cell membrane [8]. The expulsion of the nucleus in the late orthochromatic erythroblast stage in mammals results in the production of reticulocytes [8]. Reticulocytes maintain ribosomal RNA and continue synthesizing hemoglobin. In circulation, the reticulocyte acquires the biconcave discoidal shape characteristic of a terminally mature erythrocyte (Fig. 1).

1.5 Identifying Erythroid Progenitors from the Complex Milieu of the Bone Marrow

The erythroid lineage-specific surface antigen, TER119, is expressed from the early erythroblast stage through to mature circulating erythrocytes [9]. While not specific to the erythroid lineage, CD71 (transferrin receptor) is highly expressed by early stage erythroblasts [10]. CD71 is highly expressed in proerythroblasts and early basophilic erythroblasts, declines with erythroblast maturation and is absent from the surface of mature erythrocytes [10–14]. The hyaluronic acid receptor, CD44 is a useful marker in discriminating maturation stages of erythropoiesis as CD44 expression is downregulated by at least 30-fold in a stepwise manner during erythroid maturation from proerythroblast to orthochromatic erythroblast stages [8]. Flow-cytometric assays have utilized the different expression patterns of these molecules to help identify various erythroblast maturation stages.

1.6 The Distinct Erythroid Progenitor Populations

Throughout the life span of a mouse (the representative mammalian species that is best characterized for the erythroid lineage), at least five waves of erythroid populations can be identified [6, 14, 15]. These include the first red blood cell population—the primitive erythroid with a corresponding progenitor population known as the primitive erythroid colony-forming cell (EryP-CPC); a yolk sac-derived definitive (adult-type) erythroid population; a definitive erythroid population that arises in the fetal liver; the definitive erythroid cells produced under homeostatic conditions in the bone marrow; and a population that migrates from the bone marrow to the spleen (and other organs) during events of anemic stress such as trauma-related blood loss, pregnancy, or hemolysis. This last wave of erythropoiesis is termed "stress erythropoiesis" and can be induced in a nonlethal animal model using the hemolytic agent phenylhydrazine.

1.6.1 Primitive Erythroid Colony-Forming Cells

The first wave of erythroid cells to appear in the mammalian embryo is distinct from all other waves and has been termed primitive erythroid cells (EryP). Compared to adult-type or definitive erythroid cells, EryP exhibit distinct physiological characteristics including; possessing nuclei while in circulation; being six times larger in volume than circulating adult RBCs; and expressing embryonic globin genes [16]. Primitive erythroid cells mature as a semisynchronous wave and dominate the early embryonic circulation [12, 14]. Primitive erythroid cells are derived from the primitive erythroid colony-forming cell (EryP-CFC) which is present within the early yolk sac for a window of approximately 36 h. EryP-CFC was first isolated in a semisolid colony assay containing platelet-depleted serum and Epo. The erythroid cells which matured under these conditions expressed embryonic globin genes [17]. The discrete window of EryP-CFC appearance in mouse embryogenesis was mapped by Palis and colleagues [14]. Using a transgenic mouse line that used epsilon-globin regulatory elements to drive a nuclear-green fluorescent protein, EryP-CFC could be highly enriched by fluorescence-activated cell sorting (FACS) of GFP+ cells from early mouse embryos [18]. This allowed for transcriptome analysis of isolated EryP-CFC that demonstrated a unique profile of gene expression patterns compared to mature primitive erythroid cells [18].

1.6.2 Definitive Erythroid Progenitors: Burst-Forming Unit-Erythroid (BFU-E) and Colony-Forming Unit-Erythroid (CFU-E)

The two earliest identified erythroid progenitors are the BFU-E and more mature CFU-E progenitors [19, 20]. CFU-E progenitors are rapidly dividing cells that are very responsive to low concentrations of erythropoietin, and gives rise to colonies consisting of 8–32 red cells in 2–3 days in mice and 7 days in humans [21, 22]. The BFU-E cell is a more immature cell that divides more slowly, but can give rise to colonies consisting of more than 500 red cells over 15 days in humans and 7–10 days in mice [19]. Stimulating BFU-Es with a combination of erythropoietin and either stem cell factor or IL3 or GM-CSF, results in the induction of division and differentiation of these cells into CFU-E cells [23].

Maturing mouse erythroblasts can be isolated according to TER119, CD71 and other marker expression. However, it is currently not possible to definitively isolate mouse CFU-E or BFU-E by FACS. Complex flow cytometry regimens utilizing negative selection criteria have been described in an attempt to isolate these cells for further study. Lineage$^-$ Sca-1$^-$ IL7Rα$^-$ IL3Rα$^-$ CD41$^-$ c-Kit$^+$ CD71$^+$ cells have been described to account for most of the CFU-E activity in the murine bone marrow [24]. BFU-E and CFU-E cells have been isolated from the mouse fetal liver by utilizing negative selection for TER119, B220, Mac-1, CD3, Gr1, Sca-1, CD16/CD32, CD41, CD34, followed by separation based on CD71 expression levels, highlighting the difficulty of this process [25, 26]. Socolovsky and colleagues have further refined the process of purification of erythroid progenitor populations using the surface markers c-Kit, CD55, and CD49f also known as integrin alpha-6 [27]. Human BFU-E and CFU-E have been enriched according to expression of CD45, CD71 and the fatty acid translocase CD36 [28].

1.6.3 Stress Erythroblasts

Stress erythroblasts are a specialist subpopulation of erythroblasts that utilizes signals distinct from steady-state erythropoiesis [29, 30]. Stress erythroblasts are derived from a group of BFU-Es that arise in the spleen in response to hypoxic stress via a BMP4 and SMAD5-dependent mechanism [30–32]. Notch and Hedgehog signaling are also required for stress erythropoiesis [31, 33]. The splenic stress-specific BFU-E progenitor units are derived from CD34$^+$Kit$^+$Sca1$^+$Lineage$^-$ (34KSL) cells that migrate to the spleen from the bone marrow [34, 35]. Splenic stress erythroblasts are distinct from steady state erythroblasts by simultaneous expression of both immature cell markers (c-Kit and Sca1) and late erythroid markers (CD71 and TER119) [30, 34, 35].

1.7 Immortalized Erythroid Progenitor Cell Lines

Erythroblasts can be maintained in culture following transformation into an immortal cell line. Cell lines such as mouse erythroleukemic cells (MEL) or human leukemic cell lines (HEL) can be useful tools for studying erythroblast biology though do not completely resemble the untransformed erythroblast. MEL are an immortalized cell line isolated from the spleen cells of mice infected with the Friend virus [36]. MEL cells grow in suspension or are semiadherent and are highly proliferative in culture with most cell lines dividing every 12 h. MEL cells are a well-established in vitro model of erythropoiesis as they are representative of early erythroid precursors (proerythroblast or CFU-e cells) arrested in development but will undergo erythropoiesis in the presence of various differentiating agents [36].

2 Materials

2.1 Dissecting Equipment and Consumables

1. Dissection scissors.
2. Toothed forceps.
3. 25G needles.

4. 3 mL syringes.
5. 3 mL sterile plastic transfer pipettes.
6. 15 mL conical polypropylene tubes.
7. 35 mm petri dishes.
8. 60 mm petri dishes.
9. 96 mm petri dishes.
10. 70 μm filter mesh (*see* **Note 1**).
11. Heparinized blood collection tubes.
12. Tubes appropriate for flow-cytometric analysis (material and shape depend on model of flow cytometer), herein termed "FACS tubes."
13. 6-well or 96-well tissue culture quality plates.

2.2 Buffers and Chemicals

1. Phosphate buffered saline (PBS) solution: 10 mM phosphate buffer, 137 mM sodium chloride, 2.7 mM potassium chloride, pH 7.3–7.5.
2. Fluorescence-activated cell sorting (FACS) buffer: PBS, containing 0.01% weight/volume bovine serum albumin.
3. Propidium iodide (1 mg/mL) solution.
4. 4′,6-diamidino-2-phenylindole (DAPI) (1 mg/mL) solution: 1 μg/mL DAPI final concentration in FACS buffer.
5. FACS buffer with PI: FACS buffer as described above containing 1 in 1000 dilution of 1 mg/mL propidium iodide solution.
6. Phenylhydrazine (PHZ): Commercially available 97% phenylhydrazine.
7. Phenylhydrazine stock solution: Create a stock solution of 1 mg/mL PHZ in sterile PBS. Adjust to pH 7.4 using NaOH.
8. 70% ethanol.
9. Dimethyl sulfoxide.

2.3 Flow Cytometry

1. Directly conjugated fluorescent antibodies against mouse surface markers.
2. Multiparametric flow cytometer.
3. Flow-cytometric data analysis software package—FlowJo.

2.4 Cell Culture

1. Erythroid colony assays optimized for mouse erythroid progenitors (Stem Cell Technologies). Catalogue number M3334 for BFU-E; catalogue number M3434 for late BFU-E and CFU-E.
2. Methylcellulose powder.

3. Recombinant mouse growth factors: erythropoietin (Epo), stem cell factor (SCF), interleukin-3 (IL-3), interleukin-6 (IL-6).
4. Iscove's modified Dulbecco's medium (IMDM).
5. Protein-free hybridoma medium-II (PFHM-II).
6. Sodium ascorbate: prepare 5 mg/mL stock in distilled water.
7. Monothioglycerol (MTG).
8. Platelet-poor platelet-derived fetal bovine serum.
9. Primitive erythroid colony-forming cell methylcellulose: Alpha-modified Eagle's medium (alpha-MEM): 4.4% w/v methylcellulose powder melted in 480 mL sterilized double distilled water. When cooled, add 500 mL 2xIMDM/penicillin–streptomycin/MTG.
10. EryP medium: 55% Primitive erythroid colony-forming cell methylcellulose, 10% platelet-derived fetal bovine serum, 1:100 dilution GlutaMAX, 2 mM sodium ascorbate, 0.6 mM MTG, 4 U/mL recombinant Epo, IMDM with penicillin–-streptomycin to 100% final volume.
11. OP9 differentiation medium: alpha-MEM medium with 20% fetal calf serum, penicillin–streptomycin.
12. TrypLE: recombinant trypsin solution with EDTA.
13. MEL cell growth medium: Roswell Park Memorial Institute (RPMI) 1640 with L-Glutamine, 10% heat-inactivated fetal bovine serum, 1% penicillin–streptomycin.

3 Methods

3.1 Identifying Distinct Stages of Erythroid Cell Development by Flow Cytometry

3.1.1 Isolating Single Cell Suspensions from the Adult Mouse

Bone Marrow

1. Following euthanasia of mouse according to protocols approved by institutional animal ethics committee, isolate the femur, the largest long bone.
2. Wet the cadaver with 70% ethanol to control the dander during dissection.
3. Pull the skin from the hind limbs, find the knee joint and using toothed forceps, grip the quadriceps femoris muscle, which covers the front of the femur.
4. Holding the quadriceps femoris muscle in place with toothed forceps, make a cut parallel to the bone but not through the bone. Peel away the rectus femoris toward the hip joint. The femur should be exposed. Insert the scissors ventrally (underneath) the femur in closed configuration.
5. Once beneath the bone, open the scissors making a blunt dissection. This will separate the vastus lateralis from the

bone. Snip the femur at the hip joint first then from the knee joint.

6. Flush bone marrow with PBS to harvest the marrow from the murine femur (*see* **Note 2**). To do so, grip the femur with toothed forceps, use a 5 or 10 mL syringe with cell culture media.
7. Using a 25G needle, insert the needle into one end of the bone while holding the bone over a 15 mL tube with the lid off. Flush out the marrow with the media into the tube (*see* **Note 3**).

Spleen

1. To harvest spleen cells, dissect the entire spleen from the euthanized animal, trim any adipose tissue away from the organ and place onto a 70 μM mesh filter in a small petri dish.
2. Gently push the spleen through the filter material (*see* **Note 4**).
3. Collect all the cells from the 35 mm petri dish and transfer to a 15 mL conical tube.

Circulating Blood Cells

Obtain circulating blood according to the techniques allowed by institutional animal ethics committees. To obtain circulating erythrocytes by external cardiac puncture;

1. Euthanize mouse according to local institutional regulations and approvals.
2. Immediately dissect mouse chest wall and expose beating heart.
3. Aspirate blood from the beating left and right ventricle using an insulin syringe. Flush any blood obtained directly into a blood collection tube with heparin or EDTA or into a 15 mL tube with PBS containing heparin to prevent clotting (*see* **Note 5**).

3.1.2 Immunofluorescent Staining for Flow Cytometry

1. Centrifuge all single cell suspensions (bone marrow, spleen, blood) in 10 mL of PBS at 200 × *g* for 5 min to remove serum and other contaminating proteins as well as removing floating cell debris and adipose cells.
2. Aspirate supernatant and resuspend cells at a concentration of 0.5–1 × 10^6 cells in 100 μL of FACS buffer (*see* **Note 6**).
3. Incubate cells with primary conjugated antibodies for 30–60 min on ice (*see* **Note 7**).
4. Wash cells with FACS buffer to remove unbound antibodies by using 1 mL of FACS buffer per sample and centrifuging at 400 × *g* for 5 min.
5. Remove supernatant and resuspend cells in FACS buffer+PI or FACS buffer+DAPI then transfer the samples to FACS tubes. For an explanation of appropriate controls for flow cytometry (*see* **Note 8**).

3.1.3 Utilizing the Cell Surface Markers CD71 and Ter119 to Identify Mouse Erythroid Maturation Stages

1. Obtain bone marrow, spleen, or blood samples as described above and stain with fluorescently conjugated antibodies as described above.
2. Gate cells according to forward scatter (FSC) assessing size and side scatter (SSC) assessing granularity, to exclude cell debris (very small and nongranular), dead cells (very small and very granular) and doublets (very large). Doublets can also be excluded using FSC-H vs FSC-A analysis, if those parameters are available on the flow cytometer being used.
3. Nonviable cells were excluded by gating out cells that had high levels of fluorescence in the propidium iodide fluorescent channel. DAPI can be used in place of PI though this requires a violet laser for excitation and appropriate filter/photometer.
4. Use unstained controls to set the gates for fluorescence.
5. Stain sample of interest with antibodies against TER119 and CD71 that are conjugated to two independent fluorophores (*see* Subheading 3.1.2) to identify four different cell populations that correspond morphologically to proerythroblasts (1), basophilic erythroblasts (2), late basophilic and polychromatic erythroblasts (3), and orthochromatic erythroblasts (4) (*see* **Note 9**).
6. Integrate the forward scatter parameter to further delineate the stages of erythroblast maturation. This will provide five populations roughly corresponding to erythrocytes, reticulocytes, orthochromatic erythroblasts, polychromatic erythroblasts, basophilic erythroblasts, and proerythroblasts [37].
7. Plot CD71 versus forward scatter of all TER119 positive cells (populations 2, 3, 4 in Fig. 2), this will help separate the three populations into a further four populations, though with marked overlap in the profiles of CD71 between the gated clusters. The TER119^{low}/CD71^{high} population (Population 1 in Fig. 2) will correspond to proerythroblasts (*see* **Note 9**).

3.1.4 Utilizing the Cell Surface Markers TER119, CD44, and the Forward Scatter Property to Identify Erythroid Maturation Stages

1. Stain sample of interest with antibodies against TER119 and CD44 that are conjugated to two different fluorophores.
2. Gating the expression levels of CD44 as a function of the forward scatter for all TER119 positive cells enable distinction of five distinct clusters of cells that have a progressive decline in CD44 expression levels with decreasing cell size (Fig. 3). The population 4 can be further subdivided into CD44^{high} and CD44^{low} populations. The TER119^{neg} CD44^{+} population corresponds to proerythroblasts. The other remaining populations 2, 3, 4a, 4b, and 5 correspond to basophilic erythroblasts, polychromatic erythroblasts, orthochromatic erythroblasts, reticulocytes, and erythrocytes, respectively [8].

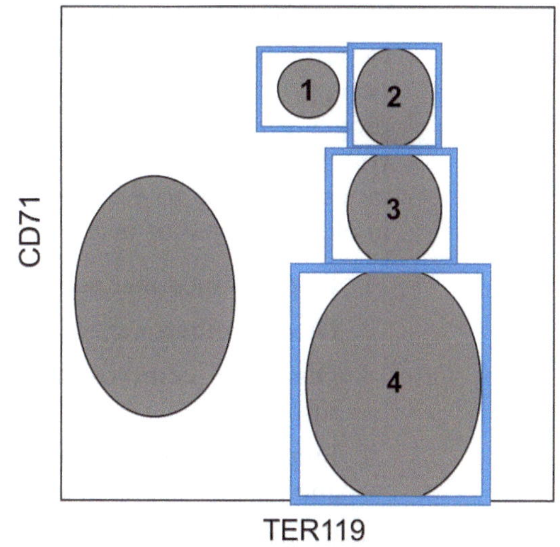

Fig. 2 Schematic representation of the gating strategy for adult mouse bone marrow cells with progressively maturing mouse erythroid cells indicated by CD71 and Ter-119 expression profiles. Gate 1 represents the least mature population that can be isolated with erythroid commitment while Gate 4 represents the phenotype of mature circulating erythrocytes

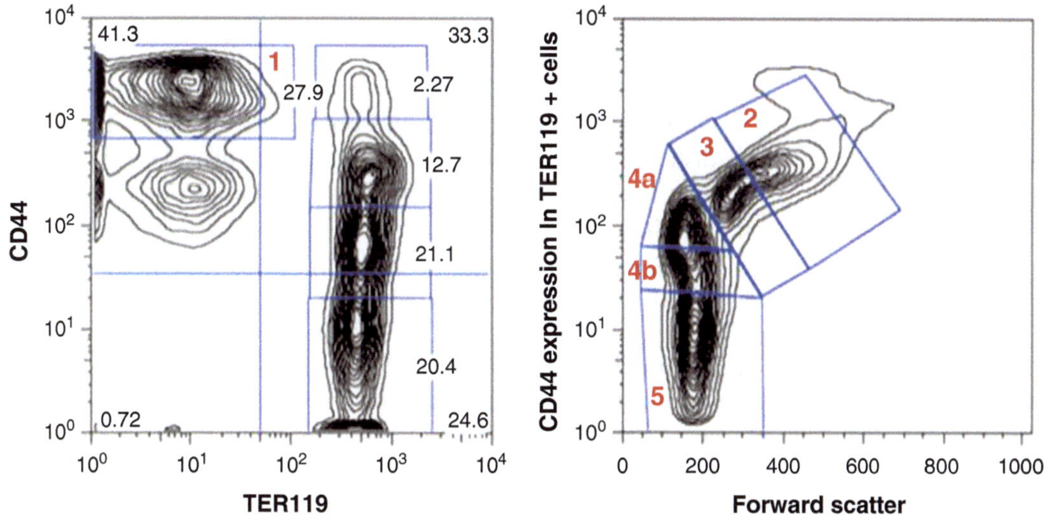

Fig. 3 Schematic representation of the gating strategy for adult mouse bone marrow cells with progressively maturing mouse erythroid cells indicated by CD44 and Ter-119 expression profiles

3.2 Erythroid Colony Assays

Erythroid colony assays are a useful means for quantifying the erythroid progenitor activity of mixture of cells. Three distinct erythroid colony types can be assayed as described below.

3.2.1 Primitive Erythroid Colony Assays

As mentioned in the introduction, primitive erythroid cells are the earliest form of red blood cell and dominate the embryonic circulation prior to the development of the fetal liver. Primitive erythroid progenitor cells (EryP-CFC) are only found in the extraembryonic yolk sac, and only for a limited window of development (in the mouse from 7.5 to 8.5 days postcoitus). EryP-CFC are not found in older embryos or in adult blood producing tissues.

1. Prepare timed matings of mice as described [38].
2. Dissect embryos at either 7.5 or 8.5 dpc.
3. Dissociate embryos using trypsin to obtain a single cell suspension.
4. Prepare the semisolid methylcellulose plating mixture with Epo and platelet-depleted serum.
5. Resuspend embryonic cells in EryP medium and add to Primitive erythroid colony-forming cell methylcellulose mixture at 1:10 ratio.
6. Vortex to mix cells. Stand for 5–10 mins until all bubbles have surfaced.
7. Plate into 3.5 cm petri dishes using a 16G needle and 3 or 5 mL syringe with 1 mL of cells/methylcellulose per dish.
8. Carefully place the 3.5 cm petri dishes into a 15 cm petri dish. Six individual 3.5 cm dishes can be placed in a rosette inside the 15 cm petri dish. Fill another 3.5 cm dish with autoclaved, distilled water and add into the middle of the 15 cm dish without the lid. Place the lid on the 15 cm dish. The water dish will humidify the 15 cm dish once placed in the 37 °C incubator and prevent evaporation from the colony assays.
9. Monitor the colony assays. EryP colonies will appear after 2.5–3 days and should be counted at this time point as EryP colonies rapidly die off after this period.

3.2.2 BFU-E and CFU-E Assays

Two distinct erythroid progenitors can be isolated from adult mouse and human bone marrow, the burst-forming units—erythroid (BFU-E) and the colony-forming units—erythroid (CFU-E). These two colony types are supported by different cytokine combinations and show distinct kinetics in appearance in methylcellulose colony assays. CFU-E require Epo alone and appear in colony assays after 2–3 days and then die off rapidly. In contrast, BFU-E represent a more immature progenitor, require Epo, Interleukin-3 (IL-3) and stem cell factor (SCF). Interleukin-6 addition assists in BFU-E culture. BFU-E appear after 7–8 days of culture. These colony types also vary in size. CFU-E are small (8–200 erythroblasts in each colony) whereas BFU-E are larger containing 200 or more erythroblasts per colony. Both develop a brown-red color due to hemoglobinization. *See* **Note 10** on choice of CFU-E and BFU-E colony assay systems.

3.2.3 OP9 Stromal Cell Cocultures

Colony assays allow for quantification of progenitor numbers with a static endpoint. In some cases, researchers may wish to investigate the generation of EryP-CFC, BFU-E, and CFU-E from the same progenitors by assessing development at different time points. For this, stromal cell cocultures can be useful for monitoring erythroid progenitor development over time (*see* **Note 11**).

1. Trypsinize a 70% confluent flask of OP9 cells. Do not allow OP9 cells to overgrow as they will lose hematopoietic-supportive activity (*see* **Note 1**).
2. Culture OP9 cells into 96-well plates until 80–90% confluent in alpha-MEM with 20% FCS and penicillin–streptomycin.
3. Change the medium in each well to; 100 units/mL SCF and 2 units/mL Epo in alpha-MEM with 20% FCS and penicillin–streptomycin.
4. Under sterile conditions, add FACS-sorted putative progenitor populations into each well. Return 96-well plate to 37 °C incubator for culture.
5. Check cultures at day 4 for BFU-E activity and day 8 for CFU-E activity and/or expression of erythroid development markers as described earlier (CD71, Ter119, CD44).

3.3 Stress Erythroblast Isolation by Phenylhydrazine Administration

3.3.1 Administration of PHZ

1. Hemolytic anemia can be induced in laboratory mice by injecting the PHZ solution intraperitoneally at a dose of 60 mg/kg over 2 days, 24 h apart. A maximum volume of 200 μL is allowed per injection per animal per day as per University of Sydney Animal Ethics Committee (*see* **Note 12**).
2. Control mice should be injected with sterile PBS utilizing the same technique. Mice not injected with vehicle or PHZ should also be used as controls for treatment.
3. Mice are visibly anemic 4 days post-PHZ treatment with pale ears, paw pads, and tails compared to the control mice (*see* Fig. 4).
4. Anemia can be further quantified via hematocrit.
5. Marked splenomegaly is seen from day 2 post-PHZ treatment onward, and is an indicator that the expected stress erythropoiesis is occurring in the spleen (Fig. 4).

3.3.2 Analyzing Bone Marrow and Splenic Tissue from PHZ-Treated Mice

1. Follow the same protocol as mentioned in Subheading 3.1 to prepare samples and to perform flow cytometry. It is important to perform a pilot experiment with single stain controls to clarify PHZ emission spectrum, as it is known to autofluoresce and therefore limits the number of possible channels able to be used during flow cytometry. We were unable to use the fluorescein (FITC) fluorophore in our experiments with PHZ due to overlapping emission spectrum.

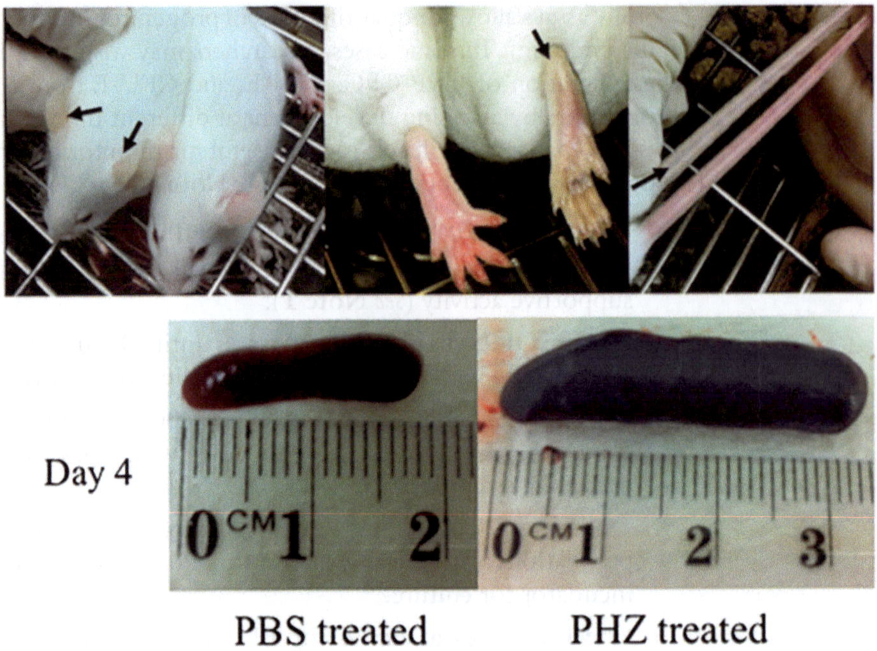

Fig. 4 The erythroblast response to chemically induced anaemia. Upper panels-Examining physical characteristics of anaemia such as ear, paw pad, and tail pallor can be utilised to confirm an adequate anaemic response to PHZ (arrows indicate PHZ-treated mouse). Lower panels-PHZ treatment induces splenomegaly within 4 days of injection due to stress erythroblast migration to the spleen

2. Cell surface markers of interest can be used and expression levels can be studied by staining in conjunction with TER119 and CD71 or CD44. Figure 5 demonstrates blood taken from 2 and 4 days post PHZ treatment and the resultant reticulocytosis seen compared with the PBS treated mice.

3.4 Murine Erythroleukemia Cell (MEL) Culture

Erythroleukemic cell lines such as MEL (mouse erythroid leukemia), HEL (human erythroid leukemia) and K562 (a human early myeloid leukemia) can be useful models of erythropoiesis in vitro. However, these cells are transformed, do not always behave in the same manner as their in vivo counterparts and can vary in the expression of surface antigens. However, these cell lines can be induced to differentiate into hemoglobinized cells in vitro and can be useful for studying aspects of red blood cell production.

3.4.1 Preparation of Cells from Frozen Cell Suspensions and Culture Maintenance

1. Warm growth media in water bath at 37 °C.
2. Add 500 μL of warm MEL growth media to vial of frozen cells to thaw and gradually transfer to T75 filter top flask with a pipette.
3. Transfer 10–15 mL of warm MEL growth media to flask.
4. Place seeded flask in an incubator at 5% CO_2 and 37 °C.

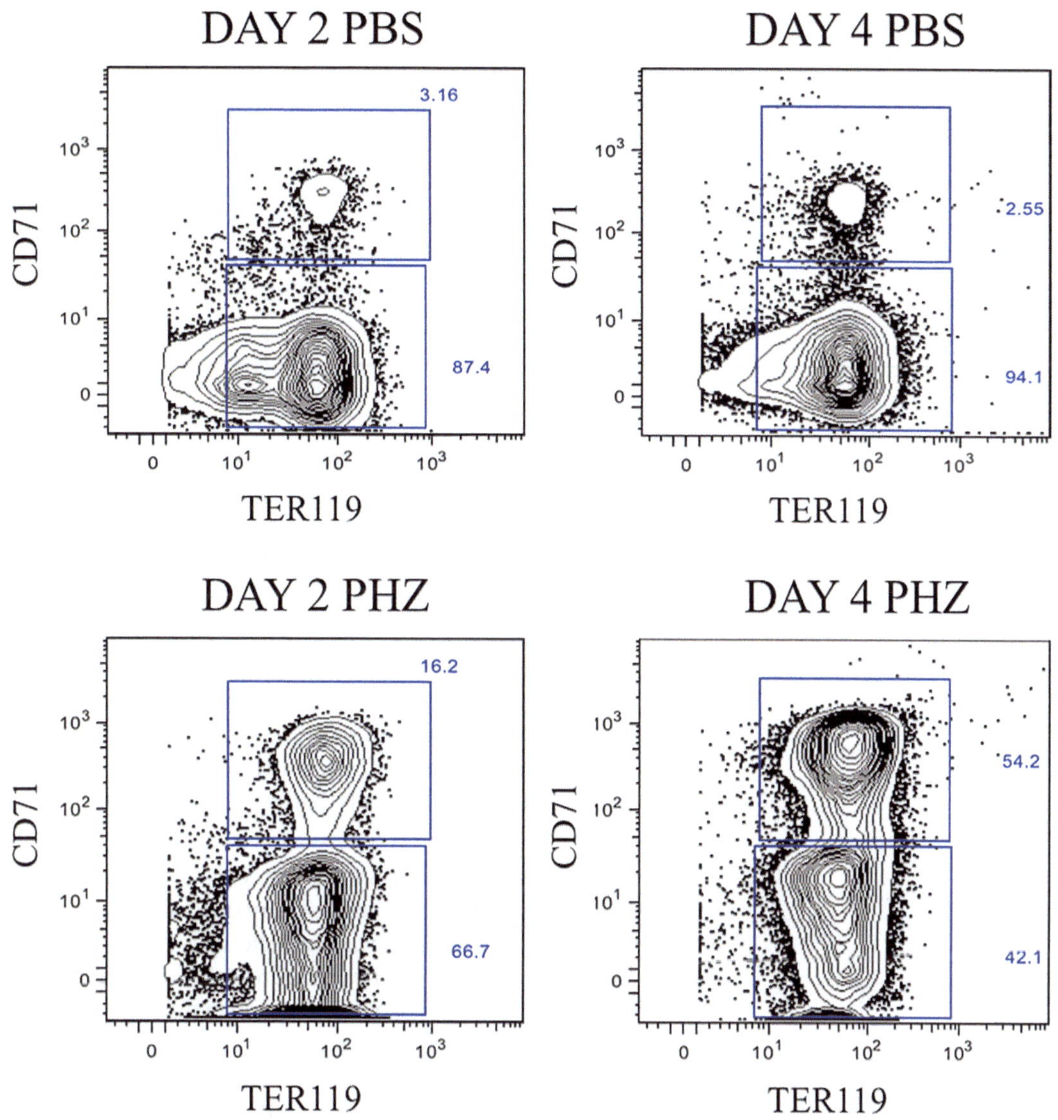

Fig. 5 Representative flow-cytometric plots demonstrating marked changes in erythroid progenitors seen in day 2 and 4 PHZ-treated mice. Samples are stained with CD71 and TER119 antibodies and demonstrate expression levels in blood taken from day 2 and day 4 PBS or PHZ-treated mice

5. Incubate cells until a density of 1×10^6 cells/mL.
6. Split cells 1:10 every 1–2 days to maintain density.

3.4.2 Differentiation Protocol and Analysis

1. Prepare aliquots of warm growth media in separate flasks or well plates with 2% (v/v) dimethylsulfoxide (DMSO).
2. Remove required number of cells from cultured flasks and transfer to media prepared above.
3. Maintain culture in an incubator for no longer than 72 h.

3.4.3 Preparation of Samples for Single Cell Analysis

1. Gently remove cells from flasks to a 50 mL tube using a transfer pipette.
2. Centrifuge cells at 100 × *g* for 5 min.
3. Remove supernatant and gently resuspend cell pellet in FACS buffer.
4. Stain with directly conjugated fluorescent antibodies against mouse surface antigens as described above in Subheading 3.1.
5. Analyze by flow cytometry as described above (*see* **Note 13**).

4 Notes

1. Cell suspensions need to be filtered to remove cell clumps, debris, and matrix, which all have the potential to cause blockages in the flow cytometer fluidics system. Cell filters can be surprisingly expensive. We purchase mesh shower curtain material from haberdashery/craft material shops to use instead of laboratory standard filters.
2. Using a 3 mL syringe with a smaller gauge and longer needle length (e.g., 25 gauge, 2 in.) can be an effective way to flush marrow from smaller mouse bones. We often used cell culture media containing FCS that is beyond the expiry date for cell culture but is still physiologically balanced and useful for this process.
3. Bone marrow flushing: A strong flush may result in most of the marrow cells becoming single cells. A gentler flush may result in a "marrow ribbon" with the marrow held together by mesenchymal cell–cell interactions. If a ribbon is generated, this can be made into a single cell suspension by inverting the tube (lid on) or gentle pipetting. Avoid dispersing the ribbon by drawing through the needle and syringe multiple times as this leads to considerable reduction in cell viability.
4. To make a single cell suspension of the spleen, as well as other immune organs such as the thymus and lymph nodes, place the organ on the filter mesh material and push the organ through the material using the flat end of the syringe plunger as this applies a broader force through the material. Round hematopoietic cells will move easily through a 70 μM filter mesh with the stromal vascular fraction will remain on the filter.
5. This technique can provide up to 1 mL of blood. The key to a high yield is to ensure that the time between euthanasia and cardiac puncture is kept to a minimum, as the yield will be much higher with a beating heart. Make sure all the necessary equipment, including a 25 gauge needle (insulin injecting needles can be very useful here), heparinized PBS and

dissection equipment, are all prepared before euthanasia. Blood should not be drawn too quickly, as this will cause the heart to collapse.

6. Resuspend the cells in FACS buffer. Always ensure reagents/ buffers are kept on ice to minimize death of cells.
7. Antibodies should be titrated first to final optimal staining concentration. Detection of antigens is dependent on antibody quality (can be optimized using monoclonal antibodies directly conjugated to fluorophores); the fluorophore itself; and the frequency of antigen expression on the cells of interest. For example, anti-mouse CD71 antibody is used at 1/tenth the concentration of anti-Ter119 antibody due to the very high fluorescence intensity of CD71 expression on developing mouse erythroid cells.
8. Flow-cytometric data is only reliable when performed with a suite of control samples. Indeed, in nearly all of our studies, the controls greatly outnumber the test samples. Controls for flow cytometry should include the following:
   (a) Unstained controls: cells not exposed to any fluorescent compound including viability stains. Unstained controls are important to ensure there is no autofluorescence of the samples, and to allow for setting of the laser strengths in each fluorescence channel.
   (b) Single stain controls: cells stained with only one conjugated antibody fluorescing in a known fluorescent channel including viability dyes. Single stained samples are important to aid with compensation to ensure that fluorescence spectra do not overlap leading to false results. Isotype controls are samples stained with the Fc portion of antibodies that are conjugated to a fluorophore. This ensures that there is no nonspecific binding of antibodies to samples.
   (c) Isotype controls: cells exposed to an antibody of the same isotype (both immunoglobulin type—IgM or IgG) as well as same isotype of antibody—IgG1, IgG2a, IgG2b, IgG3—from the same species and conjugated to the same fluorochrome. However, isotype control antibodies should completely lack reactivity to mammalian cells. These controls primarily indicate Fc receptor binding to the cells rather than antigen-specific F(ab) binding.
   (d) Fluorescence minus one (FMO) control: This type of control shows the potential spectral overlap of fluorochromes when multiple antibodies are combined into one "cocktail."

9. Four cell populations that morphologically correspond to proerythroblasts, basophilic erythroblasts, late basophilic and polychromatic erythroblasts, and orthochromic erythroblasts can be identified utilizing the cell surface expression of TER119 and CD71 in bone marrow and splenic cells. However, due to the gradual decline in CD71 expression in these cell stages, it is difficult to differentiate well resolved subpopulations using this method. The forward scatter parameter can be utilized as a function of cell size to help delineate the erythroblast maturation stage, as the erythroblasts decrease in size with maturation. There is still difficulty separating clear erythroblast maturation stages despite the use of this assay. There is marked overlap in the profiles of CD71 between the gated clusters I–III, and it is not possible to obtain pure populations of specific erythroblast subsets other than for proerythroblasts.
10. Methylcellulose assays can be challenging to work with due to the highly viscous nature of methylcellulose and the difficulty in preparing consistent, sterile batches with the appropriate growth factors and serum. The authors have chosen, for ease and consistency, to utilize commercially available colony assay kits, in particular, those developed by Stem Cell Technologies (Vancouver, Canada). These ready-made colony assay kits require the addition of the putative progenitors to the mix, vortexing and plating. Stem Cell Technologies offers detailed instructions on using these methylcellulose preparations hence we will not go into specific details here. Of note, Methocult SF M3436 is optimized for BFU-E quantification whereas CFU-E quantification is best assessed using Methocult M3334.
11. A number of stromal cell lines have proven useful in identifying the erythroid potential of isolated populations. This technique is particularly useful in determining the erythroid potential from populations that have not been well characterized such as a newly identified putative progenitor cell [39–41]. Previously, the authors have used the OP9 stromal cell line to assess erythroid progenitor activity in different embryonic populations. OP9 was derived from the *Op/Op* osteopetrotic mouse calvarium. These mice have a mutation in gene encoding macrophage colony-stimulating factor and do not secrete M-CSF [39]. This prevents overgrowth of macrophages during hematopoietic progenitor cultures, a problem encountered with other stromal cell lines.
12. Phenylhydrazine (PHZ) can result in acute toxicity from inhalation and dermal contact, resulting in serious skin and eye irritation and as well sensitization. Germ cell mutagenicity and possible carcinogenicity are major concerns with use of this chemical in high concentrations. PHZ can also result in

aquatic toxicity if disposed of inappropriately. PHZ is rapidly metabolized once injected into the mouse and urine does not contain toxic levels of PHZ. PHZ solution is prepared by dissolving the 97% PHZ in PBS in a fume hood wearing personal protective equipment including nitrile gloves (double glove), lab coat, safety glasses, and face shield. PHZ stock solution can be stored in a laboratory refrigerator at <4 °C and will continue to work effectively for at least 5 years. This procedure should be performed by a researcher with skilled mouse handling experience. Ensure that the mouse is well immobilized prior to attempting injection. The injecting procedure should be carried out in a way that eliminates any spillage of PHZ. This can be achieved by injecting the mouse with the bevel up, and holding the needle in the mouse for 2 s prior to gently twisting the needle such that the bevel is facing down prior to removing the needle from the mouse. Never resheath a used needle to avoid needle-stick injuries. Our experiments on outbred Swiss mice found the mice to have a maximal stress response at day 4 post-PHZ treatment. It is best to perform a timed analysis, day 0, day 2, day 4, and day 6 post-PHZ injections, to determine the maximal response for specific murine breeds.

13. MEL cells show differences in surface antigen expression compared to wild type adult mouse erythroblasts. In our studies, MEL cells lack expression of Ter119 possibly due to a difference in glycosylation of the antigen. MEL cells, in our studies, also lack CD45 but express the SCF receptor, c-Kit: c-Kit expression is lost during differentiation.

References

1. Maximow A (1909) Der Lymphozyt als gemeinsame Stammzelle der verschiedenen Blutelemente in der embryonalen Entwicklung und im postfetalen Leben der Säugetiere, Folia Haematologica. pp 125–134
2. Lappin TR, Rich IN (1996) Erythropoietin—the first 90 years. Clin Lab Haematol 18:137–145
3. Mizel SB, Farrar JJ (1979) Revised nomenclature for antigen-nonspecific T cell proliferation and helper factors. Cell Immunol 48 (2):433–436
4. Orkin SH, Zon LI (2008) SnapShot: hematopoiesis. Cell 132:712.e1–712.e2
5. Al-Drees MA, Yeo JH, Boumelhem BB et al (2015) Making blood: the haematopoietic niche throughout ontogeny. Stem Cells Int 2015:571893
6. Palis J (2014) Primitive and definitive erythropoiesis in mammals. Front Physiol 5:3
7. Sanchez M, Weissman IL, Pallavicini M et al (2006) Differential amplification of murine bipotent megakaryocytic/erythroid progenitor and precursor cells during recovery from acute and chronic erythroid stress. Stem Cells 24:337–348
8. Chen K, Liu J, Heck S et al (2009) Resolving the distinct stages in erythroid differentiation based on dynamic changes in membrane protein expression during erythropoiesis. Proc Natl Acad Sci 106:17413–17418
9. Kina T, Ikuta K, Takayama E et al (2000) The monoclonal antibody TER-119 recognizes a molecule associated with glycophorin A and specifically marks the late stages of murine erythroid lineage. Br J Haematol 109:280–287
10. Socolovsky M, Nam H-S, Fleming MD et al (2001) Ineffective erythropoiesis in Stat5a−/−5b−/− mice due to decreased survival of early erythroblasts. Blood 98(12):3261–3273

11. Koulnis M, Pop R, Porpiglia E et al (2011) Identification and analysis of mouse erythroid progenitors using the CD71/TER119 flow-cytometric assay. J Vis Exp e2809–e2809
12. Fraser ST, Isern J, Baron MH (2007) Maturation and enucleation of primitive erythroblasts during mouse embryogenesis is accompanied by changes in cell-surface antigen expression. Blood 109:343–352
13. Yeo JH, McAllan BM, Fraser ST (2016) Scanning electron microscopy reveals two distinct classes of erythroblastic island isolated from adult mammalian bone marrow. Microsc Microanal 22:368–378
14. Palis J, Robertson S, Kennedy M et al (1999) Development of erythroid and myeloid progenitors in the yolk sac and embryo proper of the mouse. Development 126:5073–5084
15. Fraser ST, Midwinter RG, Berger BS et al (2011) Heme oxygenase-1: a critical link between iron metabolism, erythropoiesis, and development. Adv Hematol 2011:1–6
16. Baron MH, Isern J, Fraser ST (2012) The embryonic origins of erythropoiesis in mammals. Blood 119:4828–4837
17. Wong PM, Chung SW, Reicheld SM et al (1986) Hemoglobin switching during murine embryonic development: evidence for two populations of embryonic erythropoietic progenitor cells. Blood 67:716–721
18. Isern J, He Z, Fraser ST et al (2011) Single-lineage transcriptome analysis reveals key regulatory pathways in primitive erythroid progenitors in the mouse embryo. Blood 117:4924–4934
19. Iscove NN, Sieber F (1975) Erythroid progenitors in mouse bone marrow detected by macroscopic colony formation in culture. Exp Hematol 3:32–43
20. Lodish H, Flygare J, Chou S (2010) From stem cell to erythroblast: regulation of red cell production at multiple levels by multiple hormones. IUBMB Life 62:492–496
21. Stephenson JR, Axelrad AA, McLeod DL et al (1971) Induction of colonies of hemoglobin-synthesizing cells by erythropoietin in vitro. Proc Natl Acad Sci U S A 68:1542–1546
22. Wu H, Liu X, Jaenisch R et al (1995) Generation of committed erythroid BFU-E and CFU-E progenitors does not require erythropoietin or the erythropoietin receptor. Cell 83:59–67
23. Hattangadi SM, Wong P, Zhang L et al (2011) From stem cell to red cell: regulation of erythropoiesis at multiple levels by multiple proteins, RNAs and chromatin modifications. Blood 118 (24):6258–6268
24. Terszowski G (2005) Prospective isolation and global gene expression analysis of the erythrocyte colony-forming unit (CFU-E). Blood 105:1937–1945
25. Stumpf M, Waskow C, Krötschel M et al (2006) The mediator complex functions as a coactivator for GATA-1 in erythropoiesis via subunit Med1/TRAP220. Proc Natl Acad Sci U S A 103:18504–18509
26. Flygare J, Estrada VR, Shin C et al (2011) HIF1 synergizes with glucocorticoids to promote BFU-E progenitor self-renewal. Blood 117:3435–3444
27. Tusi BK, Wolock SL, Weinreb C et al (2018) Population snapshots predict early haematopoietic and erythroid hierarchies. Nature 555:54–60
28. Li J, Hale J, Bhagia P et al (2014) Isolation and transcriptome analyses of human erythroid progenitors: BFU-E and CFU-E. Blood 124:3636–3645
29. Lenox LE, Perry JM, Paulson RF (2005) BMP4 and Madh5 regulate the erythroid response to acute anemia. Blood 105:2741–2748
30. Perry JM, Harandi OF, Paulson RF (2007) BMP4, SCF, and hypoxia cooperatively regulate the expansion of murine stress erythroid progenitors. Blood 109:4494–4502
31. Perry JM, Harandi OF, Porayette P et al (2009) Maintenance of the BMP4-dependent stress erythropoiesis pathway in the murine spleen requires hedgehog signaling. Blood 113:911–918
32. Socolovsky M (2007) Molecular insights into stress erythropoiesis. Curr Opin Hematol 14:215–224
33. Porayette P, Paulson RF (2008) BMP4/Smad5 dependent stress erythropoiesis is required for the expansion of erythroid progenitors during fetal development. Dev Biol 317:24–35
34. Harandi OF, Hedge S, Wu D-C et al (2010) Murine erythroid short-term radioprotection requires a BMP4-dependent, self-renewing population of stress erythroid progenitors. J Clin Invest 120:4507–4519
35. Xiang J, Wu DC, Chen Y et al (2015) In vitro culture of stress erythroid progenitors identifies distinct progenitor populations and analogous human progenitors. Blood 125:1803–1812
36. Antoniou M (1991) Induction of erythroid-specific expression in murine erythroleukemia (MEL) cell lines. Methods Mol Biol 7:421–434
37. Socolovsky M (2001) Ineffective erythropoiesis in Stat5a−/−5b−/− mice due to decreased

survival of early erythroblasts. Blood 98:3261–3273

38. Fraser ST, Isern J, Baron MH (2010) Use of transgenic fluorescent reporter mouse lines to monitor hematopoietic and erythroid development during embryogenesis. Methods Enzymol 476:403–427

39. Nakano T, Kodama H, Honjo T (1994) Generation of lymphohematopoietic cells from embryonic stem cells in culture. Science 265:1098–1101

40. Fraser ST, Ogawa M, Yu RT et al (2002) Definitive hematopoietic commitment within the embryonic vascular endothelial-cadherin(+) population. Exp Hematol 30:1070–1078

41. Schroeder T, Fraser ST, Ogawa M et al (2003) Recombination signal sequence-binding protein Jkappa alters mesodermal cell fate decisions by suppressing cardiomyogenesis. Proc Natl Acad Sci U S A 100:4018–4023

Chapter 12

Single-Cell Assays Using Hematopoietic Stem and Progenitor Cells

Ashwini S. Hinge and Marie-Dominique Filippi

Abstract

Hematopoietic stem cells (HSCs) undergo division, making two daughter cells with unique fate decision choices, that is, whether to self-renew to maintain stemness or differentiate to committed progenitors. Since HSCs are heterogeneous in nature understanding this phenomenon at the single cell level is important. In vitro single-cell assays like the paired-daughter cell and myeloid multilineage differentiation are useful to understand this unique stem cell process. Both assays are performed using cytokine combination which allows four-lineage myeloid differentiation—neutrophil, erythroid, macrophage/monocyte, and megakaryocyte. Paired-daughter cell assay examines symmetric or asymmetric retention of four myeloid lineages after first cell division in the paired-daughter cells. Thus, it defines asymmetric versus symmetric division patterns in the paired daughter cells. Thus, this assay may provide HSC fate decision cues. Myeloid multilineage differentiation assay examines the ability of a single cell to form multipotent clones containing four or less myeloid lineages. Here, we discuss in detail methodology of these assays.

Key words Hematopoietic stem cell, Single cell, Paired-daughter cell assay, Myeloid multilineage differentiation assay, Fate decision, Self-renewal, Differentiation

1 Introduction

Hematopoietic stem cells (HSCs) are characterized by their ability to undergo self-renewal and commitment to differentiation to produce mature blood cells throughout one's life span [1]. HSCs are mostly quiescent [2] but when undergo cell division, the decision whether to produce stem cell (self-renewal ability is retained) or committed progenitor (undergo differentiation), that is, fate decision is crucial. The fate decision choices are pivotal for sustained hematopoiesis and clinically important for efficient engraftment following transplantation [3]. Understanding the factors and/or signals which control this mechanism remains the central focus of HSC research. HSCs can undergo asymmetric division (producing one stem cell and one progenitor) or symmetric division (producing two stem cells or two progenitors) to meet the

Mugdha V. Joglekar and Anandwardhan A. Hardikar (eds.), *Progenitor Cells: Methods and Protocols*, Methods in Molecular Biology, vol. 2029, https://doi.org/10.1007/978-1-4939-9631-5_12,

physiological demand. It has been reported that these divisions occur in vitro as well as in vivo [4–7]. Using single-cell analysis, few reports have provided valuable information on HSC self-renewal, revealing stem cell factor (SCF) signaling intensity [4], Lnk signaling pathway [8] and lipid metabolism are important in HSC fate [5]. Through these unique patterns of cell division, some studies demonstrated asymmetric segregation of endocytic marker Ap2a2 [9], numb [10] and numb along with p-p38MAPK [11] and have been implicated in HSC fate changes.

Increasing evidences suggest heterogeneity of the HSC compartment [12–15], making each cell possess a unique cellular and molecular signature and thus carry a unique fate decision ability. Therefore, assessment of HSCs at the single cell level is essential. Drs. Suda and Nakauchi demonstrated at the single cell level the existence of different HSC division patterns using the in vitro paired-daughter cell assay [6, 16, 17]. It quantitatively assesses myeloid lineage commitment of hematopoietic stem and progenitor cells (HSPCs) after first division. In this assay, a single cell can give rise to daughter cells that retain—or not—multilineage differentiation potential (i.e., producing a clone that contains neutrophil, erythroid, macrophage/monocyte, and megakaryocyte, nemM). When both daughter cells produce nemM clones, it is considered as symmetric division. When only one daughter cell produces nemM clone, it is asymmetric division. When both daughter cells fail to produce nemM clones, it is committed division. Thus, assessment of each daughter cell reveals asymmetric versus symmetric cell division patterns possibly giving clues for HSC fate decision.

Single-cell myeloid multipotential differentiation assay examines multilineage differentiation ability of the cell. In this assay, daughter cells are not separated after cell division. Clones forming from single cells are identified as multipotent (presence of four myeloid lineages), tripotent (presence of three lineages), bipotent (presence of two lineages), and unipotent (presence of only one lineage). By assessing the differentiation ability of different cell populations in the HSC hierarchy (such as long-term HSC, short-term HSC, and HSPC), the cascade of differentiation progression of individual cells can be studied. Here, we discuss each assay in detail (*see* **Note 1**).

2 Materials

2.1 Isolation of Whole Bone Marrow (BM) Cells by Flushing the Bones

1. 1 mL tuberculin syringe
2. 23G needle
3. 5 mL round bottom polystyrene tube
4. Petri dish.

5. 40 μm sterile cell strainer
6. Medium of preference—Iscove's Modified Dulbecco's medium (IMDM) or Hank's Balanced Salt Solution (HBSS) with or without fetal bovine serum (FBS).
7. Hemocytometer or automated cell counter.
8. Blunt forceps and pair of scissors.

2.2 Isolation of BM Cells by Crushing the Bones

1. Autoclaved mortar pestle.
2. 40 μm sterile cell strainer
3. Medium of preference—IMDM or HBSS with or without FBS.

2.3 Optional: Density Gradient Separation of Whole BM Cells

The reagents used for this procedure should be kept at room temperature before use.

1. Histopaque (density 1.083 g/mL).
2. Whole BM cell suspension in medium of choice.
3. Medium with or without FBS.
4. 15 mL conical tubes.

2.4 Enrichment of Whole BM Cells or Density Gradient Separated Cells

The reagents used in this protocol are commercially available such as from Miltenyi biotech.

1. c-Kit enrichment or lineage depletion kits,
2. Buffers—running and wash buffers (commercially available).
3. Magnetic cell separation columns and magnetic stand for manual cell separation or.
4. autoMACS Pro cell separation machine for automated cell separation.

2.5 Immunostaining for Cell Surface Markers Specific for HSCs

1. Antibodies—anti-mouse Ly-6G/Ly-6C (Gr1; clone RB6-8C5), CD11b (clone M1/70), CD45R/B220 (clone RA3-6B2), CD3e (clone 145-2C11), TER-119 (clone Ter-119), Ly-6A/E (Sca-1, clone D7), CD117 (c-Kit, clone 2B8), CD48 (clone HM48-1), CD150 (clone 9D1), and CD45.2 (clone 104).
2. 2% bovine serum albumin (BSA) in phosphate buffered saline (PBS).
3. 5 mL round bottom polystyrene tube.

2.6 For Single-Cell Assays

1. Terasaki plates or 96 well U bottom plates.
2. StemSpan medium (commercially available).
3. Cytokines-Stem cell factor (SCF), thrombopoietin (TPO), interleukin-3 (IL3), erythropoietin (EPO), granulocyte colony-stimulating factor (G-CSF).

4. Single cytology funnel.
5. Microscope slides.
6. Wright's and Giemsa stains or eosin and methylene blue stains.
7. Myeloid multilineage differentiation medium: IMDM medium containing 10% FBS, 20 ng/mL SCF, 20 ng/mL TPO, 20 ng/mL G-CSF, 50 ng/mL IL3, 4 U/mL EPO.

3 Methods

For all the steps described, maintain sterile conditions to avoid contamination during cell culture (Fig. 1).

3.1 Isolation of Whole BM Cells

1. Sacrifice mice as per the institutional approved animal protocol. Collect tibia, femur, and iliac crest bones from hind legs of the mouse in a tube containing medium of choice (IMDM or HBSS). Make sure the bones are immersed in the medium.

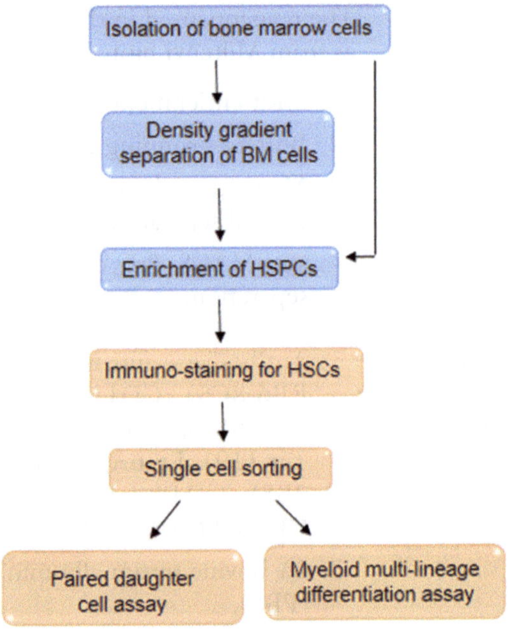

Fig. 1 *Experimental schema for the assays*: Mouse whole bone marrow cells were isolated. Either cells were further enriched by density gradient separation or directly used for enrichment of hematopoietic stem and progenitor cells by c-Kit enrichment or lineage depletion method. Enriched cells were immunostained for specific cell surface markers which defines long-term HSC. Single HSC was sorted using flow sorter machines in Terasaki plate or 96-well plate with U bottom. The single cell was used to perform paired daughter cell assay or myeloid multilineage differentiation assay

Mouse whole bone marrow (BM) cells can be isolated either by crushing or flushing the bones with medium.

2. Attach 23G needle to 1 mL tuberculin syringe and fill the syringe with medium. Cut very small top portion of the bone and insert needle of the syringe into cut bone portion using sterile forceps.
3. Flush 2 mL of medium through the bone at least 5 to 6 times. Make sure marrow from the bone is flushed out of the bone. Repeat the process for all collected bones in the same tube containing medium (*see* **Note 2**).
4. Filter single-cell suspension through a 40 μm cell strainer. Count cells either using a hemocytometer or an automatic cell counter machine.
5. After whole BM cells isolation either proceed to density gradient cell separation protocol (optional) or directly perform cell enrichment protocol.

3.2 Density Gradient Separation of Whole BM Cells

All reagents and centrifuge should be at room temperature before use.

1. In a tapering bottom 15 mL conical tube, take 3 mL of Ficoll Histopaque. Layer 4 mL of whole BM single-cell suspension (isolated from one mouse) such that Ficoll layer is at the bottom (*see* **Note 3**).
2. Centrifuge the tube immediately at speed of 700 × *g* with slow acceleration and off/no deceleration. Spin for 30 min at 24–28 °C. After centrifugation, cells will be separated as per their densities.
3. Collect middle buffy coat using pipette in a separate tube. Add 8 mL of plain medium to ensure excess Ficoll is washed off.
4. Centrifuge at 524 × *g* at maximum acceleration and deceleration for 10 min at 4 °C.
5. Aspirate medium and resuspend cell pellet in running buffer used for cell enrichment protocol. Count density gradient separated cells using hemocytometer or using automated cell counter. Proceed to cell enrichment protocol.

3.3 Hematopoietic Stem and Progenitor Cell (HSPCs) Enrichment

This protocol is used to enrich HSPCs and subsequently to minimize cell sorting time. All reagents used should be at 4 °C. Commercially available kits (lineage depletion or c-Kit enrichment) can be used. Cell enrichment can be done by using manual magnetic separating columns or automated machine with magnetic columns. Follow company protocol as mentioned.

3.3.1 c-Kit Enrichment Using Automated Cell Separator Instrument

1. Resuspend ten million cells in 80 μL of running buffer and add 20 μL of c-Kit/CD117 microbeads.
2. Incubate at 2–8 °C for 15 min. Add 2 mL of running buffer to wash the cells. Centrifuge at 524 × *g* for 10 min at 4 °C.
3. Aspirate the supernatant and resuspend the cell pellet in 500 μL of running buffer.
4. Select program on automated machine for positive selection.
5. Place the tube containing cell suspension and two empty tubes to collect separated cell fractions on tube rack provided by company.
6. After cell separation, centrifuge tubes containing c-Kit enriched cells at 524 × *g* for 10 min at 4 °C. Resuspend cells in PBS buffer containing 2% BSA; henceforth called buffer (0.5 to six million cells in 100 μL of buffer).
7. Count cells. Proceed to Subheading 3.4.

3.3.2 Lineage Depletion Using Automated Cell Separator Instrument

1. Resuspend ten million cells in 40 μL of running buffer and add 10 μL of biotin labeled antibody cocktail. Incubate at 2–8 °C for 10 min.
2. Add 30 μL of running buffer and 20 μL of anti-biotin microbeads. Incubate at 2–8 °C for 15 min.
3. Wash cells with 2 mL of running buffer. Centrifuge at 524 × *g* for 10 min at 4 °C.
4. Aspirate the supernatant and resuspend the cell pellet in 500 μL of running buffer.
5. Select program for negative selection—deplete on an automated cell separator instrument.
6. Place the tube containing cell suspension and two empty tubes to collect separated cell fractions in tube rack provided by company.
7. After cell separation, centrifuge tubes containing lineage depleted fraction at 524 × *g* for 10 min at 4 °C. Resuspend cells in buffer (0.5 to six million cells in 100 μL of buffer).
8. Count cells. Proceed to Subheading 3.4.1 (*see* **Note 4**).

3.4 Immunostaining for Long-Term HSCs (LT-HSCs) for Flow Cytometry

LT-HSCs are characterized by specific cell surface markers such as lineage^{-} Sca-1^{+}c-Kit^{+}CD48^{-}CD150^{+} (LSK-SLAM) using combination of specific anti-mouse antibodies.

Keep all the reagents at 4 °C for immunostaining protocol.

If direct tagged antibodies are used, all antibodies can be added together, if not, a two-step immunostaining protocol is required.

3.4.1 One-Step Staining Using Direct Tagged Antibodies

1. Add all antibodies together at concentration—1 μL of each antibody in 100 μL of buffer for 0.5 to five million HSPCs. Fluorochromes of each antibody can be selected as per lasers available on flow cytometry machines. For example typically for 5 laser machine, this combination can be used—lineage (Gr-1, CD11b, B220, CD3e, Ter119)-APCCy7, Sca-1–PECy7, c-Kit-APC, CD48-PE, CD150-FITC.
2. Incubate on ice for 1 h. Proceed to common protocol for both the methods (*see* Subheading 3.4.3).

3.4.2 Two-Step Staining

1. For lineage staining, add biotin labeled Gr-1, CD11b, B220, CD3e, Ter119 antibodies; 1 μL of each antibody in 100 μL of buffer (primary antibody staining).
2. Incubate on ice for 30 min. Wash the cells using 1 mL of buffer, spin down at 524 × *g*, 10 min at 4 °C and aspirate the supernatant.
3. Resuspend cell pellet in 100 μL of buffer. Add 1 μL of each antibody with fluorochrome of choice. For example, typically for 5 laser machine, this combination can be used—lineage Streptavidin-APCCy7 (secondary antibody), Sca-1–PECy7, c-Kit-APC, CD48-PE, CD150-FITC. Incubate on ice for 1 h.

3.4.3 Common Protocol for Both the Methods

1. Wash the cells using 1 mL of buffer, spin down at 524 × *g* for 10 min at 4 °C and aspirate the supernatant.
2. Resuspend cell pellet in flow sorting buffer; 300–400 μL of buffer for two to three million cells. (Typically, 2% BSA in PBS or 2% FBS in PBS is used.) (*see* **Note 5**).

3.5 Single-Cell Sorting

1. Add 17 μL of StemSpan medium in Terasaki plate containing murine SCF and murine TPO (both100 ng/mL) or 150 μL of medium + cytokines if 96 well U bottom plate is used (*see* **Note 6**).
2. Set up compensation controls on flow sorting machine for the fluorochromes used in the immunostaining (*see* **Note** 7).
3. Set up gates and subgates to sort LSK-SLAM cell population (Fig. 2).
4. Adjust the liquid stream of the sorter machine such that liquid droplet falls in the center of the well. Select 4-way purity or single-cell options for single-cell sorting (*see* **Note 8**).
5. For Terasaki plates: After sorting keep Terasaki plates in a carrier plate with 1-2 petri plates with distilled water to avoid evaporation of the medium in the plates. Incubate plates in incubator at 37 °C and 5% CO_2 settings. For 96-well U bottom plates: Incubate plates in incubator directly.

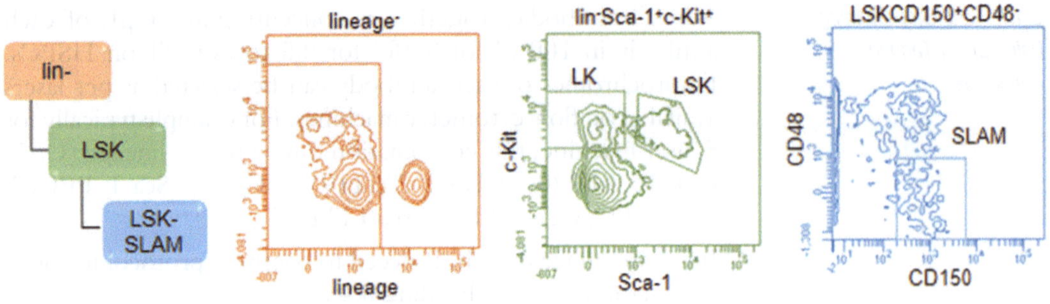

Fig. 2 *Gating strategy to sort LSK-SLAM*: c-Kit enriched cells were immunostained for cell surface markers specific for LSK-SLAM. Lineage negative cells were gated first. Under this gate Sca-1$^+$ and c-Kit$^+$ (LSK) cell population was gated. Further CD150$^+$CD48$^-$ population was gated under LSK gate (LSK-SLAM). Single LSK-SLAM cell was sorted to perform single-cell assays

6. To observe a single cell in the well immediately, incubate at least for 30 min (if required spin down the plate at 524 × *g* for 1 min), otherwise incubate overnight.
7. Observe each plate and each well for a single cell under microscope. Note down wells having a single cell. Do not consider wells with more than one cell or with no cell for further experiment (*see* **Note 9**).
8. For Terasaki plates: After overnight incubation, add 7–10 μL of StemSpan medium containing SCF and TPO (100 ng/mL of each) to wells containing a single cell. For 96-well plate, addition of more medium is not required. Again, incubate the plates overnight in the incubator. Proceed to either myeloid multilineage differentiation single-cell assay or paired daughter cell assay.

3.6 Single-Cell Assays

3.6.1 Myeloid Multilineage Differentiation Single-Cell Assay

In this single-cell assay, multilineage differentiation of each single cell is studied. Here, the daughter cells after first cell division are not separated. Therefore, differential differentiation pattern of each daughter cell cannot be studied (Fig. 3).

1. After 48 h of incubation, transfer a single cell into U bottom 96-well plate if sorting in Terasaki plate, otherwise add myeloid multilineage differentiation medium in the same plate containing 200 μL of myeloid multilineage differentiation medium.
2. Incubate plates in the incubator (at 37 °C and 5% CO_2) for 12 days. This cytokine combination allows for four-lineage myeloid differentiation (**n**eutrophil, **e**rythroid, **m**acrophage/**m**onocyte, **M**egakaryocyte, nemM).
3. After incubation, observe plates under microscope. Mark wells with clones, that is, those cells that have undergone myeloid differentiation. Proceed to Subheading 3.6.3, common steps for both the protocols—Cytospin preparation and staining.

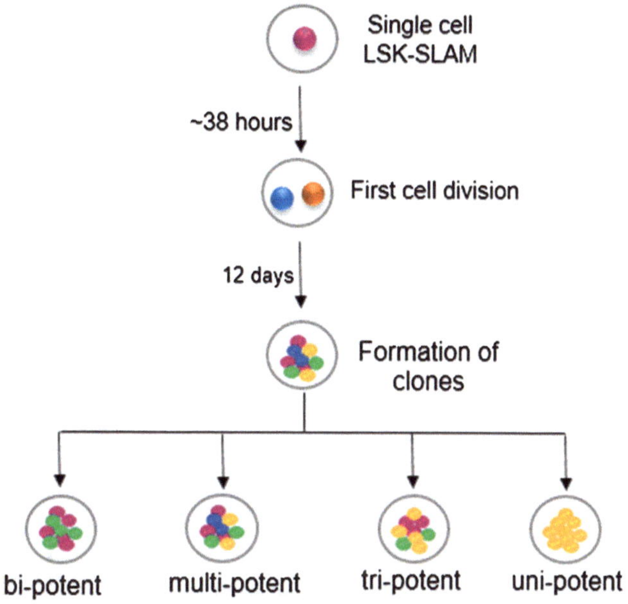

Fig. 3 *Myeloid multilineage differentiation assay schema*: Single LSK-SLAM cell was sorted in multiwell plate containing StemSpan medium and SCF, TPO cytokines. After first cell division, medium with fetal bovine serum and combination of cytokines which allows for four-lineage myeloid differentiation was added to each well. After 12 days, clones were formed and scored based on the presence of 4 lineages or the absence of either of the lineage as multipotent (presence of 4 lineages), tripotent (presence of 3 lineages), bipotent (presence of 2 lineages), or unipotent (presence of only 1 lineage)

3.6.2 Paired Daughter Cell Assay

In this single-cell assay, after first cell division two daughter cells are separated and are paired during identification of clones. Therefore, differential pattern of myeloid differentiation symmetric or asymmetric division of each daughter cell can be studied (Fig. 4).

1. LSK-SLAM cells undergo first cell division after 38–40 h of incubation. Observe each well which was marked for a single cell under microscope.
2. Check for wells containing two cells, that is, those cells that have undergone first cell division producing two daughter cells. Mark all the wells with two cells.
3. Use pipette to separate daughter cell from each well. Pipet medium up and down 5–6 times. Split medium into half into two separate wells. Add the medium into 2 wells of new Terasaki plate. Pair the daughter cells from each well after separating into new wells (*see* **Note 10**).
4. Incubate the plates in the incubator for 30–40 min. Observe under the microscope to ensure separation of the daughter cells.

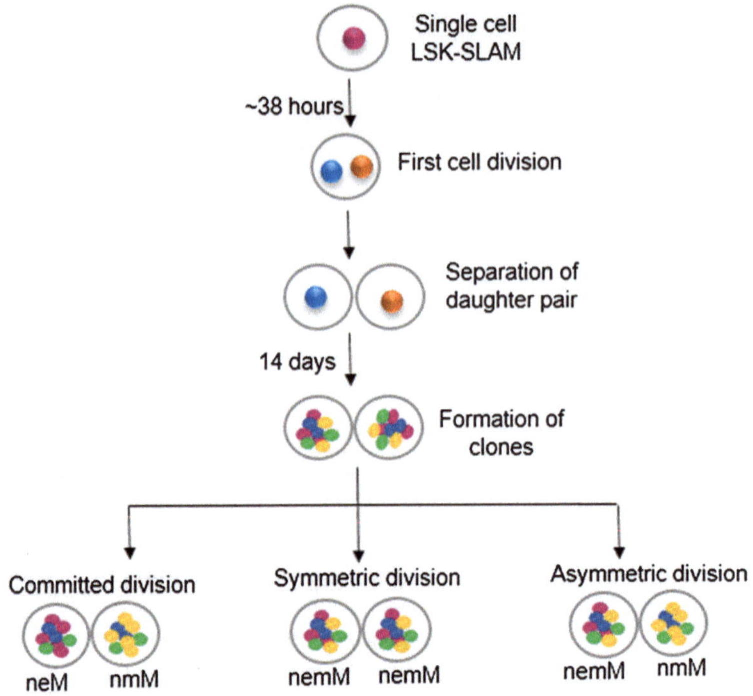

Fig. 4 *Paired daughter cell assay schema:* Single LSK-SLAM cell was sorted in multiwell plate containing StemSpan medium and SCF, TPO cytokines. After first cell division, daughter cells were separated and cultured into two separate wells containing medium with fetal bovine serum and combination of cytokines which allows four-lineage myeloid differentiation (neutrophil, erythroid, monocyte/macrophage, and Megakaryocyte; nemM). After 14 days, clones were formed and scored based on the presence of 4 lineages or the absence of either of the lineage. The clones were paired and identified for the type of the division occurred

5. Transfer the separated cells to U bottom 96-well plate. Add 200 μL of myeloid multilineage differentiation medium, which allows for differentiation into four myeloid lineages (nemM).
6. Incubate plates in the incubator for 2 weeks. Ensure that there is no medium evaporation.
7. After incubation, observe plates under microscope. Mark wells of each paired daughter cells which formed clones, that is, those cells that have undergone differentiation. There will be a few wells with no clones or only one well has clones, do not use these wells to prepare cytospins. Proceed to Subheading 3.6.3, common steps for both the protocols—Cytospin preparation and staining.

3.6.3 Common Steps for Both Single-Cell Assays: Cytospin Preparation and Staining

1. Use disposable cytology funnel with thin filter pads to have good Cytospin smears on the slide with low cell number.
2. Keep the slide on Cytospin clip and place cytology funnel on the top.
3. For myeloid differentiation assay: Collect the clone from the well after mixing the medium.

   For pair daughter cell assay: Mix medium from each paired wells with clones to avoid clumping of cells. Add them separately into two cytology funnels. At each step mark the paired daughters for correct interpretation of the results.
4. Place cytology funnels in the centrifuge holder and spin at 500 rpm for 5 min at medium speed at room temperature. Remove cytology funnels from the holder carefully and let Cytospin smear air-dry.
5. Fix Cytospin smears using diff quick fixation solution or methanol for 1 min. Air-dry Cytospin smears after fixation.
6. To stain nucleus and cytoplasm of the Cytospin smears of clones, stain with Wright's and Giemsa or eosin and methylene blue stains for 30 s to 1 min for each. Wash the slides with distilled water. Observe air-dried slides under the microscope.

3.6.4 Identification of Clones

1. Identify the clones from myeloid multilineage differentiation assay as **Multipotent:** Presence of 4 lineages, **Tripotent:** Presence of 3 lineages, **Bipotent:** Presence of 2 lineages, **Unipotent:** Presence of only 1 lineage (Figs. 3 and 5).

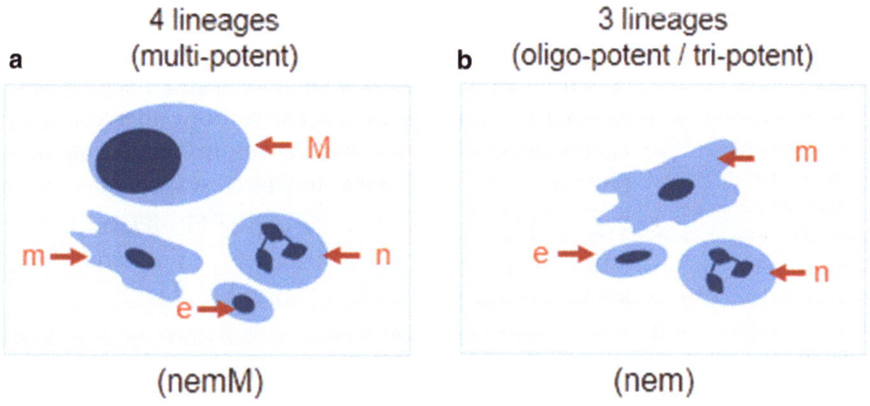

Fig. 5 *Identification of clones*: Cytospin smears were prepared after the single-cell assay. The clones were stained with either Wright's–Giemsa or eosin–methylene blue stains. Myeloid lineages were identified as per the cell morphology. Schematic representation of clones: Clone A shows the presence of 4 myeloid lineages (neutrophil, erythroid monocyte/macrophage, and Megakaryocyte) while clone B shows the presence of 3 lineages (neutrophil, erythroid monocyte/macrophage). In paired daughter cell assay, these types of paired clone suggest asymmetric division. In myeloid multilineage differentiation assay, each clone is identified separately as multipotent; clone A (presence of 4 myeloid lineages) and tripotent; clone B (presence of 3 myeloid lineages). For stained clones' images refer to Hinge et al., Nat Com, 2017 (ref 11)

2. Identify the clones from the paired daughter wells as-

   Symmetric division—Presence of 4 lineages in both clones of the same daughter pair.

   Asymmetric division—Presence of 4 lineages only in one clone of the same daughter pair.

   Committed division—Both the clones of the same daughter pair have less than 4 lineages (Figs. 4 and 5).

3. After scoring of each paired daughter clone, calculate percent of symmetric, asymmetric and committed divisions for the experimental data set. Perform statistical analysis on the data set. To compare two groups, use 2×2 contingency table using Fisher's exact test or to compare three groups, use 3×2 contingency table using chi-square test.

4 Notes

1. Though these assays are useful to study murine HSC division patterns and multipotential differentiation ability at the single cell level, results should be interpreted with caution. Since the assays are performed using cytokine combination in vitro, the results obtained cannot be used to assess HSC self-renewal. It is only an indication of daughter cells having similar or differential fate choices to form a myeloid multipotent clone in vitro. However, results must be validated using the gold standard assay for hematopoietic stem cell function—that is, transplantation assays. Also, these assays do not measure lymphoid differentiation potential of HSCs nor do they provide information on the role of the microenvironment on HSC fate decision.
2. It is important to make a single-cell suspension while flushing the bones. Sometimes the entire bone marrow is not dissociated completely; flush the medium containing this marrow few more times to have a single-cell suspension.
3. While adding the cells on the top of Ficoll layer, maintain slow constant speed during addition of medium containing cells to avoid mixing of the cells with the Ficoll layer.
4. In lineage depletion protocol, cells are already labeled with lineage biotin labeled antibodies. Therefore, proceed to one-step staining protocol.
5. Adjust final volume of resuspending buffer as per the cell concentration. Do not make sorting sample very concentrated.
6. Due to low volume of medium, avoid bubbles while pipetting the medium. It is easy to locate a single cell in the medium not containing any bubbles.

7. It is always a good practice to have single stained compensation controls for each fluorochrome that has been used in the multicolor immunostaining. If automatic compensation is used, always check the compensation manually to ensure correct compensation setup.
8. Before sorting cells of interest in the plate, run test sort to set up correct positioning of the wells. Check the test plate for liquid droplet to make sure it is getting sorted in the center or middle of the well.
9. Have sheet of paper with plate layout. Mark the wells with a single cell on the paper. Go back to these marked wells the next day to observe whether the cell has divided or not.
10. While pipetting medium to separate daughter cells, avoid frothing. Do not use micro tips; instead, use 2–20 μL tips to be gentle on the single cell.

References

1. Seita J, Weissman IL (2010) Hematopoietic stem cell: self-renewal versus differentiation. Wiley Interdiscip Rev Syst Biol Med 2 (6):640–653
2. Wilson A, Laurenti E, Trumpp A (2009) Balancing dormant and self-renewing hematopoietic stem cells. Curr Opin Genet Dev 19 (5):461–468
3. Dahlberg A, Delaney C, Bernstein ID (2011) Ex vivo expansion of human hematopoietic stem and progenitor cells. Blood 117 (23):6083–6090
4. Kent DG, Dykstra BJ, Cheyne J, Ma E, Eaves CJ (2008) Steel factor coordinately regulates the molecular signature and biologic function of hematopoietic stem cells. Blood 112 (3):560–567
5. Ito K, Carracedo A, Weiss D, Arai F, Ala U, Avigan DE, Schafer ZT, Evans RM, Suda T, Lee CH, Pandolfi PP (2012) A PML–PPAR-δ pathway for fatty acid oxidation regulates hematopoietic stem cell maintenance. Nat Med 18(9):1350–1358
6. Takano H, Ema H, Sudo K, Nakauchi H (2004) Asymmetric division and lineage commitment at the level of hematopoietic stem cells: inference from differentiation in daughter cell and granddaughter cell pairs. J Exp Med 199(3):295–302
7. Yamamoto R, Morita Y, Ooehara J, Hamanaka S, Onodera M, Rudolph KL, Ema H, Nakauchi H (2013) Clonal analysis unveils self-renewing lineage-restricted progenitors generated directly from hematopoietic stem cells. Cell 154(5):1112–1126
8. Seita J, Ema H, Ooehara J, Yamazaki S, Tadokoro Y, Yamasaki A, Eto K, Takaki S, Takatsu K, Nakauchi H (2007) Lnk negatively regulates self-renewal of hematopoietic stem cells by modifying thrombopoietin-mediated signal transduction. Proc Natl Acad Sci U S A 104(7):2349–2354
9. Ting SB, Deneault E, Hope K, Cellot S, Chagraoui J, Mayotte N, Dorn JF, Laverdure JP, Harvey M, Hawkins ED, Russell SM, Maddox PS, Iscove NN, Sauvageau G (2012) Asymmetric segregation and self-renewal of hematopoietic stem and progenitor cells with endocytic Ap2a2. Blood 119(11):2510–2522
10. Wu M, Kwon HY, Rattis F, Blum J, Zhao C, Ashkenazi R, Jackson TL, Gaiano N, Oliver T, Reya T (2007) Imaging hematopoietic precursor division in real time. Cell Stem Cell 1 (5):541–554
11. Hinge A, Xu J, Javier J, Mose E, Kumar S, Kapur R, Srour EF, Malik P, Aronow BJ, Filippi MD (2017) p190-B RhoGAP and intracellular cytokine signals balance hematopoietic stem and progenitor cell self-renewal and differentiation. Nat Commun 8:14382
12. Dykstra B, Kent D, Bowie M, McCaffrey L, Hamilton M, Lyons K, Lee SJ, Brinkman R, Eaves C (2007) Long-term propagation of distinct hematopoietic differentiation programs in vivo. Cell Stem Cell 1(2):218–229
13. Benz C, Copley MR, Kent DG, Wohrer S, Cortes A, Aghaeepour N, Ma E, Mader H, Rowe K, Day C, Treloar D, Brinkman RR, Eaves CJ (2012) Hematopoietic stem cell subtypes expand differentially during development

and display distinct lymphopoietic programs. Cell Stem Cell 10(3):273–283

14. Challen GA, Boles NC, Chambers SM, Goodell MA (2010) Distinct hematopoietic stem cell subtypes are differentially regulated by TGF-beta1. Cell Stem Cell 6(3):265–278
15. Sieburg HB, Cho RH, Dykstra B, Uchida N, Eaves CJ, Muller-Sieburg CE (2006) The hematopoietic stem compartment consists of a limited number of discrete stem cell subsets. Blood 107(6):2311–2316
16. Ema H, Takano H, Sudo K, Nakauchi H (2000) In vitro self-renewal division of hematopoietic stem cells. J Exp Med 192(9):1281–1288
17. Suda T, Suda J, Ogawa M (1984) Disparate differentiation in mouse hemopoietic colonies derived from paired progenitors. Proc Natl Acad Sci U S A 81(8):2520–2524

Chapter 13

Isolation and Characterization of Cardiac Progenitor Cells

Parul Dixit and Rajesh G. Katare

Abstract

Cardiac progenitor cells (CPCs) are gaining interest as a therapeutic option for the treatment of the heart. Due to the limited pool of CPCs residing in the heart, it is essential to isolate and expand the CPCs in vitro. Here we describe the protocol for isolation and culture of the heterogeneous population of CPCs from right atrial appendage and left ventricular tissue collected from patients undergoing on-pump coronary artery bypass graft surgery for the treatment of ischemic heart disease. Our protocol is developed to simultaneously isolate, culture, and characterize the CPCs from both atrial and ventricular tissues. We also describe the protocol for flow cytometry and immunohistochemical characterization of the isolated CPCs.

Key words Cardiac progenitor cells, Heterogeneous CPC population, Isolation and culture, Characterization, Flow cytometry

1 Introduction

The most common form of heart disease is ischemic heart disease (IHD) which results from a failure to meet the oxygen demands of the myocardium. Current strategies for the treatment of IHD are either pharmacological treatment or surgical revascularization of the occluded coronary artery. Marked advance in both these strategies has made them vital for the life-saving rescue after a myocardial infarction (MI). Nonetheless, the pharmacological and surgical therapies cannot principally make up for the loss of myocytes in the heart.

Cell therapy refers to the transplantation of externally administered stem cell/progenitor cells or mature cells for inducing tissue regeneration and repair. Various cell types at different developmental stages, including embryonic, fetal and adult stem cells/progenitor cells and mature cells, have been considered for transplantation into the heart. Among different types of cells, cardiac progenitor cells (CPCs) are the front runners among the candidate cell types, which can be translated to the clinic for cardiac regenerative therapy.

Mugdha V. Joglekar and Anandwardhan A. Hardikar (eds.), *Progenitor Cells: Methods and Protocols*, Methods in Molecular Biology, vol. 2029, https://doi.org/10.1007/978-1-4939-9631-5_13, © Springer Science+Business Media, LLC, part of Springer Nature 2019

We and others have used the heterogeneous population of CPCs. These cells express the mesenchymal stem cell (MSC) markers CD105, CD90 along with a small population of cells expressing C-Kit [1, 2].

This chapter describes the isolation procedure optimized in our laboratory for isolation of CPCs from the right atrial appendage (RAA) and left ventricle (LV) sampled from the same hearts of the patients undergoing on-pump coronary artery bypass graft surgery. Using this protocol, we have demonstrated the functional difference between the RAA and LV CPCs in terms of surface markers, growth, expansion potential, and differentiation capacity.

2 Materials

2.1 Tissue Collection

1. Cold Krebs Ringer HEPES (KRH) buffer solution: Recipe in Table 1.
2. 5 mL screw capped glass bottles to collect the tissues.
3. Right atrial appendage tissue—100–300 mg.
4. Left ventricular biopsy tissue, 20–40 mg.
5. Hanks Balance Salt Solution (HBSS): Recipe in Table 1.

2.2 CPC Isolation

1. 10 cm culture dish.
2. 15 mL conical tubes.
3. T25 culture flasks.
4. Forceps, scissors, and scalpel.
5. Temperature controlled shaker-incubator.
6. Temperature controlled centrifuge.
7. Collagenase II solution—Dissolve 1357 unit collagenase II in 1 mL of HBSS (Recipe in Table 1). Prepare 30 mL in total, just before the isolation.
8. 10 mL of donor bovine serum (DBS).

2.3 CPC Culture

1. 6-well culture dish, T25, and T75 culture flask.
2. Ham's F12 complete medium: 10% DBS, 1% antibiotic–antimycotic, 10 ng/mL human basic fibroblast growth factor (b-FGF), 0.005 units/mL human erythropoietin (h-epo) in Ham's F12 medium.

2.4 CPC Subculture and Freezing

1. 2 mL cryovial tubes.
2. Phosphate buffered saline (pH 7.4) (PBS) Recipe in Table 1.
3. TrypLE dissociation reagent.
4. Mesenchymal stem cells (MSC) freezing medium.

Table 1
List of buffers used in the study and their composition

No.	Solution	Reagents
1	Phosphate Buffer Saline (PBS)	10 mM Sodium Phosphate Monobasic 1.8 mM Potassium Phosphate Dibasic 137 mM Sodium Chloride 2.7 mM Potassium Chloride (adjusted to pH 7.4)
2	Hanks Balanced Salt Solution (HBSS)	1.26 mM Calcium Chloride Dihydrate 0.49 mM Magnesium Chloride Hexahydrate 0.41 mM Magnesium Sulphate Hepta Hydrate 5.3 mM Potassium Chloride 0.22 mM Mono Potassium Phosphate 4.17 mM Sodium Bicarbonate 138 mM Sodium Chloride 0.34 mM Sodium Phosphate Dibasic 5.55 mM D-Glucose (adjusted to pH 7.4)
3	Krebs Ringer Hienslet Buffer (KRH)	116 mM Sodium Chloride 4 mM Potassium Chloride 1mM Magnesium Chloride 1.8 mM Calcium Chloride 25 mM Glucose 10 nM HEPES Acid (adjusted to pH 7.4) 10 nM HEPES Acid (adjusted to pH 7.4)
4	Flow Buffer	3 mM HEPES in HBSS buffer 1% Fetal Bovine Serum (adjusted to pH 7.4)
5	4% Paraformaldehyde solution	4% (w/v) Paraformaldehyde in PBS

2.5 Differentiation of CPCs

1. Cardiomyocyte differentiation medium (Merck Millipore, USA).
2. 8 well chamber slides
3. 20 μg/mL poly-L-ornithine.

2.6 Flow Cytometry Characterization of CPCs

1. Vortex mixer.
2. 5 mL polystyrene tubes
3. Flow cytometer.
4. FlowJo 9.6 software.
5. Flow cytometry buffer freshly prepared on the day of the experiment according to the recipe in Table 1.
6. Primary antibodies against CD34, CD105, CD90, CD45, c-Kit, and CD31.
7. Propidium iodide.
8. Antibody capture beads.

2.7 Immuno-cytochemistry Characterization of CPCs

1. 4% formaldehyde. Recipe in Table 1. Prepare just before use.
2. 10% goat serum
3. Primary antibodies against FSP-1 (fibroblast marker), myosin heavy chain (cardiomyocytes marker).
4. DAPI.
5. Immunofluorescence microscope.
6. Fluorescent mounting medium.

3 Methods

3.1 Sample Collection and Isolation of CPCs

1. Collect the RAA and the LV biopsy samples in bottles containing ice-cold sterile KRH buffer solution. The isolation procedure should start ideally within 60 min after the collection of the tissue samples.
2. Transfer the tissues to sterile HBSS (Fig. 1).
3. Remove the epicardial fat using scissors and scalpel.
4. Weigh the tissue and wash with sterile HBSS.
5. Mince the RAA and LV tissues and suspend in 15 mL of fresh HBSS in a 15 mL conical tube. As a further washing step, allow the tissue to settle down at the bottom of the tube by gravity, remove the supernatant buffer and replace it with fresh HBSS. Repeat this process twice to ensure efficient washing of the tissue.

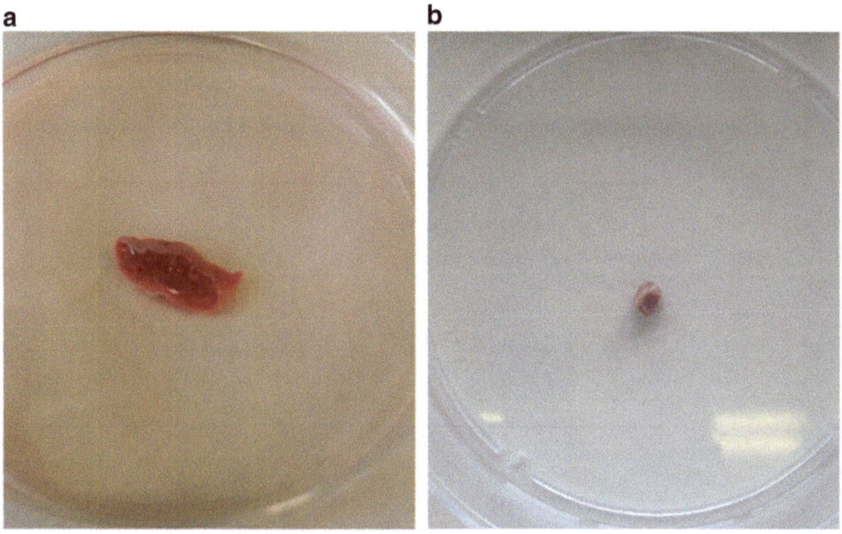

Fig. 1 Tissue samples. The RAA (**a**) and LV biopsy (**b**) received from one of the Patients undergoing on-pump coronary artery bypass graft surgery

6. Resuspend the minced tissue pieces in 7 mL of warm 0.04% collagenase II solution kept at 37 °C. Transfer the suspension to T25 flasks and place in the shaker-incubator for 1 h at 37 °C.
7. At the end of the incubation, place the T25 Flasks on ice allowing the tissue pieces inside the flask to settle down by gravity. Remove the supernatant consisting of collagenase solution and isolated cells in a fresh sterile 15 mL conical tube containing 0.5 mL DBS to neutralize collagenase activity. Store the tube on ice until remaining tissues are digested.
8. Add fresh collagenase solution to the remaining tissue pieces in T25 Flask and mechanically digest using repeated pipetting (10 times). Leave the undigested pieces to settle on ice for 3 min before removing the supernatant into the same conical tube containing the previously collected supernatant collagenase solution and isolated cells.
9. Centrifuge the collected supernatant solution for 5 min at 300 × *g* and resuspend the cell pellet in warm Ham's F12 complete medium. Transfer the cells to culture flasks (details mentioned in Table 2) (*see* **Notes 1** and **2**) and culture in 5% CO_2/95% humidified air at 37 °C.

3.2 The Maintenance, Subculturing, and Expansion of CPCs

1. Replenish with fresh Ham's F12 complete medium after 24 h of isolation and henceforth change the medium every 2 days.
2. As soon as either LV or the RAA flasks reach 90% confluence, passage both the cell types together on the same day so that they remained in culture for the same length of time and were paired for passages (*see* **Note 3**).

Table 2
Criteria for selection of the size of the culture flask based on the weight of the tissue received

Weight of the tissue	Type/size of the flask plated in
100 -300 mg	T25 Flask
30-100 mg	6 well plate
<30 mg	24 well plate

3. For passaging, aspirate the culture medium from the cell culture flasks, wash with prewarmed PBS at 37 °C (80 μL/cm^2 of the culture dish). After removing PBS, add TrypLE dissociation reagent to the culture dishes (80 μL/cm^2) and incubate at 37 °C for 5 min.
4. Once the cells are detached add fresh Ham's F12 complete medium to the culture flasks in the ratio of 2:1 (culture medium: TrypLE) and centrifuge this mixture at 300 × *g* for 5 min.
5. Resuspend the resultant pellet in the fresh Ham's F12 complete medium.
6. Count the cells using a hemocytometer and transfer the required fraction of cells to fresh culture flasks Passage one onwards, plate both the cell types at the seeding density of 10^4 cells/mL in similarly sized flasks so that the seeding density remained constant between the two tissue types. Maintain the cells as described above.
7. The human CPCs show a mesenchymal cell type morphology, with a spread phenotype and frequent protrusions (Fig. 2) (*see* **Notes 4** and **5**).

3.3 Freezing Aliquots of RAA and LV CPCs at Paired Passages

1. Passage two onward; freeze the aliquots of both the cell types to be used later for future experiments.
2. After trypsinizing cells (*see* Subheading 3.2, **steps 3** and **4**), freeze an aliquot by suspending the pellet of dissociated cells in 500 μL of the ice cold MSC freezing medium, then transferring it to a freezing vial.
3. To minimize the damage occurring during freezing, the freezing vial can be wrapped with paper towels to facilitate slow freezing and transferred to −80 °C for 24 h.
4. At the end of 24 h transfer the cells to −150 °C liquid nitrogen storage canister.

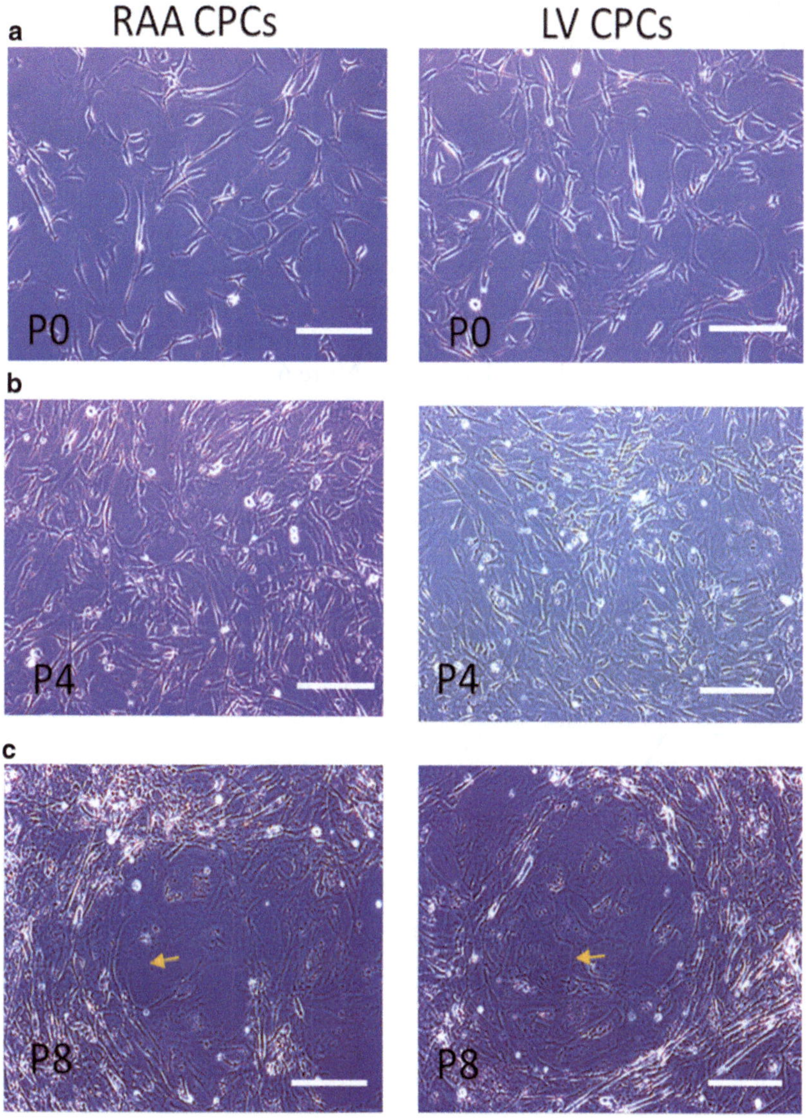

Fig. 2 Images of RAA and LV CPCs at different passages in culture. Representative phase contrast microscopic images of CPCs from RAA and LV of the same patient at passage 0 (**a**), 4 (**b**), and 8 (**c**). It can be seen that at passage 8 (**c**) the cells assemble in peculiar circular arrangement (yellow arrows) in both CPCs. Scale bars are 100 μm

3.4 Initiating Cells from Frozen Aliquots

1. To initiate frozen aliquots of CPCs, thaw the frozen aliquots in the water bath at 37 °C. Stop this process when you see the last piece of ice.
2. Add this cell suspension to 9 mL of the warm Ham's F12 complete medium in drops using a pipette.
3. Centrifuge the suspension at 300 × *g* for 5 min and plate in Ham's F12 complete medium. Change the medium after 24 h.

3.5 Calculation of Cumulative Population Doublings (CPD)

1. Calculate CPD for each cell type at each trypsinization using the formula

$$\text{CPD for } P_n = \frac{\ln(\text{cell count at the end of } P_n) - \ln(\text{cell count at the beginning of } P_n)}{\ln(2)} + \text{CPD for } P_{n-1}$$

   where P_n is the current passage number, CPD for P_n is the CPD for the most recent passage number, and CPD for P_{n-1} is the CPD for the previous passage number.

2. Plot a CPD versus time graph for all the samples displaying the CPD at each passage with respect to the time it had been in culture (Fig. 3) (*see* **Note 6**).
3. Create a new interpolated data set on this graph by modeling straight lines joining adjacent data points and manually reading off (the red dots in Fig. 3) the CPD at day 7, 14, 21, and so on up till day 56 for all the samples.

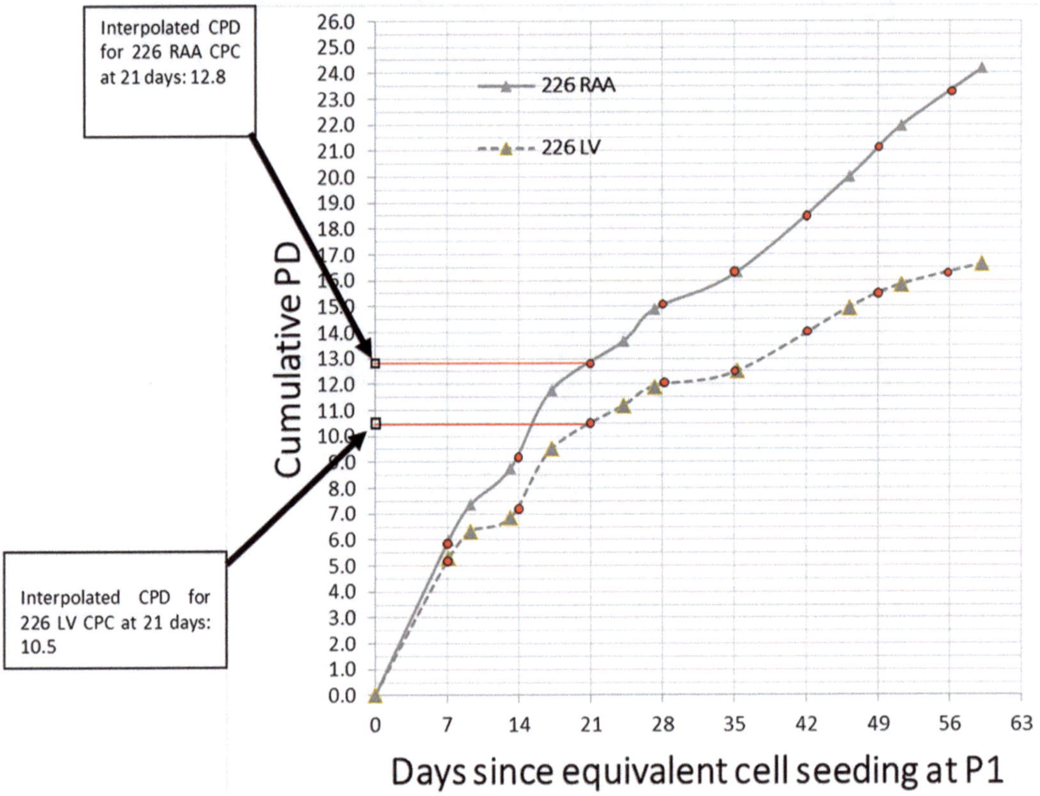

Fig. 3 Method for calculating CPDs. The graph shows the procedure for modeling the weekly CPDs using adjacent straight lines to manually interpolate and read off the CPD at weekly intervals. The red dots show the points where the CPDs were interpolated using straight lines. The triangular points show the actual CPD reading at that passage. The red lines show the interpolation to the y-axis to read off the CPD at the mentioned time point (21 days' time point in this graph is displayed as an example). The CPD vs. time graph shown here is from the CPD data of one of the sample isolated in our laboratory

The resolution of the data will therefore conservatively approximated in this manner, and the CPDs for all the samples can be predicted at the same time intervals (*see* **Note 7**).

3.6 Characterization of RAA and LV CPCs for Progenitor Cell Surface Markers Using Low Cytometry

3.6.1 Staining of Cells

1. Use the RAA and LV CPCs at the same passage numbers (between 2 and 4) for characterization experiments (*see* **Note 8**).
2. Dissociate the cells as described in Subheading 3.2 **step 3**, centrifuge at 300 × *g* for 5 min and suspend in the flow buffer such that the concentration is not more than 1–2 × 10^6 cells/mL.
3. Transfer 50 μL of the cell suspension to 5 mL polystyrene tubes.
4. Add 50 μL of the fluorochrome-conjugated antibody solution (2.5 μL of the antibody diluted in 47.5 μL of the flow buffer) to it and incubate on ice in the dark for 30 min. Incubate the cells used as unstained/control samples for the same time with 50 μL of flow buffer.
5. Wash the cells twice with 2 mL of the flow buffer by centrifuging them at 400 × *g* for 5 min.
6. Suspend the final cell pellet in 400 μL of flow buffer and add 5 μL of 1 mg/mL propidium iodide (PI) to the tubes containing the test samples and the PI compensation controls, just before taking the samples to the flow cytometry facility for analysis (*see* **Note 9**).

3.6.2 Staining of Compensation Controls

1. For staining, add two drops of the capture beads to the polystyrene tubes and incubate with 2 μL of the respective antibodies in the dark for 15 min (*see* **Note 10**).
2. Subsequently, wash the beads by adding 3 mL of PBS and centrifuging at 300 × *g* for 5 min.
3. Resuspend the final bead pellet in 300 μL of PBS and two drops of the negative bead (which is used as an unstained compensation control) to the final bead suspension before the acquisition.
4. To have a positive compensation control for the PI channel, stain the dead cells (killed by incubating at −80 °C for 30 min) with PI.
5. Prepare the fluorescence minus one (FMO) controls by excluding one of the fluorophore-conjugated antibodies from the complete antibody cocktail for that particular fluorophore. For example, the FMO control for APC fluorochrome contains all the other fluorophore-conjugated antibodies except for APC.

3.6.3 Instrument Setup, Data Acquisition, and Analysis

The protocol described here describes the procedure for cell acquisition using the Gallios flow cytometer (Beckman Coulter Inc., USA).

1. Set the flow cytometer voltages using a combination of unstained, single stained bead or FMO cell controls in such a way that the spill-over was minimum in the adjacent channels.
2. Discriminate the doublets by visualizing the FSA (forward scatter area) and FSW (forward scatter width). Doublets are formed when two or more cells clump together and give a false signal.
3. Exclude the dead cells by gating the negative cells in the PI channel [3]. Respective unstained samples were used as controls to define the gating to identify the positive staining.
4. Post-acquisition compensation and data analysis is performed using FlowJo 9.6 software.

3.7 Differentiation of CPCs into Cardiomyocytes

1. Seed 5×10^5 CPCs on poly-L-ornithine (20 μg/mL) coated 24-well culture dishes for gene expression experiments (*protocol not described in this chapter*) and 5×10^3 CPCs on 8-well chamber slides for immunocytochemistry (ICC) experiments (*see* **Note 8**).
2. Culture the cells in the complete Ham's F12 complete medium for 1 day to allow them to form a confluent monolayer, after which change the medium to the cardiomyocyte differentiation medium. Culture the control cells in the Ham's F12 complete medium.
3. At the end of 15 days, isolate the RNA for gene expression analysis (protocol not described here) or fix the cells with freshly prepared 4% paraformaldehyde to validate the expression of cardiomyocyte lineage markers at the protein level.

3.8 Immuno-cytochemistry Characterization of CPCs and Differentiated Cardiomyocytes

1. The undifferentiated cells will be used as negative controls. Seed the undifferentiated CPCs in fibronectin-coated 8-well chamber slides at a seeding density of 2×10^3 cells/well in Ham's F12 complete medium and culture for 24 h until a monolayer is formed. After 24 h, fix the cells by incubating them with freshly prepared 4% paraformaldehyde in PBS at room temperature for 10 min and wash three times with ice-cold PBS after fixing. Proceed with staining of both the differentiated and undifferentiated CPCs as below.
2. Block the nonspecific binding sites by incubating the cells in 10% goat serum (in PBS) for 30 min.
3. Treat the cells with the primary anti-FSP1 antibody or MHC antibody diluted in 1% goat serum, for 2 h in a humidified chamber.

4. Following this, remove the antibody solution and wash the cells three times with PBS with a 5-min incubation after each wash.
5. Next, incubate the cells with the fluorophore-conjugated secondary antibody (Alexa Fluor-488 or -568) at 1:100 dilution in 1% goat serum in PBS for 1 h in a humidified chamber in the dark.
6. Decant the antibody and wash the cells three times with PBS (in dark) with a 5-min incubation after each wash.
7. Finally, add 0.1 μg/mL DAPI to the cells for 1 min to stain the nucleus, and rinse the cells with PBS.
8. Finally, detach the glass walls/chamber walls of the slides and mount a coverslip on the slides by adding a drop of Fluoromount G/any fluorescent mounting medium.
9. Allow the slides to dry for 24 h before visualization under a fluorescence microscope.
10. To test the specificity of the secondary antibody, add a negative control group where the cells undergo the same immunohistochemical and immunofluorescent labeling procedures with the exclusion of primary antibodies.
11. Analyze the images using Image J software (NIH, USA).

4 Notes

1. The size of the RAA tissue is usually much larger than the LV tissue, and hence the size of the pellet attained at the last centrifugation step in the isolation process is much larger for the RAA tissue. A previous study by Davis et al. showed that even when normalized for the size of the tissue, the yield of CPCs from the atria is much higher compared to other locations in the heart [4].
2. Although it would be ideal to plate the same number of cells for both RAA and LV tissue after isolation to maintain consistency between culture conditions, this is not possible because, on the day of isolation, the isolates also contain debris, endothelial cells, cardiomyocytes, smooth muscle cells, and dead cells. So the only way to maintain a similar cell number to surface area ratio was to plate them in appropriately sized flask which is decided according to the weight of the starting tissue sample, which was standardized in our laboratory after multiple isolation trials (Table 2).
3. It is important to make sure that both the cell types from the same patient were passaged on the same day at the same time. Subculturing them at the same time, ensure that both the cells

remained in culture for the same length of time. Our preliminary trials showed that the proliferation rate of RAA and LV CPCs remained the same until the first 5–6 passages; hence, we could assume the population doublings to be the same for both the cell types for early passages. Hence when the two cells types from the same passage number from the same patients were used for comparing their functional properties, it was a fair comparison because both cell types had undergone similar population doublings.

4. We can observe changes in the morphology of both the cell types at higher passages with longer duration of culture. At later passages, the cells usually acquire a senescent phenotype which could be seen with the presence of a larger flattened cytoplasm containing vacuoles and cytoplasmic filaments (Fig. 2c). The increase in cell size was noticeable when compared to passage zero (Fig. 2a). This type of change in morphology has been previously reported in primary cells in various studies [5, 6].
5. Based on our experience, the senescent RAA and LV CPCs assembled in a characteristic circular arrangement (as can be seen in the images from P8 of RAA and LV CPCs marked by yellow arrows, Fig. 2c). However, there is no evidence of such a phenomenon in the literature to our knowledge.
6. To aggregate the CPD versus time data, it is required to have the CPDs of all the samples at the same time points during culture. However, as mentioned previously, the cell count reading could only be taken when the cells were subcultured at the confluence, making it impossible for CPD reading to be recorded at universally fixed weekly intervals for all the patients. Moreover, it was highly unlikely that the cells reach confluence at the same fixed time points for all the patients. Hence we devised a techniques described here to predict the weekly CPD for all the samples using a simple method.
7. The last time point for recording CPD was decided as 8 weeks (56 days) of culture. However, this can go further depending on the potential of the isolated CPCs.
8. All the experiments for assessing functional properties with CPCs should be usually conducted between passages 3 and 6. The in vitro experiments required large cell numbers, which could only be reached after passage two. Although the passage numbers at which the proliferation capacity of CPCs started to decline was dependent on the patient source of the sample, we identified that for all the patients the cells remained in the log phase growth until 6 weeks.
9. PI is a dye used to identify dead cells [7].

10. After incubation with a fluorochrome-conjugated mouse antibody, the two bead components provide distinct positive and negative populations of beads that can be used to set the compensation for a combination of fluorochrome-labeled mouse antibodies.

References

1. Mishra R, Vijayan K, Colletti EJ, Harrington DA, Matthiesen TS, Simpson D, Goh SK, Walker BL, Almeida-Porada G, Wang D, Backer CL, Dudley SC Jr, Wold LE, Kaushal S (2011) Characterization and functionality of cardiac progenitor cells in congenital heart patients. Circulation 123:364–373
2. Dixit P, Donnelly H, Edamatsu M, Galvin I, Bunton R, Katare R (2017) Progenitor cells from atria, ventricle and peripheral blood of the same patients exhibit functional differences associated with cardiac repair. Int J Cardiol 228:412–421
3. Herzenberg LA, Tung J, Moore WA, Herzenberg LA, Parks DR (2006) Interpreting flow cytometry data: a guide for the perplexed. Nat Immunol 7:681–685
4. Davis DR, Kizana E, Terrovitis J, Barth AS, Zhang Y, Smith RR, Miake J, Marban E (2010) Isolation and expansion of functionally-competent cardiac progenitor cells directly from heart biopsies. J Mol Cell Cardiol 49:312–321
5. Rodier F, Campisi J (2011) Four faces of cellular senescence. J Cell Biol 192:547–556
6. Toussaint O, Medrano EE, von Zglinicki T (2000) Cellular and molecular mechanisms of stress-induced premature senescence (SIPS) of human diploid fibroblasts and melanocytes. Exp Gerontol 35:927–945
7. Cummings BS, Schnellmann RG (2004) Measurement of cell death in mammalian cells. Curr Protoc Pharmacol Chapter 12:Unit 12.8

Chapter 14

Mitochondrial Assays Using Cardiac Stem Cells

Ayeshah A. Rosdah, Lea M. D. Delbridge, and Shiang Y. Lim

Abstract

Ischemic heart disease is the leading cause of death worldwide. Stem cell therapy to repair and regenerate the infarcted myocardium is a promising approach to address this unmet medical need. However, the poor survival of transplanted cells in the hostile ischemic myocardium has been a major hurdle in achieving an effective cell therapy against myocardial infarction. As such, novel strategies to promote the survival of transplanted cells are highly sought after. Mitochondria are intimately involved in cell survival and have been the main organelles being targeted for cytoprotection. Mitochondrial morphology is linked to mitochondrial function and cell viability. Therefore, quantitative methodologies to obtain reliable and reproducible results of mitochondrial morphology and function are essential for identifying and developing new cytoprotective strategies to enhance the survival of stem cells post-transplantation. Here, we describe methods for assessing mitochondrial morphology, mitochondrial membrane potential, and mitochondrial reactive oxygen species production.

Key words Cardiac stem cells, Mitochondria, Mitochondrial morphology, Mitochondrial membrane potential, Reactive oxygen species, Cell survival

1 Introduction

Mitochondria are dynamic organelles that undergo fission and fusion events to produce fragmented and elongated mitochondrial phenotypes, respectively. These two opposing processes are regulated by a group of mitochondrial fusion proteins (mitofusin-1, mitofusin-2, and optic atrophy-1) as well as mitochondrial fission proteins (dynamin-related protein-1 (Drp1), mitochondrial fission 1 protein, mitochondrial fission factor, mitochondrial dynamics proteins of 49 and 51 kDa) [1–3]. These processes play central roles in quality control of mitochondria and are important for maintaining cell viability and various cellular functions. Mitochondrial fusion is required for appropriate distribution of mitochondrial DNA, lipids, and proteins among mitochondria, which in turn ensure optimal conditions for mitochondria to carry out key cellular processes such as energy metabolism, cellular differentiation and calcium homeostasis [3]. The opposite process, mitochondrial

Mugdha V. Joglekar and Anandwardhan A. Hardikar (eds.), *Progenitor Cells: Methods and Protocols*, Methods in Molecular Biology, vol. 2029, https://doi.org/10.1007/978-1-4939-9631-5_14, © Springer Science+Business Media, LLC, part of Springer Nature 2019

fission, plays an important role in mitochondrial proliferation following mitosis and is involved in removing damaged mitochondria from the cell through the process of mitophagy [4].

Mitochondrial fission occurs during apoptosis and necrosis, where the loss of the mitochondrial network facilitates cell death [5–7]. Using cell culture models of oxidative stress and simulated ischemia–reperfusion injury, we have reported that inhibition of mitochondrial fission protein Drp1 with a small molecule, Mdivi-1, improved survival of human W8B2$^+$ cardiac stem cells [8]. Contradictory to other studies [9–12], the prosurvival effect of Mdivi-1 is not dependent of effects on recovery of mitochondrial membrane potential and reduction in reactive oxygen species production [8]. In this chapter, we describe three mitochondrial assays that are applicable to most adherent cells: mitochondrial morphology, mitochondrial membrane potential, and mitochondrial reactive oxygen species.

2 Materials

2.1 Reagents

1. **Explant Media:** 80% Iscove's Modified Dulbecco's Medium (IMDM), 10% fetal bovine serum, 0.1 mM β-mercaptoethanol, 2 mM L-glutamine, 100 U/mL penicillin, 100 μg/mL streptomycin, 0.25 μg/mL amphotericin B (*see* **Note 1**).
2. **Growth Media:** Endothelial Cell Growth Medium-2 (EGM-2):M199 media in 1:3 ratio, 10% fetal bovine serum, 0.1 mM nonessential amino acids, 100 U/mL penicillin, 100 μg/mL streptomycin, 0.25 μg/mL amphotericin B.
3. PBS with (PBS+/+) and without (PBS−/−) calcium and magnesium.
4. Gelatin from porcine skin: 0.1% gelatin in distilled water and autoclave to solubilize. Store at 4 °C for 3–4 months.
5. 3 mM hydrogen peroxide in growth media. Prepare fresh when required.
6. 100 μM EUK-134 (chloro[[2,2′-[1,2-ethanediylbis[(nitrilo-κN)methylidyne]]bis[6-methoxyphenolato-κO]]]-manganese; Cayman Chemicals) in growth media. Prepare fresh when required.
7. 50 μM carbonyl cyanide m-chlorophenyl hydrazone (CCCP) in growth media. Prepare fresh when required.
8. 100 nM MitoTracker Red (Thermo Fisher Scientific) in growth media. Prepare fresh when required.
9. 1 μM JC-1 (5,5′,6,6′-tetrachloro-1,1′,3,3′-tetraethyl-imidacarbocyanine iodide, 5,5′,6,6′-tetrachloro-1,1′,3,3′-tetraethylbenzimidazolocarbocyanine iodide, Thermo Fisher Scientific) in PBS+/+. Prepare fresh when required.

10. 1 μM MitoSOX Red in PBS+/+. Prepare fresh when required.
11. 4% Paraformaldehyde in PBS. Store at −20 °C.
12. Acetone. Store at −20 °C.
13. Collagenase A: Prepare aliquots of stock solution 50 mg/mL in PBS−/− for long-term storage at −20 °C. Prepare a final concentration of 1 mg/mL with PBS−/− before use.
14. 0.25% trypsin–EDTA.
15. 10 μg/mL fibronectin in PBS−/−. Prepare fresh when required.
16. Fluorescent mounting agent.
17. W8B2 microbead (mouse monoclonal anti-human).
18. W8B2 PE-conjugated antibody (mouse anti-human IgG).

2.2 Instruments

1. MACS Cell separator.
2. Flow cytometer and sorter.
3. Fluorescence microscope such as BX60.
4. Inverted fluorescence microscope such as IX71.

2.3 Plasticware and Glassware

1. Black 96-well plate with clear flat bottom, black polystyrene.
2. Clear 24-well plate with clear flat bottom, clear polystyrene.
3. Clear 100 mm diameter dish.
4. Round cover glass (diameter 12 mm).

2.4 Software

1. ImageJ software.

3 Methods

3.1 Isolation of Human W8B2$^+$ Cardiac Stem Cells

1. Isolate human W8B2$^+$ cardiac stem cells from freshly incised human atrial appendage specimens, obtained after informed consent as per the institutional ethics approvals [13].
2. Mince tissue manually into small fragments and digest with 1 mg/mL of collagenase A for 20 min at 37 °C.
3. Plate tissue fragments onto fibronectin coated dishes (100 mm diameter) and culture in explant media (*see* **Note 2**). Change media every 3–4 days.
4. After 2–3 weeks, collect monolayer outgrowth cells from the adherent tissue fragments by enzymatic digestion using trypsin–EDTA.
5. Expand cells in growth media. When confluent, label cells with W8B2 microbeads and enrich by MACS cell separator. Expand enriched W8B2$^+$ cells in growth media.

6. After confluent, label cells with W8B2 PE-conjugated antibody and further enrich by fluorescence-activated cell sorting. Expand enriched W8B2$^+$ cells in growth media.

3.2 Mitochondrial Morphology Assay

1. Assess mitochondrial morphology with the cell permeant mitochondrial selective fluorescent dye, MitoTracker Red.
2. Seed W8B2$^+$ cardiac stem cells onto gelatin-coated cover glass in 24-well plates at 211 cells/mm^2 (*see* **Notes 2** and **3**).
3. Stain cells with 100 nM MitoTracker Red for 30 min at 37 °C in a humidified 5% CO_2 incubator and wash cells once with PBS−/−.
4. To induce oxidative stress and mitochondrial fission, treat cells with 3 mM of H_2O_2 for 3 h at 37 °C in a humidified 5% CO_2 incubator (*see* **Note 4**). Leave some cells without H_2O_2 as controls.
5. Gently remove media (*see* **Note 5**) and fix cells with 4% paraformaldehyde for 15 min at room temperature followed by ice cold acetone for 1 min at room temperature (*see* **Notes 6** and **7**).
6. Mount cover glass onto glass slides with a fluorescent mounting agent and dry overnight at 4 °C before visualization.
7. Capture images at 600× magnification with the fluorescence microscope (Fig. 1) and analyze mitochondrial morphology with the ImageJ software.

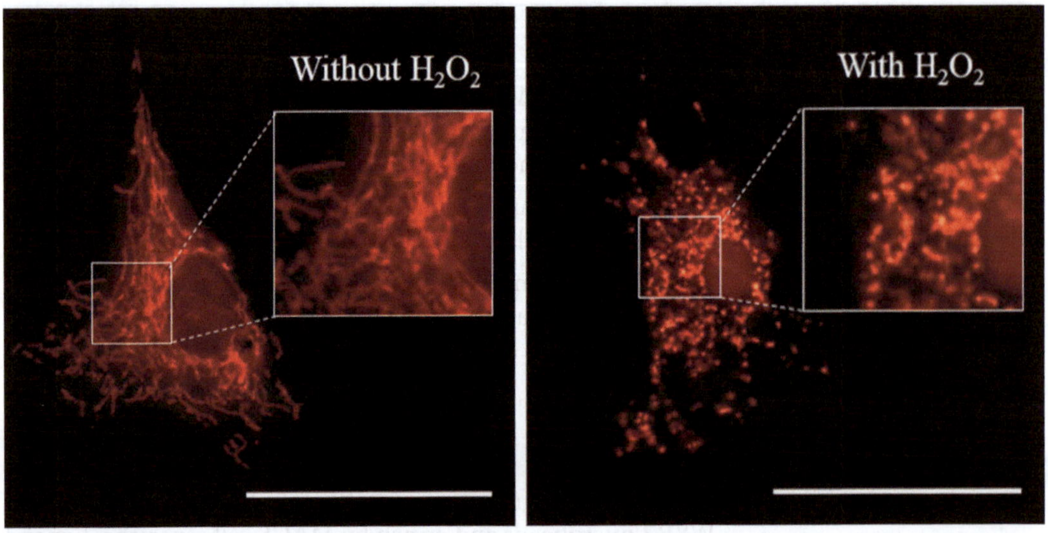

Fig. 1 Representative images of mitochondrial morphology of W8B2$^+$ cardiac stem cells treated with and without H_2O_2. Scale bar = 50 μm

3.3 Mitochondrial Membrane Potential Assay

1. The dual fluorescent cationic dye, JC-1, was used to assess mitochondrial membrane potential in live W8B2$^+$ cardiac stem cells (*see* **Note 8**).
2. Seed cells onto gelatin-coated 96-well black plates with clear flat bottom at 156 cells/mm^2 (*see* **Note 2**).
3. To induce oxidative stress, treat cells with 3 mM of H_2O_2 for 3 h at 37 °C in a humidified 5% CO_2 incubator. Leave some cells without H_2O_2 as controls.
4. To induce dissipation of mitochondrial membrane potential, treat cells with 50 μM of CCCP, a mitochondrial respiratory uncoupler, for 5 min at 37 °C in a humidified 5% CO_2 incubator.
5. Gently remove media and replace with PBS+/+ containing 1 μM of JC-1 dye, and incubate for 10 min at 37 °C in a humidified 5% CO_2 incubator.
6. Wash cells with PBS−/− twice prior to imaging in PBS−/− (*see* **Notes 5** and **9**).
7. Visualize J-aggregates (orange to red fluorescence) and monomers (green fluorescence) of JC-1 in cells with the fluorescence microscope (Fig. 2) (*see* **Notes 10** and **11**). Capture subsequent images from both channels under 200× magnification. Quantify fluorescence intensity using ImageJ software by subtracting the fluorescence intensity of cells with that of background (a cell-free region). Express the data as red-to-green fluorescence ratio for individual cells.

3.4 Mitochondrial Reactive Oxygen Species Assay

1. Mitochondrial reactive oxygen species, specifically superoxide, can be assessed using the fluorescent dye MitoSOX Red. This dye selectively targets the mitochondria and is specifically oxidized by superoxide.
2. Seed cells onto gelatin-coated 96-well black plates with clear flat bottom at 156 cells/mm^2 (*see* **Note 2**).
3. To induce oxidative stress, treat cells with 3 mM of H_2O_2 for 3 h at 37 °C in a humidified 5% CO_2 incubator.
4. For positive control, treat cells with 100 μM of EUK-134, an antioxidant, together with 3 mM of H_2O_2 for 3 h at 37 °C in a humidified 5% CO_2 incubator.
5. Gently remove media and replace with PBS+/+ containing 1 μM of MitoSOX Red, and incubate for 10 min at 37 °C in a humidified 5% CO_2 incubator.
6. Wash cells with PBS−/− prior to imaging in PBS−/− (*see* **Notes 5** and **9**).
7. Visualize cells with the fluorescence microscope (Fig. 3) (*see* **Notes 10** and **11**). Capture bright field and subsequent red

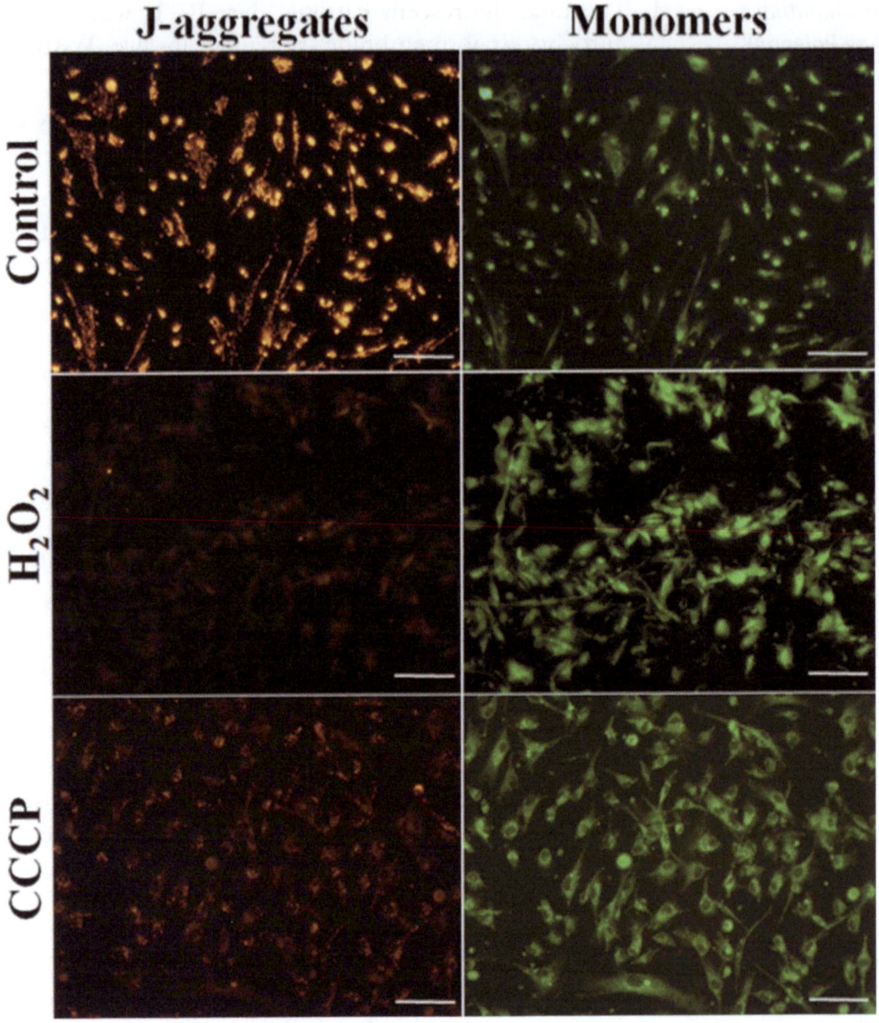

Fig. 2 Representative images of cells stained with 1 μM of JC-1. Scale bar = 50 μm

channel fluorescence images under 200× magnification. Quantify fluorescence intensity using ImageJ software by subtracting cell fluorescence intensity of the whole cell to background (a cell-free region).

4 Notes

1. It is recommended to use the same batch of fetal bovine serum and same lot of chemicals for explant and growth media to ensure consistency and reproducibility of experimental results.
2. Coat plasticware and cover glasses with fibronectin or gelatin for at least 30 min at 37 °C in a humidified 5% CO_2 incubator. Aspirate fibronectin or gelatine before use.

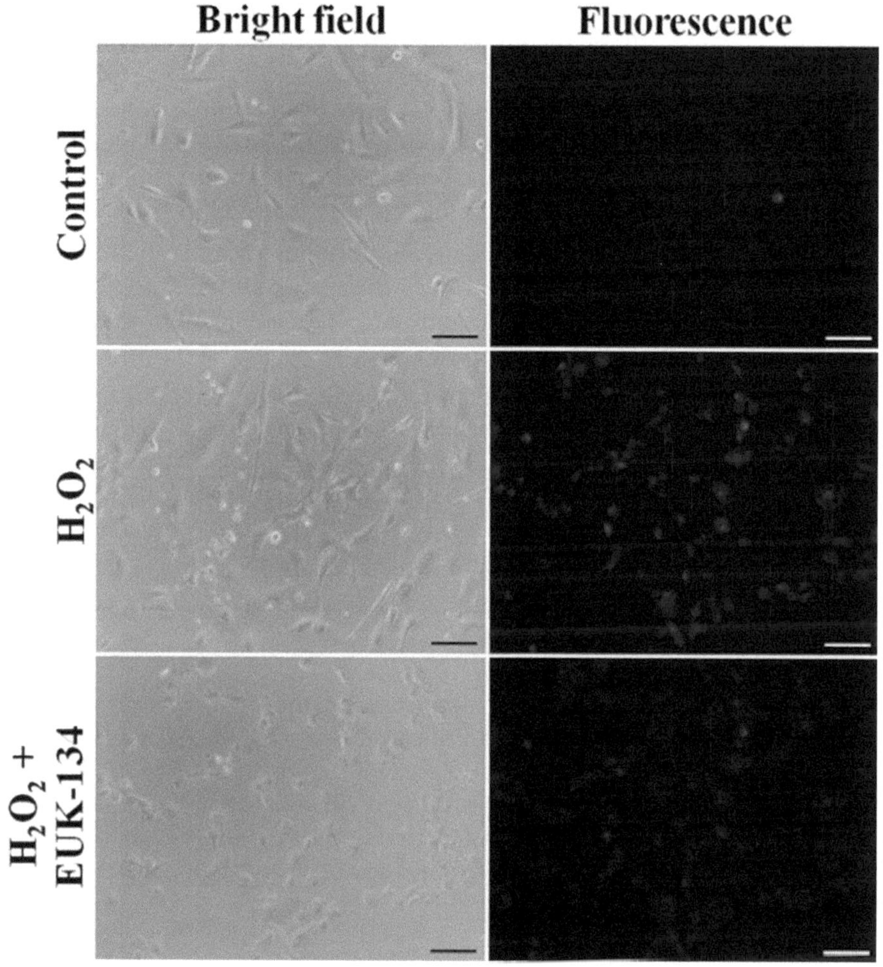

Fig. 3 Representative images of cells stained with 1 μM of MitoSOX Red. Scale bar = 50 μm

3. Ensure no air bubble is trapped in between the cover glass and the tissue culture plate during coating with gelatine.
4. Keep hydrogen peroxide away from light as it is light-sensitive. It is advisable to keep the containers containing the hydrogen peroxide wrap in aluminium foil and store at 4 °C.
5. Care must be taken when removing media and during washing with PBS as cells are loosely attached after hydrogen peroxide-induced oxidative damage.
6. Avoid leaving ice cold acetone for more than 1 min inside the wells, as it can dissolve polymer of the plasticware.
7. Care must be taken when removing cover glass from tissue culture plate with a pair of tweezer. Avoid scratching the surface of the cover glass with cells attached.

8. JC-1 exists as monomers inside the cytoplasm and yield green fluorescence. Highly negative mitochondrial membrane potential attracts large uptake of the dye into the mitochondria and form J-aggregates which yield orange to red fluorescence. These characteristics of JC-1 enable assessment of mitochondrial membrane potential in single cells by quantifying the ratio of red to green fluorescence intensity [14–16].
9. The washing step is important to remove all traces of JC-1 and MitoSOX Red dyes during imaging. Traces of the dye will cause high background fluorescence and result in variation between experiments.
10. Do not overexpose the cells to light as it can affect the fluorescence intensity and confound the image analysis.
11. Use the same fluorescence light intensity throughout imaging of JC-1 and MitoSOX fluorescence across all experimental groups.

Acknowledgments

This study was supported by the St Vincent's Hospital (Melbourne) Research Endowment Fund, CASS foundation, and Stafford Fox Medical Research Foundation. The O'Brien Institute Department and St Vincent's Institute of Medical Research receive Operational Infrastructure Support from the Victorian State Government of Innovation, Industry and Regional Development.

References

1. Da Silva AF, Mariotti FR, Maximo V, Campello S (2014) Mitochondria dynamism: of shape, transport and cell migration. Cell Mol Life Sci 71:2313–2324
2. Liesa M, Palacín M, Zorzano A (2009) Mitochondrial dynamics in mammalian health and disease. Physiol Rev 89:799–845
3. Palmer CS, Osellame LD, Stojanovski D, Ryan MT (2011) The regulation of mitochondrial morphology: intricate mechanisms and dynamic machinery. Cell Signal 23:1534–1545
4. Otera H, Mihara K (2012) Mitochondrial dynamics: functional link with apoptosis. Int J Cell Biol 2012:821676
5. Brooks C, Cho SG, Wang CY, Yang T, Dong Z (2011) Fragmented mitochondria are sensitized to Bax insertion and activation during apoptosis. Am J Physiol Cell Physiol 300: C447–C455
6. Wang Z, Jiang H, Chen S, Du F, Wang X (2012) The mitochondrial phosphatase PGAM5 functions at the convergence point of multiple necrotic death pathways. Cell 148:228–243
7. Whelan RS, Konstantinidis K, Weld AC et al (2012) Bax regulates primary necrosis through mitochondrial dynamics. Proc Natl Acad Sci U S A 109:6566–6571
8. Rosdah AA, Bond ST, Sivakumaran P et al (2017) Mdivi-1 protects human W8B2+ cardiac stem cells from oxidative stress and simulated ischemia-reperfusion injury. Stem Cells Dev 26:1771–1780
9. Grohm J, Kim SW, Mamrak U et al (2012) Inhibition of Drp1 provides neuroprotection in vitro and in vivo. Cell Death Differ 19:1446–1458
10. Lim S, Lee SY, Seo HH et al (2015) Regulation of mitochondrial morphology by positive feedback interaction between PKC and Drp1 in vascular smooth muscle cell. J Cell Biochem 116:648–660

11. Liu JM, Yi Z, Liu SZ, Chang JH, Dang XB, Li QY, Zhang YL (2015) The mitochondrial division inhibitor Mdivi-1 attenuates spinal cord ischaemia-reperfusion injury both in vitro and in vivo: involvement of BK channels. Brain Res 1619:155–165
12. Wang J, Wang P, Li S, Wang S, Li Y, Liang N, Wang M (2014) Mdivi-1 prevents apoptosis induced by ischemia–reperfusion injury in primary hippocampal cells via inhibition of reactive oxygen species–activated mitochondrial pathway. J Stroke Cerebrovasc Dis 23:1491–1499
13. Zhang Y, Sivakumaran P, Newcomb AE et al (2015) Cardiac repair with a novel population of mesenchymal stem cells resident in the human heart: W8B2^{+} cardiac stem cells for heart repair. Stem Cells 33:3100–3113
14. Cottet-Rouselle C, Ronot X, Leverwe X, Mayol JF (2011) Cytometric assessment of mitochondria using fluorescent probes. Cytometry A 79:405–425
15. Griffiths EJ (2000) Mitochondria - potential role in cell life and death. Cardiovasc Res 46:24–27
16. Mathur A, Hong Y, Kemp BK, Barrientos AA, Erusalimsky JD (2000) Evaluation of fluorescent dyes for the detection of mitochondrial membrane potential changes in cultured myocytes. Cardiovasc Res 46:126–138

Chapter 15

Two-Dimensional Electrophoresis and Mass Spectrometry for Protein Identification

Amaresh K. Ranjan and Anil Gulati

Abstract

Two-dimensional electrophoresis is useful for separation of individual proteins that can be easily isolated and processed with mass spectrometry for their identification. Here we describe a simplified method of 2D electrophoresis with aim to help beginners in optimization of protein separation protocol, which can be further modified to enhance protein spot resolution. Current protocol can be used in different fields of biology including progenitor cells.

Key words 2D Electrophoresis, Proteins, Proteomics, Mass spectrometry, Isoelectric focusing, SDS-PAGE

1 Introduction

The two-dimensional electrophoresis (2-DE) is considered as the core technology of proteomics, with the capability of resolving hundreds to thousands of proteins in one go. The technique for 2D electrophoresis has evolved over the past few decades with the advancement of technology in the field of proteomics. The history of 2-D electrophoresis is often traced back to the early 1970s, when laboratories experimented with linking the techniques of isoelectric focusing (IEF) and SDS-PAGE (sodium dodecyl sulfate–polyacrylamide gel electrophoresis), the first and second dimensions of 2D gel electrophoresis, respectively [1–5]. At first, the methodology of groups was difficult to replicate as McGillivray et al. in 1974 were only able to demonstrate a few spots on gels using Coomassie staining, resulting in a limited appeal [6]. However, in the following year Patrick O'Farrell's group published the molecular weight-based separation of radiolabeled proteins from an *E. coli* sample using native gel isoelectric focusing (IEF) [7]. The method used by the O'Farrell group was successful and became widely employed in the next years of what would be known as the proteomics revolution of the 1980s. However, in his reposted 2D technique, the first

Mugdha V. Joglekar and Anandwardhan A. Hardikar (eds.), *Progenitor Cells: Methods and Protocols*, Methods in Molecular Biology, vol. 2029, https://doi.org/10.1007/978-1-4939-9631-5_15,

dimension gel run was performed in carrier ampholyte containing polyacrylamide gels. These gels were very elongated and cylindrical and emerged from a glass tube, scarcely resembling the gels that are known today to researchers worldwide. Moreover, this system was constrained by cathodic drift, variable pH gradients, and deformations of the gel, which were significant barriers to reproducibility.

Later, the development of immobilized pH gradient gel and Immobiline reagent (ampholytes) revolutionized the technique with superior resolution and reproducibility [8–10]. The IEF is the most important step of 2D electrophoresis, which helps in separation of proteins with even slight difference in their net charge according to their isoelectric points (pI). The isoelectric point is the specific pH at which the effective total charge of the protein is zero. Proteins are amphoteric molecules, where the net charge (positive, negative, or zero) is dependent on the pH of their surroundings. They have net positive charge at pH below their pI and negative charge above their pI. Therefore, when they are loaded in the gel with a pH gradient they have certain charge depending upon the pH of the gel at that position (Fig. 1a). In this type of pH gradient

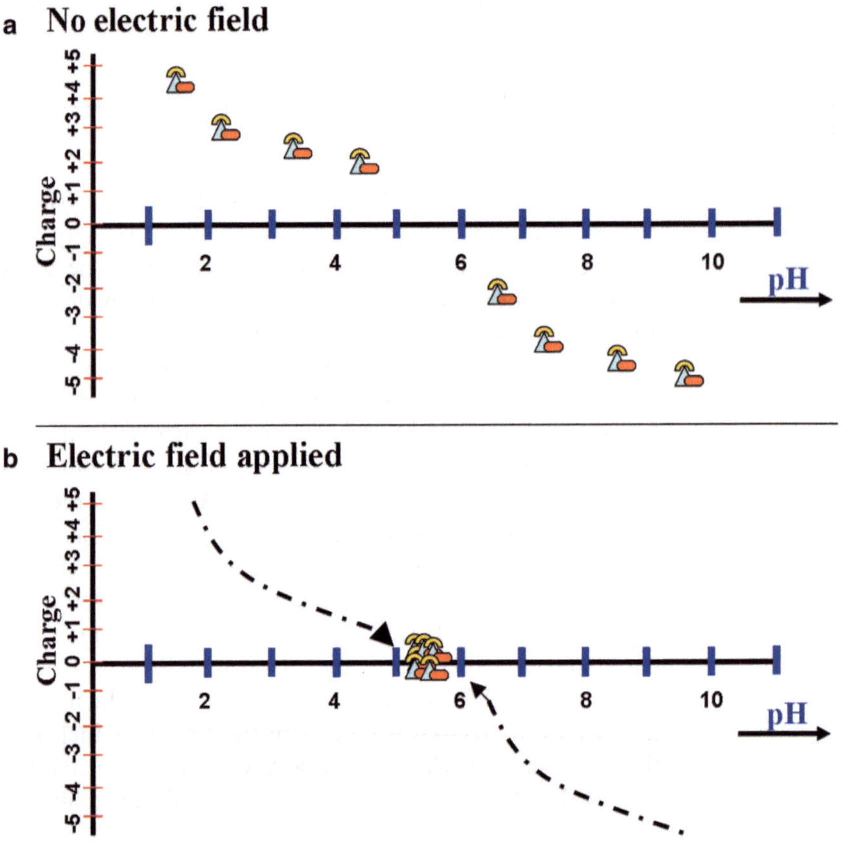

Fig. 1 Isoelectric point focusing principle. (**a**). Diagrammatic representation of a group of proteins with different net charge at different pH. (**b**) Diagrammatic representation of isoelectric focusing of a group of proteins in the presence of electricity

1. Isolation of protein
2. Precipitation and solubilization of protein in urea buffer
3. Protein estimation by using the Nanodrop
4. Loading of protein on gel strip by the method of rehydration
5. Iso electric focusing (IEF) of protein as 1st dimension electrophoresis
6. Equilibration of gel strip to SDS and DTT containing buffer
7. Equilibration of gel strip to SDS and IAA containing buffer
8. Placement of gel strip to SDS polyacrylamide gel and sealing with 0.5% agarose
9. SDS polyacrylamide gel run as 2nd dimension electrophoresis
10. Staining of SDS polyacrylamide gel with either Coomassie brilliant blue or silver stain
11. Downstream processing of gel including spot analysis, protein identification by using MALDI-TOF, etc.

Fig. 2 Schematic representation of the steps of 2-DE gel electrophoresis

gel these proteins start moving under the influence of electric field until they reach to pH gradients where their net charge becomes zero (Fig. 1b). Thus, the movement of proteins and their focusing (stopping at pI) depend upon the pH gradient and electric field strength. Therefore, the resolution of proteins depends mainly on these two factors; hence, IEF is performed at high voltage (generally >1000 V, with low current ~1 mA).

We performed IEF with IPG strips, consisting of immobilized pH gradient acrylamide gel casted on a plastic backing (Fig. 2; flowchart of 2D electrophoresis). The IPG strips were rehydrated with protein samples in rehydration buffer for overnight. After rehydration, IEF was performed with placing the strip horizontally on a flatbed first dimension electrophoresis unit (such as Ettan IPGphor Isoelectric Focusing System), equipped with proper cooling (Fig. 3a). After IEF, the strips were equilibrated in SDS containing buffer and second dimension SDS-PAGE was performed (Fig. 3b).

This chapter discusses in detail the methodology of 2D electrophoresis that we have adapted and carry out in our laboratory for different tissues and cells including endothelial progenitor cells.

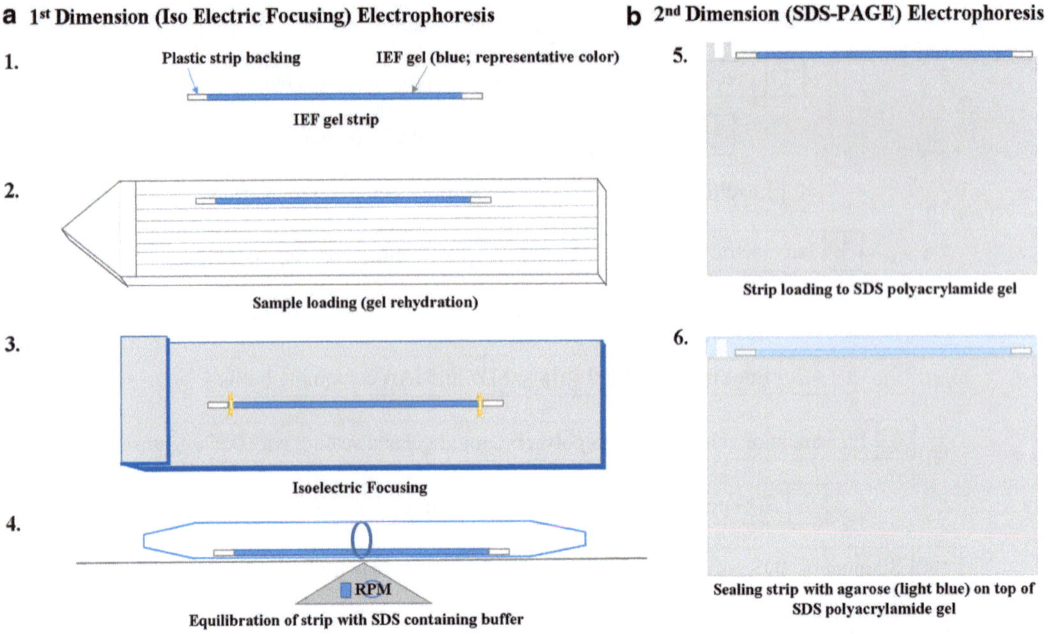

Fig. 3 Diagrammatic representation of the first dimension (**a**) and the second dimension (**b**) of 2-DE gel electrophoresis

2 Materials

2.1 First Dimension Gel Electrophoresis

1. *Urea lysis buffer*: 6 M urea, 2 M thiourea, 4% CHAPS, 2% Pharmalyte (3–10)

   Add 14.4 g urea, 6.1 g thiourea, 1.6 g CHAPS, and 800 μl Pharmalyte 3–10 to 40 ml of sterile protease and nuclease-free molecular grade water. Dissolve and store at room temperature (RT).

2. *Trichloroacetic acid (TCA)*: Make 20% (W/V) TCA by dissolving 5 g of TCA in 25 ml of nuclease-free molecular grade water.

3. *Acetone.*

4. *Sample buffer*: 8 M urea, 2% CHAPS, 0.5% Pharmalyte, and 0.002% Bromophenol Blue. Add 19.2 g urea, 0.8 g CHAPS, 200 μl Pharmalyte, and 0.8 mg Bromophenol Blue to 40 ml sterile protease and nuclease-free molecular grade water. Dissolve and store at −20 °C.

5. *Rehydration stock solution*: 8 M urea, 2% CHAPS, 0.5% Pharmalyte, and 0.002% Bromophenol Blue. Add 19.2 g urea, 0.8 g CHAPS, 200 μl Pharmalyte, and 0.8 mg Bromophenol blue to 40 ml sterile protease and nuclease-free molecular grade water. Dissolve and store at −20 °C. For *rehydration working solution*

(2.5 ml for each IEF strip) add 7 mg DTT (dithiothreitol) to 2.5 ml of rehydration stock solution just before use.

6. *Molecular grade mineral oil.*
7. *SDS equilibration stock solution*: 50 mM Tris–HCl, pH 8.8, 6 M urea, 30% glycerol, 2% SDS, 0.002% Bromophenol Blue. Add 10 ml of 1.5 M Tris–HCl pH 8.8, 72.07 g urea, 69 ml glycerol, 4.0 g SDS, and 4.0 mg Bromophenol Blue to 200 ml sterile protease and nuclease-free molecular grade water. Dissolve and store at −20 °C. For making 10 ml of SDS equilibration buffer I—add 100 mg DTT to 10 ml stock solution just prior to use, and for making 10 ml SDS equilibration buffer II—add 250 mg IAA (Iodoacetamide) to 10 ml stock solution just prior to use.

2.2 Second Dimension Gel Electrophoresis (SDS-PAGE)

1. *1.5 M Tris–HCl pH 8.8*: Dissolve 181.7 g Tris base to 750 ml of double distilled water and adjust the pH to 8.8 with concentrated HCl. After pH adjustment, add double distilled water to make the final volume 1 l. Store at 4 °C.
2. *SDS electrophoresis buffer*: 25 mM Tris–HCl, pH 8.3, 192 mM glycine, 0.1% SDS.

   Add 30.3 g Tris base, 144.0 g glycine, and 10 g SDS to 10 l of double distilled water. The pH of the solution will be ~8.3; therefore, there is no need of adjusting the pH. Store at 4 °C.
3. *Acrylamide–bis-acrylamide*: 30% acrylamide, 0.8% *N*, *N′*-methylbisacrylamide

   Add 60 g of acrylamide and 1.6 g of *N*,*N′*-methylbisacrylamide to 200 ml double distilled water. Store at 4 °C.
4. *10% SDS solution*: Add 10 g SDS to 100 ml double distilled water. Store at RT.
5. *10% APS solution*: Add 1.0 g ammonium persulfate to 10 ml double distilled water. Store at −20 °C
6. *N,N,N′,N′-Tetramethylethylenediamine (TEMED).*
7. *Agarose sealing solution*: Add 0.5 g agarose and 2 mg Bromophenol to 100 ml SDS electrophoresis buffer and heat slowly (microwave) with stirring to dissolve agarose completely. Store at 4 °C and reheat (microwave) to bring it to liquid state before use.

2.3 Downstream Processing

1. *Gel fixing solution*: 40% methanol and 10% acetic acid. Mix 400 ml methanol and 100 ml acetic acid in 500 ml deionized water to prepare 1 l of gel fixing solution. Store at room temperature.
2. *Coomassie Blue dye*: (0.1% Coomassie Brilliant Blue R-250, 40% methanol, and 10% acetic acid. Add 0.2 g Coomassie Brilliant Blue R-250 in 200 ml gel fixing solution.

3. *Destaining solution for Coomassie Blue*: 40% Methanol and 10% Acetic acid. Same as gel fixing solution.
4. *Gel storage solution*: 5% acetic acid. For 500 ml of total volume add 25 ml of acetic acid in 475 ml deionized water, mix, and store at room temperature.
5. *Silver nitrate ($AgNO_3$) stock solution*: *20% $AgNO_3$*. Add 20 g of $AgNO_3$ in a brown bottle and add deionized water to make a final volume of 100 ml. Store at 4 °C in dark.
6. *Paraformaldehyde (PFA) stock solution*: 40% paraformaldehyde. For 100 ml volume add 70 ml 1× PBS (phosphate buffer saline) in a beaker, place it on stir hot plate in a fume hood and heat it at 50 °C, add 40 g paraformaldehyde slowly with stirring, add 1 N sodium hydroxide solution dropwise until paraformaldehyde dissolves completely, adjust the pH to 7.4, and store at −20 °C.
7. *Silver stain*: 0.2% AgNO3 and 0.0075% paraformaldehyde. (Prepare fresh just prior to use.) Add 12 ml of 20% $AgNO_3$ stock solution and 9 μl of 40% PFA to 120 ml deionized water.
8. *Sensitization solution*: 0.025% sodium dithionite. (Prepare fresh just prior to use.) Add 25 mg of sodium dithionite in 100 ml water and mix.
9. *Developer solution*: 6% sodium carbonate, 0.4% sodium thiosulfate, and 0.01% paraformaldehyde. (Prepare fresh just prior to use.) Dissolve 12 g sodium carbonate to 150 ml deionized water, add 0.8 g sodium thiosulfate and 50 μl of 40% paraformaldehyde solution, mix, and make the final volume to 200 ml with deionized water.
10. *Stop solution*: 5% acetic acid. Add 10 ml of acetic acid to 190 ml deionized water and mix. Store at room temperature.
11. *Destaining solution for silver stain*: 0.2% potassium ferricyanide and 0.1% sodium thiosulfate. Add 0.2 g potassium ferricyanide to 80 ml deionized water and dissolve, add 0.1 g sodium thiosulfate and make final volume to 100 ml with deionized water. Store at 4 °C.
12. *Gel trypsin digestion buffer*: 2.5% trypsin and 50 mm ammonium bicarbonate. Add 250 mg sequencing grade trypsin and 39.5 mg ammonium bicarbonate to 10 ml sterile deionized water and dissolve with slow mixing. Store at −20 °C.
13. *5% formic acid*: Add 2.5 ml of formic acid to 47.5 ml sterile deionized water to make 50 ml of 5% formic acid. Store at 4 °C.
14. *Acetonitrile.*
15. *0.1% trifluoroacetic acid*: Add 50 μl of trifluoroacetic acid to 50 ml sterile deionized water to make 0.1% trifluoroacetic acid solution. Store at 4 °C.

3 Methods

The steps of 2D gel electrophoresis are schematically represented in Fig. 2.

3.1 Protein Sample Preparation

Protein samples can be prepared from different tissues or cells. We isolated proteins from adult rat tissues and cultured adult human endothelial progenitor cells and processed the tissue and cells by using following described method:

1. Dissect out tissues with sterile scissors, put them into sterile Eppendorf's tubes and snap freeze in liquid nitrogen. (In case of cultured endothelial progenitor cells, isolate cells after trypsinization and wash them three times with sterile 1× PBS with centrifugation at 1000 × *g* at 4 °C. Resuspend the cell pellet in 1 ml of 1× PBS and count the cells by using automated cell counter or Neubauer's chamber. Centrifuge at the same settings, remove PBS and snap freeze the cell pellet in liquid nitrogen.)
2. Use about 5–10 mg of tissues or 1×10^6 to 5×10^6 cultured cells as starting sample quantity for 2-DE.
3. Add 500–1000 μl of urea lysis buffer for tissues or 250–500 μl of buffer for cells.
4. Place the sample tubes in ice and lyse by using the ultrasonicator homogenizer at settings of 55–125 W and 20 kHz frequency at 4 °C with total of 6–7 pulses (each pulse of 3–5 s) (*see* **Note 1**).
5. Centrifuge samples at 16,000 × *g* at 4 °C for 30 min.
6. Collect supernatant in new sterile tubes and discard pellet.
7. Precipitate proteins using TCA method of precipitation as follows—mix equal volume of cold 20% (w/v) TCA to supernatant from **step 6**, store them at −20 °C for 1 h, and centrifuge at 16,000 × *g* at 4 °C for 15 min. Discard supernatant and add ice-cold acetone (500 μl) to pellet and centrifuge at 16,000 × *g* at 4 °C. Discard the supernatant and air-dry the pellet.
8. Dissolve the pellet in sample buffer (vol. 200–500 μl) (*see* **Note x2**) and estimate protein by using the NanoDrop spectrophotometer.

3.2 Isoelectric Focusing (IEF)

The methodology of IEF is similar for most of the IEF strips, however following detailed protocol is described for IEF strip of 11 cm, pH 3–10.

1. Determine the volume of sample buffer for loading total protein (100–500 μg).

2. Load protein to IEF gel strips by incubating the strips with protein sample in sample buffer and 200 μl rehydration buffer at room temperature for overnight (*see* **Note 3**). Cover the strip with molecular grade mineral oil after 5 min of incubation to avoid sample drying.
3. Run IEF for strip 11 cm pH 3–10 at the following settings: Phase 1—300 V for 1 min (Step and hold). Phase 2—1000 V for 1 h (Step and hold). Phase 3—7500 V for 2 h 30 min (Step and hold).
4. After isoelectric focusing the strips are equilibrated in SDS containing buffer as follows:

   Strips are placed in 2.5 ml equilibration buffer I containing DTT in tubes and incubated at room temperature at a rocker for 15 min. The buffer is changed with 2.5 ml equilibration buffer II with IAA and incubated for 15 min at the same settings.

3.3 SDS PAGE

1. Use a suitable SDS-PAGE apparatus to accommodate the 11 cm IEF strip and cast 12% polyacrylamide gel. For 20 ml of 12% polyacrylamide gel mix 8 ml of acrylamide–bis-acrylamide solution, 5 ml of 1.5 M Tris–HCl pH 8.8, 200 μl of 10% SDS solution, 100 μl of 10% APS solution, 30 μl of TEMED, and 6.7 ml of double distilled water.
2. Use a suitable comb of 14–16 cm long and 3–4 cm depth to make a well for protein molecular weight markers and space to accommodate 11 cm long IEF gel on the top of the gel.
3. After the gel gets solidified, remove the comb and place the equilibrated gel strip horizontally with back of the strip facing the glass plate of the polyacrylamide gel (as shown in Fig. 3b, **step 5**)
4. Seal the gel strip with 0.5% agarose gel containing 0.002% Bromophenol Blue (Fig. 3b, **step 6**).
5. Perform SDS-PAGE in SDS electrophoresis buffer at constant voltage of 80–100 V until the tracking dye reaches at the end of the gel toward the cathode of the electrophoresis apparatus.

3.4 Downstream Gel Processing

3.4.1 Staining and Imaging

1. The gel is either stained with Coomassie blue dye or silver stain.
   (a) Coomassie Blue staining—fix the gel in fixing solution for 2 h with gentle shaking at RT, cover the gel with Coomassie staining solution and incubate for 45 min with gentle shaking at RT, destain the gel by using and changing destaining solution several times to remove the background staining, and store the gel in storage solution.
   (b) Silver staining—fix the gel in fixing solution for 2 h with gentle shaking at RT, wash the gel with 40% methanol

twice with incubation of 10 min each, wash the gel with deionized water twice with incubation of 10 min each, cover the gel with sensitizer solution and incubate for 1 min with gentle shaking at RT, wash twice with deionized water with incubation of 2 min each, incubate the gel with silver staining solution for 30 min at RT with gentle shaking, wash gel with deionized water with incubation of 2 min, discard water and add developer solution immediately, develop the stained spots with gentle shaking and stop it with adding stopping solution after desired appearance of protein spots, and store the gel in storage solution.

2. After staining, scan the gels and analyze protein spots by using PDQuest 2-D analysis software. Determine the Isoelectric points and molecular weights of protein spots with mapping of the gel images.
3. The gel analysis software is also used for identifying differentially expressed proteins among different treatment groups.

3.4.2 Protein Identification

1. To identify proteins, the stained spots on the gels are selected and excised carefully (*see* **Note 4**).
2. The excised gel pieces are destained with incubating them with destaining solution at 37 °C for 20 min (*see* **Note 5**).
3. After destaining, the gel pieces are submerged in gel trypsin digestion buffer.
4. The gels are crushed with handheld motorized pestle and incubated at 37 °C for 12–16 h with minimum required volume of gel trypsin digestion buffer.
5. Tubes are centrifuged at 11,000 × *g* at RT for 5 min and the supernatant is discarded.
6. Pelleted gel pieces are washed twice with 250 μl of 50 mM ammonium bicarbonate solution.
7. After washing, submerge the gel pieces into 20–30 μl 5% formic acid–acetonitrile 1:1 (v/v) and incubate at 30 °C for 20 min.
8. The tubes are centrifuged at 11,000 × *g* at RT for 5 min and the supernatant is collected in sterile tubes. **Steps 6** and 7 are repeated twice for maximum peptide elution.
9. The collected supernatant is dried by speed-vac and peptides are resuspended in 5 μl of 0.1% trifluoroacetic acid.
10. The digested peptides are subjected to mass spectrometry (MALDI-TOF; matrix-assisted laser desorption ionization—time of flight) followed by peptide mass fingerprint (PMF) searching of databases to identify protein (*see* **Note 6**).

4 Notes

1. Sonication of the sample should be done with proper care, wearing ear protective equipment. Also, avoid heat generation during sonication with short pulses of 3–5 s.
2. The sample loading buffer must contain ampholytes compatible to the gel strips that are being used in the experiment.
3. The amount of loading protein and volume of the samples should be determined according to the gel strips being used in the experiment. For example: use 100–500 μg of protein and the total volume (sample buffer + rehydration buffer) should not exceed 300 μl for a gel strip of 11 cm pH 6–11.
4. The excised gels can be stored in sterile Eppendorf's tubes in −80 °C until further processing.
5. The step of destaining should be repeated until complete destaining is achieved.
6. MALDI-TOF PMF method is useful to identify proteins from species with available genome sequences, otherwise tandem mass spectrometry (MS/MS) followed by peptide sequencing is required to identify proteins.

Acknowledgments

We acknowledge Dipavo Banerjee and Divya Khandekar for manuscript preparation. We thank National Centre for Cell Science, Pune, India, for supporting the related original diabetes research.

References

1. Macko V, Stegemann H (1969) Mapping of potato proteins by combined electrofocusing and electrophoresis identification of varieties. Hoppe Seylers Z Physiol Chem 350 (7):917–919; Epub 1969/07/01
2. Margolis J, Kenrick KG (1969) 2-dimensional resolution of plasma proteins by combination of polyacrylamide disc and gradient gel electrophoresis. Nature 221(5185):1056–1057; Epub 1969/03/15
3. Mets LJ, Bogorad L (1974) Two-dimensional polyacrylamide gel electrophoresis: an improved method for ribosomal proteins. Anal Biochem 57(1):200–210; Epub 1974/01/01
4. Orrick LR, Olson MO, Busch H (1973) Comparison of nucleolar proteins of normal rat liver and Novikoff hepatoma ascites cells by two-dimensional polyacrylamide gel electrophoresis. Proc Natl Acad Sci U S A 70 (5):1316–1320; Epub 1973/05/01
5. Kaltschmidt E, Wittmann HG (1970) Ribosomal proteins. VII. Two-dimensional polyacrylamide gel electrophoresis for fingerprinting of ribosomal proteins. Anal Biochem 36(2):401–412; Epub 1970/08/01
6. MacGillivray AJ, Rickwood D (1974) The heterogeneity of mouse-chromatin nonhistone proteins as evidenced by two-dimensional polyacrylamide-gel electrophoresis and ion-exchange chromatography. Eur J Biochem 41(1):181–190; Epub 1974/01/03
7. O'Farrell PH (1975) High resolution two-dimensional electrophoresis of proteins. J Biol Chem 250(10):4007–4021; Epub 1975/05/25
8. Bjellqvist B, Ek K, Righetti PG, Gianazza E, Gorg A, Westermeier R et al (1982) Isoelectric

focusing in immobilized pH gradients: principle, methodology and some applications. J Biochem Biophys Methods 6(4):317–339; Epub 1982/09/01

9. Gorg A, Postel W, Weser J, Patutschnick W, Cleve H (1985) Improved resolution of PI (alpha 1-antitrypsin) phenotypes by a large-scale immobilized pH gradient. Am J Hum Genet 37(5):922–930; Epub 1985/09/01

10. Gorg A, Postel W, Gunther S (1988) The current state of two-dimensional electrophoresis with immobilized pH gradients. Electrophoresis 9(9):531–546. https://doi.org/10.1002/elps.1150090913; Epub 1988/09/01

Chapter 16

High-Efficiency Lentiviral Gene Modification of Primary Murine Bone-Marrow Mesenchymal Stem Cells

Dario Gerace, Binhai Ren, Rosetta Martiniello-Wilks, and Ann M. Simpson

Abstract

Lentiviral vectors are the method of choice for stable gene modification of a variety of cell types. However, the efficiency with which they transduce target cells varies significantly, in particular their typically poor capacity to transduce primary stem cells. Here we describe the isolation and enrichment of murine bone-marrow mesenchymal stem cells (MSCs) via fluorescence-activated cell sorting (FACS); the cloning, production, and concentration of high-titer second generation lentiviral vectors via combined tangential flow filtration (TFF) and ultracentrifugation; and the subsequent high-efficiency gene modification of MSCs into insulin-producing cells via overexpression of the furin-cleavable human insulin (INS-FUR) gene.

Key words Lentivirus, Tangential flow filtration, Mesenchymal progenitor cells, Protein tyrosine phosphatase receptor type C, Stem cell antigen 1, Fluorescence-activated cell sorting, Gene modification

1 Introduction

MSCs reside within a number of adult tissues and organs including the placenta [1], cord blood [2], umbilical cord Wharton's jelly [3], lung [4], and adipose tissue [5] at varying frequencies; however the major source of MSCs is the bone marrow [5, 6]. A number of techniques commonly used to isolate and enrich for MSCs include antibody-based cell sorting [7], low and high-density culture [8], positive and negative selection [9], frequent defined-media changes [10], and enzymatic digestion [11]. For the purposes of this work, antibody-based positive and negative selection cell sorting was used to enrich for a primitive population of MSCs from the stromal fraction of the bone marrow of nonobese diabetic (NOD) mice [12–15].

MSCs were subsequently gene-modified utilizing high-titer lentivirus produced by combined TFF and ultracentrifugation [16]. Second-generation lentiviral vector particles were produced

Mugdha V. Joglekar and Anandwardhan A. Hardikar (eds.), *Progenitor Cells: Methods and Protocols*, Methods in Molecular Biology, vol. 2029, https://doi.org/10.1007/978-1-4939-9631-5_16,

by calcium phosphate transfection of three plasmids (pVSV-G, envelope plasmid; pCMVΔR8.2, packaging plasmid; and pHMD or pHMD-INS-FUR, transfer plasmids) in HEK293T cells cultured in cell factories, then purified and concentrated using a benchtop Minimate™ TFF machine. To further increase viral titers, ultracentrifugation of TFF filtered viral supernatants was performed, resulting in lentiviral titers of 10^7–10^8 TU/ml [17, 18]. Thus, primary MSCs were transduced with an efficiency of 50–70%, of which INS-FUR-expressing MSCs demonstrated a >900 fold-increase in insulin secretion in comparison to parental and empty vector transduced MSCs.

2 Materials

2.1 Animal Work

1. Twenty female NOD mice at 6 weeks of age (19–23 g) (Animal Resources Centre, WA, Australia).
2. Corncob bedding.
3. Irradiated rat and mouse feed.
4. 80% (v/v) ethanol.
5. Phosphate-buffered saline (PBS) 1×.
6. 3 ml syringe.
7. Scalpel and scalpel blade (No. 10).
8. Forceps (Medshop Australia®).
9. Surgical scissors (Medshop Australia®).
10. 23 Gauge PrecisionGlide™ needle.
11. 10 cm petri dishes.
12. 50 ml falcon tubes.
13. 75 cm^2 culture flasks.
14. Culture medium: 20% (v/v) 0.2 μm-filtered fetal bovine serum, qualified, heat-inactivated, US origin and 1% (v/v) 100× penicillin–streptomycin–glutamine (PSG) added to 500 ml of α-MEM, no nucleosides. Store at 4 °C.

2.2 FACS Sorting and Culture of MSCs

1. Hanks' Buffered Salt Solution.
2. TrypLE Express Enzyme (1×), no phenol red.
3. 15 ml falcon tubes.
4. Rat anti-mouse CD45 (0.2 mg/ml) conjugated to allophycocyanin (APC).
5. Rat anti-mouse Ly6 (Sca-1) (0.2 mg/ml) conjugated to phycoerythrin (PE).
6. Wash/sorting buffer: 1× PBS and 5% FBS. Store at 4 °C.

7. 5 ml Round Bottom Polystyrene Test Tubes with Cell Strainer Snap Caps
8. 25 cm^2 culture flasks.
9. Propagation medium: Culture medium (*see* Subheading 2.1, **item 14**) + 10 ng/ml (final concentration) Recombinant Human Basic Fibroblast Growth Factor.
10. Dimethyl sulfoxide (DMSO).

2.3 MSC Characterization

2.3.1 Clonogenicity Assays

1. 75 cm^2 culture flasks.
2. 10 cm^2 tissue-culture treated plates.
3. Propagation medium (*same as in* Subheading 2.2, **item 9**).
4. PBS 1×.
5. 0.4% (w/v) methylene blue solution: First dilute 100% methanol in autoclaved water to create a 60% (v/v) methanol solution. 0.4 g methylene blue powder is subsequently dissolved in 100 ml of 60% (v/v) methanol solution (*see* **Note 1**).

2.3.2 Adipogenic Differentiation

1. 75 cm^2 culture flasks.
2. 24-well tissue culture plates.
3. PBS 1×.
4. Adipogenesis control medium: Add 2% (v/v) 0.2 μm-filtered FBS and 1% (v/v) 100× PSG to 500 ml of α-MEM, no nucleoside. Aliquot into 50 ml falcon tubes and store at 4 °C for use within 2 weeks, or at −20 °C for long-term storage.
5. Adipogenesis differentiation medium: Add 2% (v/v) 0.2 μm-filtered FBS, 1% (v/v) 100× PSG, 1 μM dexamethasone, 0.5 μM 3-isobutyl-1-methyl-xanthine, 100 μM indomethacin, and 10 μg/ml insulin to a 500 ml of α-MEM, no nucleosides. Aliquot into 50 ml falcon tubes and store at 4 °C for use within 2 weeks, or at −20 °C for long-term storage (*see* **Note 2**).
6. 10% neutral buffered formalin.
7. 0.2% (w/v) Oil Red O, commercially available.
8. 0.2% sodium azide solution prepared with 1× PBS.

2.3.3 Osteogenic Differentiation

1. 75 cm^2 culture flasks.
2. 24-well tissue culture plates.
3. PBS 1×.
4. Osteogenesis control medium: Add 1% (v/v) 0.2 μm-filtered FBS and 1% (v/v) 100× PSG to a bottle of DMEM. Aliquot into 50 ml falcon tubes and store at 4 °C for use within 2 weeks, or at −20 °C for long-term storage.
5. Osteogenesis differentiation medium: Add 10% (v/v) 0.2 μm-filtered FBS, 1% (v/v) 100× PSG, 0.1 μM dexamethasone,

0.2 mM ascorbic acid, and 10 mM β-glycerophosphate to a bottle of DMEM. Aliquot into 50 ml falcon tubes and store at 4 °C for use within 2 weeks, or at −20 °C for long-term storage.

6. 10% neutral buffered formalin.
7. 2% (w/v) (pH 4.1) Alizarin Red S, commercially available.
8. 0.2% sodium azide solution prepared with 1× PBS.

2.3.4 Chondrogenic Differentiation

1. 75 cm^2 culture flasks.
2. 24-well tissue culture plates.
3. PBS 1×.
4. Chondrogenic control medium: Add 2 mM L-Glutamine to a bottle of MesenCult™-ACF Chondrogenic Differentiation Basal Medium (STEMCELL Technologies®). Store at 4 °C for use within 2 weeks.
5. Chondrogenic differentiation medium: Add MesenCult™--ACF 20× Chondrogenic Differentiation Supplement (STEMCELL Technologies®) and 2 mM L-glutamine to a bottle of MesenCult™-ACF Chondrogenic Differentiation Basal Medium. Store at 4 °C for use within 2 weeks.
6. 10% neutral buffered formalin.
7. 1% (v/v) Alcian Blue 8× solution made in 3% (v/v) acetic acid (pH 2.5). The pH was adjusted using 10 M NaOH.

2.4 Lentivirus Cloning, Production, and Titration

1. One Shot® Stbl3 TOP10 Chemically competent E. coli (LifeTechnologies™).
2. S.O.C medium (Super Optimal Broth with Catabolite Repression), commercially available.
3. Luria–Bertani (LB) broth: Dissolve 12 g of Luria–Bertani powder in 1 l of distilled H_2O, followed by autoclaving. Prepare LB agar by adding 2.3 g of agar, bacteriological per 100 ml of LB broth, followed by autoclaving.
4. PureLink® Quick Plasmid Maxiprep Kit.
5. 100 mg/ml ampicillin solution: Dissolve 1 g of ampicillin powder in 10 ml of distilled H_2O, followed by sterilization through a 0.2 μm syringe filter.
6. 10 cm petri dishes.
7. pVSV-G envelope plasmid, pCMVΔR8.2 packaging plasmid and LV-HIV/MSCV (pHMD) or pHMD-INS-FUR transfer plasmids.
8. HEK293T cells.
9. NIH3T3 cells.

10. Cell line culture medium: Add 2% (v/v) 0.2 μm-filtered FBS and 1% (v/v) 100× PSG to 500 ml of OptiMEM Store at 4 °C.
11. For calcium phosphate transfection: One 50 ml falcon tube contained 1600 μg pHMD or pHMD-INS-FUR, 800 μg pVSV-G, 800 μg pCMVΔR8.2, 2 M $CaCl_2$ and distilled H_2O. A second 50 ml falcon tube contained 1 M 4-2-hydroxyethyl-1-piperazineethanesulfonic acid (HEPES), 2 M NaCl, 150 mM Na_2HPO_4, and distilled H_2O.
12. PBS 1×.
13. Ten-tray cell factory (CF10) (Corning®).
14. 0.45 μm filter (SFCA).
15. Benchtop Minimate™ TFF machine (Pall®) with the Centramate™ LV holder and a 100 kDa Medium Screen channel-cassette membrane.
16. Ultracentrifuge.
17. 24-well tissue-culture treated plates.
18. 8 mg/ml Polybrene solution: Dissolve 80 mg of Polybrene in 10 ml of distilled H_2O, followed by sterilization through a 0.2 μm syringe filter.
19. 5 ml Round Bottom Polystyrene Test Tubes with Cell Strainer Snap Caps.
20. 4% (v/v) paraformaldehyde solution: Dilute a glass ampule containing 10 ml of 16% paraformaldehyde aqueous solution in 30 ml of distilled H_2O, followed by sterilization through a 0.2 μm syringe filter (*see* **Note 3**).
21. Any flow cytometer capable of fluorescent cell sorting such as BD LSR cytometer and BD FACSDiva™ software (Version 6.1.3).

2.5 Transduction of MSCs

1. Cell culture medium as described in Subheading 2.1, **item 14**.
2. Passage 15 MSCs.
3. 75 cm^2 and 25 cm^2 tissue-culture treated flasks.
4. 8 mg/ml Polybrene solution.
5. HMD and HMD-INS-FUR lentiviral vectors produced in Subheading 2.4.
6. HBSS 1×.
7. Wash/sorting buffer as described in Subheading 2.2, **item 6**.
8. 5 ml Round Bottom Polystyrene Test Tubes with Cell Strainer Snap Caps.
9. Freezing medium: 10% DMSO in FCS.
10. Cryovials.

2.6 Insulin Secretion

1. 6-well tissue-culture treated plates.
2. Cell culture medium (Subheading 2.1, **item 14**).
3. Untransduced and transduced MSCs.
4. 1.5 ml microfuge tubes.
5. Antibody-based insulin detection kits such as ARCHITECT™ i4000SR Immunoassay Analyzer (Abbott Diagnostics©) and ARCHITECT™ Insulin Reagent Kit (8K41).

3 Methods

All animal work was approved by the UTS Animal Care and Ethics Committee (ACEC 2011-447A; ACEC 2009-244A) and complied with the Australian code for the care and use of animals for scientific purposes [19]. All reagents are stored at room temperature, unless indicated otherwise. All steps are to be performed within a Class II Biosafety Cabinet, unless otherwise specified.

3.1 Isolation of MSCs from Murine Bone Marrow

1. House 20 female NOD mice in a polycarbonate green line individually ventilated cage (IVC) system, with corncob bedding, irradiated rat and mouse feed, and unlimited reverse osmosis water. Autoclave all bedding and housing equipment prior to use. Maintain the temperature and humidity of the animal rooms within 22–24 °C and 40–60%, respectively. House a maximum of four animals per cage and monitor daily until they are required for experimental work.
2. Euthanize mice within a CO_2 chamber, followed by a cervical dislocation to confirm death and ensure no postdeath sensation. Submerge euthanized mice in 80% ethanol to ensure sterility and transfer to a biosafety hood for further manipulation.
3. Make a circumferential incision in the skin below the ankles of the hind legs using a surgical scalpel, and a subsequent perpendicular incision ~1 cm in length proximal to the initial incision on the lateral side of the tibia using surgical scissors. Pull back the skin the entire length of limb up to the hip and remove using surgical scissors, revealing the muscle belly beneath. Wash the hind limbs with 80% ethanol to remove any contaminating fur, and subsequently remove the hind limbs from the animals using surgical scissors and transfer to a 50 ml falcon tube containing 1× PBS (*see* **Note 4**).
4. Transfer an individual hind limb to a petri dish and proceed to excise the muscle tissue from the femurs. Once the muscle tissue is removed, detach the femurs from the entirety of the limb and place the clean femurs in a new 50 ml tube containing 1× PBS. Repeat this process until all hind limbs have been cleaned (see **Note 5**).

5. Transfer the cleaned femurs to a new petri dish and remove the epiphyses from the proximal and distal joints using a scalpel, and retain the epiphyses in a separate 50 ml tube containing 1× PBS for downstream culture of the stromal cell fraction (*see* **Note 6**).
6. Connect a 23 Gauge PrecisionGlide™ needle to a 3 ml syringe and collect 1 ml of 1× PBS. Using forceps, remove a femur from the falcon tube containing the collected and cleaned femurs. While working over a new 50 ml falcon tube, place the end of the needle into the opening of the femur and flush the bone marrow into the falcon tube. Repeat this process until the bone marrow has been flushed from all cleaned femurs, and centrifuge the cell suspension at 1200 × *g* for 5 min (*see* **Note 7**).
7. Aspirate the PBS from the tube containing the epiphyses and evenly distribute the epiphyses between two 75 cm^2 flasks. Aspirate the supernatant and resuspend the cell pellet in 20 ml of culture media, and transfer 10 ml of the cell suspension to the two 75 cm^2 tissue-culture flasks containing epiphyses.
8. Incubate the flasks at 37 °C/5% CO_2 for an initial period of 6 weeks in the presence of epiphyses (*see* **Note 8**).
9. During the initial culture period, subculture and expand the adherent stromal cells for two passages (with epiphyses) prior to sorting of the unique population of MSCs. Collect the epiphyses from the existing flasks in a 50 ml falcon tube, aspirate the media, and distribute the epiphyses between new T75 cm^2 flasks. Wash the adherent stromal cells with 3 ml of 1× HBSS and then add 3 ml of 1× TrypLE Express and incubate the flasks at 37 °C/5% CO_2 for 5–10 min. Once the cells have detached, add an equal volume of culture medium to the flask, transfer the cells to a 15 ml falcon tube and centrifuge at 500 × *g* for 5 min (*see* **Note 9**).

3.2 FACS Enrichment of MSCs

1. Remove the epiphyses from the flasks and wash the plastic adherent stromal cells (passage 2) with 3 ml of 1× HBSS. Incubate the flasks with 3 ml of 1× TrypLE Express at 37 °C/5% CO_2 for 5–10 min, and add one volume of culture medium to the detached cells. Transfer the cells to a 15 ml falcon tube and centrifuge at 500 × *g* for 5 min. Aspirate the medium and wash the cell pellet with 1× HBSS, and recentrifuge at 500 × *g* for 5 min.
2. Aspirate the medium and resuspend the cell pellet in washing/sorting buffer (1× PBS containing 5% FBS) to a concentration of 5×10^6 cells/ml. Aliquot 100 μl (5×10^5 cells) into four 5 ml Round bottom polystyrene test tubes with cell strainer caps and transfer to ice.

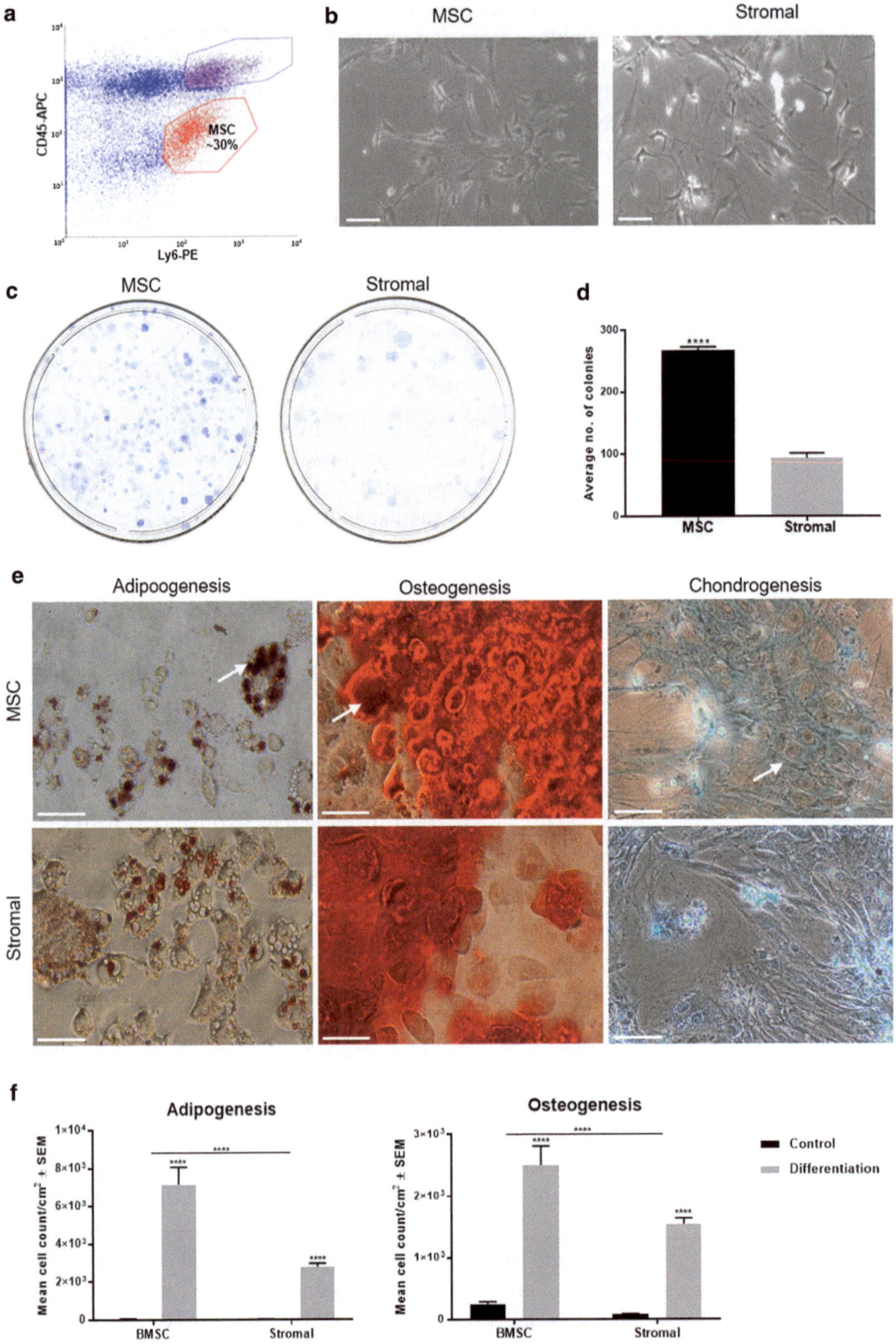

Fig. 1 Isolation and characterization of primary murine MSCs. (**a**) FACS analysis and enrichment of NOD BMSCs. Following culture for two passages, NOD bone marrow stromal cells were stained with nil antibody

3. Prepare single monoclonal antibody (mAb) control stains by adding 2 μl of 0.2 mg/ml rat anti-mouse CD45 conjugated to allophycocyanin (APC) and 0.2 mg/ml rat anti-mouse Ly6 (Sca-1) conjugated to phycoerythrin (PE) to individual tubes containing stromal cells. To prepare the experimental mAb cell stain, add 2 μl of the CD45 and Ly6 mAbs to the same tube containing stromal cells. The fourth tube serves as an unstained control. Stain the cells on ice for 30 min, followed by two washes with washing/sorting buffer (*see* **Note 10**).
4. Resuspend the cells in 100 μl of washing/sorting buffer and prepare sorting collection tubes by adding 2 ml of FBS to a 15 ml falcon tube. Sort the $CD45^{-}/Ly6^{+}$ population (BMSC) (Fig. 1a) into the collection tubes containing FBS (*see* **Note 11**).
5. Centrifuge the sorted cells at 500 × *g* for 5 min, and resuspend the cells in 5 ml of culture medium containing bFGF (Propagation medium). Transfer the cells to individual T25 cm^2 flasks and incubated at 37 °C/5% CO_2. At this point, images of sorted MSCs and unsorted stromal cells can be acquired using a light microscope (Fig. 1b).
6. Feed the cells every 3 days with culture medium, and split every 5–7 days at a 1:3 ratio. Expand until passage 15 and cryopreserve at every passage in freezing medium.

3.3 Functional Characterization of MSCs

3.3.1 Clonogenicity Assays

1. Culture MSCs in T75 cm^2 flasks and detach the cells using 1× TrypLE Express. Following detachment, resuspend the MSCs in culture medium to a concentration of 5×10^2 cells/ml, and add 1 ml of cells to 10 cm^2 tissue-culture treated plates containing 9 ml of culture medium in triplicate. Incubate the plates at 37 °C/5% CO_2 for 10 days.

Fig. 1 (continued) (Unstained), CD45 mAb-conjugated to fluorochrome APC (CD45-APC), Ly6 MAb conjugated to fluorochrome PE (Ly6-PE) and both mAbs (CD45-APC/Ly6-PE). Fluorescence dot plots of CD45-APC (y-axis) and Ly6-PE (x-axis) were used to identify MSC ($CD45^{-}/Ly6^{+}$; red) and double positive ($CD45^{+}/Ly6^{+}$; purple) cell subpopulations for cell sorting using the BD FACSAria™ II instrument. Representative of three individual FACS sorting experiments. (**b**) Morphological analysis of stromal and sorted MSCs. Cells were cultured in vitro up to passage 15 and imaged on a Leica DM Microscope. Fibroblast-like MSCs are indicated by arrowheads. Images were obtained at 10× magnification, 100 μm scale. (**c**) Comparative analysis of clonogenicity by methylene blue staining. 500 cells were seeded in 10 cm^2 tissue-culture treated plates and incubated for 10 days, after which colonies were stained with 40% methylene blue–60% methanol. Images were obtained using the Epson® Perfection 4870 Photo Scanner. (**d**) Interpopulation analysis of clonogenicity. Data are represented as mean number of colonies per 5000 cells ± SEMs ($n = 3$). A one-way ANOVA and Tukey's post hoc were performed, $* \ p < 0.05$. (**e**) Morphological analysis of trilineage differentiation. Adipocytes were stained with Oil Red O; osteocytes were stained with Alizarin Red and chondrocytes were stained with Alcian Blue. Images were acquired on a Leica DM Microscope at 20× magnification, 100 μm scale. Arrows indicate areas of adipogenesis, osteogenesis, and chondrogenesis. (**f**) Semiquantitative analysis of adipogenesis and osteogenesis under defined conditions. Data are presented as mean cell count/cm^2 ± SEM ($n = 3$). A two-way ANOVA and Tukey's post hoc tests were performed, $* \ p < 0.05$, $**** \ p < 0.0001$

2. Aspirate the media and wash the MSC colonies with 1× PBS. Stain the colonies by adding 4 ml of 0.4% (w/v) methylene blue solution for 1 h, and destain in a series of three washes using tap water. Leave the plates to dry overnight and count the total number of colonies per plate by eye (Fig. 1c) (*see* **Note 12**). Count the colonies as a measure of clonogenicity (Fig. 1d).

3.3.2 Adipogenic Differentiation

1. Culture MSCs in T75 cm^2 flasks and detach the cells using 1× TrypLE Express. Following detachment, resuspend the MSCs in culture medium to a concentration of 2.5×10^4 cells/ml. Add 1 ml of cells in triplicate to respective wells of a 24-well plate for control and differentiation assays, and incubate the cells at 37 °C/5% CO_2 until the reach 85–90% confluency.
2. Aspirate the medium and replenish with 1 ml of either adipogenic control or adipogenic differentiation medium. Incubate the cells at 37 °C/5% CO_2 for 10 days, with media replenishment every 3 days.
3. On day 10, aspirate the media from the plates and wash the cells with 1× PBS. Fix the cells with 500 μl of 10% neutral buffered formalin for 30 min. Aspirate the formalin and wash the cells twice with 1× PBS. Aspirate the 1× PBS and stain the cells with 1 ml of 0.2% (w/v) Oil Red O for 15 min. Following staining, remove the Oil Red O and wash the cells twice with 1× PBS prior to imaging and scoring (Fig. 1e, f) (*see* **Note 13**).

3.3.3 Osteogenic Differentiation

1. Culture MSCs in T75 cm^2 flasks and detach the cells using 1× TrypLE Express. Following detachment, resuspend the MSCs in culture medium to a concentration of 12.5×10^4 cells/ml. Add 1 ml of cells in triplicate to respective wells of a 24-well plate for control and differentiation assays, and incubate the cells at 37 °C/5% CO_2 until the reach 90–100% confluency.
2. Aspirate the medium and replenish with 1 ml of either osteogenic control or osteogenic differentiation medium. Incubate the cells at 37 °C/5% CO_2 for 35 days, with media replenishment every 3 days.
3. On day 35, aspirate the media from the plates and wash with 1× PBS. Fix the cells with 500 μl of 10% neutral buffered formalin for 30 min. Aspirate the formalin and wash the cells twice with 1× PBS. Aspirate the 1× PBS and stain the cells with 1 ml of 2% (w/v) (pH 4.1) Alizarin Red S for 45 min. Following staining, remove the Alizarin Red S and wash the cells twice with 1× PBS prior to imaging and scoring (Fig. 1e, f) (*see* **Note 13**).

3.3.4 Chondrogenic Differentiation

1. Culture MSCs in T75 cm^2 flasks and detach the cells using 1× TrypLE Express. Following detachment, resuspend the MSCs in culture medium to a concentration of 2.5×10^4 cells/ml. Add 1 ml of cells in triplicate to respective wells of a 24-well plate for control and differentiation assays, and incubate the cells at 37 °C/5% CO_2 until the reach 85–90% confluency.
2. Aspirate the medium and replenish with 1 ml of either chondrogenic control or chondrogenic differentiation medium. Incubate the cells at 37 °C/5% CO_2 for 18 days, with media replenishment every 3 days.
3. On day 18, aspirate the media from the plates and wash with 1× PBS. Fix the cells with 500 μl of 10% neutral buffered formalin for 30 min. Aspirate the formalin and wash the cells twice with 1× PBS. Aspirate the 1× PBS and stain the cells with 1 ml of Alcian Blue solution (8×, pH 2.5) for 1 h. Following staining, remove the Alcian Blue solution and wash the cells twice with 1× PBS prior to imaging (Fig. 1e).

3.4 Lentivirus Production

3.4.1 Cloning and TFF of Lentiviral Vectors

1. The human furin-cleavable insulin (INS-FUR) cDNA sequence [20] was cloned into the EcoRI site of the LV-HIV/MSCV (pHMD) transfer plasmid [21] to generate the pHMD-INS-FUR transfer plasmid as previously described [22]. Each lentiviral transfer plasmid contains the eGFP fluorescent reporter for downstream analysis (Fig. 2a).
2. All lentiviral plasmids are propagated in Stbl3 TOP10 E. coli according to the manufacturer's instructions (*see* **Note 14**). Maxiprep plasmid isolation should be performed for each individual lentiviral plasmid according to the manufacturer's instructions.
3. Seed a ten-layer cell factory (CF10) with 3×10^8 293 T cells/cell factory in cell line culture media as described in Subheading 2.4 and incubate at 37 °C/5% CO_2 until having reached 70–85% confluency.
4. The lentiviral vectors are produced by calcium phosphate precipitation cotransfection of three plasmids: 800 μg pCMVΔR8.2 packaging plasmid, 400 μg pVSV-G envelope plasmid and 1600 μg of either pHMD or pHMD-INS-FUR in HEK293T cell factories for 16 h (*see* **Note 15**).
5. Harvest and pool the viral supernatants at 48 and 72 h post-transfection. Prefilter the pooled supernatants using a 0.45 μm filter (SFCA).
6. Concentrate the supernatants via TFF using the Bench-top Minimate™ TFF machine (Pall®) with the Centramate™ LV holder and a 100 kDa Medium Screen Membrane. Prior to concentration of the cell-culture supernatants, wash the cassette twice with 1500 ml 1× PBS for 1 h, followed by a 30 min

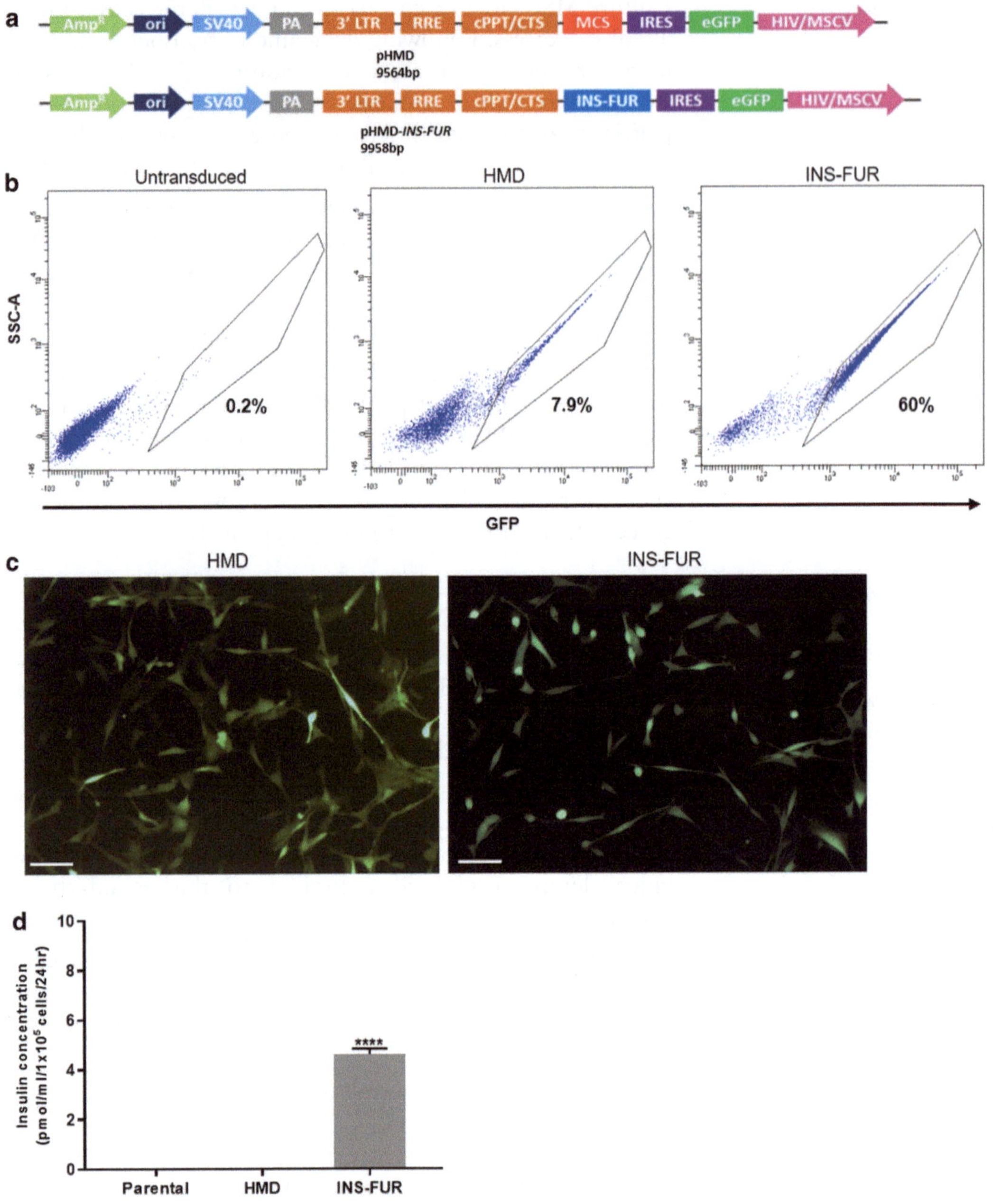

Fig. 2 Analysis of lentivirus gene-modified MSCs. (**a**) Schematic representation of pHMD and pHMD-*INS-FUR* lentiviral vectors. (**b**) FACS analysis of primary MSCs transduced to express INS-FUR. Following culture for 5 days, primary MSCs transduced with HMD and HMD-INS-FUR at an MOI of 10 were sorted into $eGFP^+$ populations. Untransduced suicide BMSCs were analyzed as a negative control. Fluorescence dot plots of SSC-A (y-axis) and GFP (x-axis) were used to identify $eGFP^+$ cells for cell sorting using the BD FACSAria™ II instrument. (**c**) Fluorescence imaging of MSCs transduced with HMD and HMD-INS-FUR. Transduced cells were returned to culture postsorting and imaged for GFP expression (day 7 posttransduction) on the Leica DM microscope (Leica Microsystems) at 10× magnification using GFP fluorescence filter sets, 100 μm scale. (**d**) Quantitation of chronic insulin secretion from transduced MSCs. Human insulin was quantified using the ARCHITECT™ i4000SR Immunoassay Analyzer. Data are represented as means ± SDs ($n = 4$). A one-way ANOVA with Sidak's post hoc test was performed, $* \ p < 0.05$

equilibration with 500 ml OptiMEM. Pooled 4–8 l supernatants are subsequently filtered through the TFF machine, which concentrates and purifies the lentiviral vectors in a final volume of 150–200 ml. To increase vector titers further following TFF, additional ultracentrifugation may be performed at 50,000 × *g* for 2 h at 4 °C.

7. The pelleted lentiviral plasmids are subsequently resuspended in 2 ml of OptiMEM and stored at −80 °C until required for downstream experiments.

3.4.2 Lentivirus Titration

1. Seed NIH3T3 cells (5×10^4 cells/well) in culture medium in a 24-well plate 24-h prior to lentiviral titration to obtain a confluency of 40–50% on the day of transduction.
2. On the day of transduction, detach and count the cells from three wells using a hemocytometer (*see* **Note 16**).
3. Prepare a tenfold serial dilution of each lentiviral vector in culture media containing 8 μg/ml Polybrene. Remove the medium from the remaining wells of the 24-well plate and add 200 μl of diluted lentiviral vector in triplicate to respective wells. Be sure to setup three wells containing medium without virus as a control. Incubate the plate at 37 °C/5% CO_2 for 4 h.
4. Following transduction, remove the virus from the 24-well plate and wash the cells twice with 250 μl of 1× PBS. Replenish the wells with 1 ml of culture media and incubate at 37 °C/5% CO_2 for 72 h.
5. On the day of FACS analysis, detach the cells and fix with 200 μl of 4% (v/v) paraformaldehyde for 15 min on ice in 5 ml round-bottom polystyrene tubes. Remove the paraformaldehyde and wash the cells twice with 1× PBS. Resuspend the cells in 200 μl of PBS and analyze the percentage GFP^+ cells immediately on a flow cytometer.
6. Following FACS analysis, determine the concentration of individual lentiviral vectors using the formula $T = (F \times C \times D)/V$; where T titer, F frequency of GFP^+ cells (FACS % divided by 100), C average number of cells per well on the day of transduction, D coefficient of lentivirus dilution (e.g., 100 for 1/100), and V volume of lentivirus used for transduction (0.2 ml).

3.5 MSC Gene-Modification

3.5.1 Lentiviral Transduction of MSCs

1. Seed MSCs at a concentration of 5×10^5 cells per T75 flasks in culture medium and incubate overnight at 37 °C/5% CO_2. Ensure to setup a flask of MSCs as an untransduced control.
2. The following day transduce MSCs (MOI = 10) with HMD and HMD-INS-FUR lentiviral vectors in culture medium containing 8 μg/ml Polybrene and incubate at 37 °C/5% CO_2 for 16 h (*see* **Note 17**).

3. Remove the lentiviral vectors from the cells and wash the cells twice with 1× PBS. Replenish the cells with culture medium and incubate at 37 °C/5% CO_2 for at least 72 h.
4. Passage the cells as required in the event that they become overconfluent during the 72 h incubation period.

3.5.2 FACS Sorting of Gene-Modified MSCs

1. At 3–5 days posttransduction, detach untransduced and transduced MSCs, resuspend the cells in 1 ml of sorting buffer, and filter through a 5 ml round-bottom polystyrene tube with cell strainer cap (*see* **Note 18**).
2. Sort the transduced MSCs on a flow cytometer based on the positive expression of GFP into 15 ml falcon tubes containing 1 ml FCS (Fig. 2b).
3. Centrifuge the sorted cells at 300 × *g* for 5 min, resuspend in culture medium and transfer to either a 6-well plate or T25 flask containing prewarmed culture medium. Incubate the cells at 37 °C/5% CO_2 (Fig. 2c).
4. Subpassage and expand the untransduced and transduced MSCs in tissue-culture treated plasticware, and cryopreserve 1×10^6 cells per vial in 1 ml of freezing medium as required.

3.6 Chronic Insulin Secretion

1. Seed untransduced and transduced MSCs at a concentration of 1×10^5 cells/well in a 6-well plate (triplicate) ($n = 4$) in culture medium and incubate for 24 h at 37 °C/5% CO_2.
2. Harvest cell culture supernatants at $t = 24$ h and transfer to new 1.5 ml centrifuge tubes. Proceed immediately with preparation of samples for insulin quantitation, or store at −20 °C until required.
3. Prepare a one in ten dilution of cell culture supernatants obtained from untransduced and transduced MSCs.
4. The concentration of human insulin in harvested cell supernatants is analyzed using the ARCHITECT™ i4000SR Immunoassay Analyzer (Abbott Diagnostics©). Briefly, vortex the samples and transfer 200 μl of each sample to an assay tube. Load the samples on the ARCHITECT™ i4000SR Immunoassay Analyzer precalibrated with the ARCHITECT™ Insulin Reagent Kit (8K41). Once loaded, combine samples, anti-insulin-coated paramagnetic microparticles, and anti-insulin acridinium-labeled conjugate. Insulin present in the sample binds to anti-insulin-coated microparticles and anti-insulin acridinium-labeled conjugate. Wash the sample, incubate with pretrigger and trigger solutions, and measure the resulting chemiluminescent reaction as relative light units (RLUs). A direct relationship exists between the amount of insulin in the samples and the RLUs detected by the ARCHITECT™ Optical system, with resulting concentrations expressed as μU/ml

converted to pmol/l. The analytical sensitivity of the assay is <1.0 μU/ml. (Fig. 2d) Any other commercially available human insulin ELISA kits can be used to measure the insulin content from the supernatants.

4 Notes

1. Methylene blue solution should be stored at room temperature, and may be filtered following the staining of MSC colonies for reuse in other experiments using filter paper.
2. Stocks of individual chemicals should be prepared and stored at −20 °C. The solubility of Indomethacin in α-MEM is poor; therefore, it should be first diluted in a small volume (10–15 ml) of α-MEM prior to addition to a full bottle of medium.
3. Dilution of the stock 16% (v/v) paraformaldehyde from glass ampules should be performed within a laminar flow hood. To break the glass ampules, hold the base of the glass ampule firmly and place the top of the ampule between your thumb and index finger. Using minimum force, break the top off the ampule and dispose of appropriately.
4. A small amount of force may be required to remove the skin off the muscle belly of the hind limb. Care should be taken when performing this step to ensure excessive force is not applied, as the limbs are fragile and may break. The skin should detach from the muscle belly as a single flap and then be cut away at the hip.
5. It is critical to remove as much of the muscle tissue as possible, as poor muscle removal can lead to an increased frequency in the number of contaminating muscle cells in the downstream stromal cell isolation. It is recommended that a combination of bulk muscle tissue removal using scissors and scraping of the femurs with a scalpel be performed to remove all muscle tissue.
6. Do not use the scalpel as a saw to cut the epiphyses off as this may create friction and cause the femurs to break. Place the scalpel blade above the joint where the epiphysis begins and firmly press down until the epiphysis is removed.
7. It may require greater than 1 ml of 1 × PBS to completely flush the bone marrow from the femurs. A good indicator of complete flushing is to observe a change in color of the femur from pale red to pale white.
8. During the initial 6-week culture period, it is critical to replenish the culture medium every 3–4 days. Nonadherent hematopoietic cells are eliminated with media replenishment, while plastic adherent stromal cells remain attached to the flasks.

9. For subpassaging of epiphyses, orientate the flasks vertically and collect the epiphyses at the base of the flask. The epiphyses may then be poured and collected in a 50 ml falcon tube. If all the epiphyses are not transferred at the first attempt, additional culture medium may be added to the flasks to assist with removing the remaining epiphyses until all the epiphyses have been collected. For subpassaging of adherent stromal cells, ensure that sufficient time is provided for the cells to detach from the culture flask. Do not incubate the cells in TrypLE Express for greater than 15 min.
10. Ensure that mAb staining is performed on ice and in the dark to avoid photobleaching of the conjugated fluorophores. A cell viability dye such as the LIVE/DEAD Fixable Violet stain (Thermofisher™) may be used to identify and sort live cells.
11. Use unstained and single mAb-stained cells to adjust the voltage settings, prepare compensation and set gates for viable (SSC-A vs FSC-A) and single cells (FSC-A vs FSC-W).
12. When removing the methylene blue solution, refilter the used solution through filter paper into a new 500 ml Schott bottle for reuse in other experiments. When destaining, colonies may become fragile and begin to detach from the plates. This can be avoided by filling a tub with tap water and slowly submerging the plates into the water rather than directly pipetting water into the plates.
13. Qualitative scoring of positively stained cells was performed according to the method described by Wang et al. [23]. For adipogenic differentiation, the number of positively stained cells in ten high power fields (40× magnification) per well of a 24-well plate were counted and scored according to the number and intensity of stained vesicles within individual cells. For osteogenic differentiation, the number of positively stained cells in ten high power fields (40× magnification) per well of a 24-well plate were counted and scored according to staining intensity correlating to immature (brown) or mature (red) osteogenesis.
14. Rather than using 50 μl of Stbl3 TOP10 E. coli cells per transformation, the bacterial volume can be halved (25 μl/transformation) to double the number of transformation reactions that can be performed from a single kit.
15. Briefly, two 50 ml falcon tubes are prepared for calcium phosphate precipitation cotransfection. Tube 1 contains the three lentiviral plasmids 2 M $CaCl_2$ and distilled H_2O. Tube 2 contains 1 M HEPES, 2 M NaCl, 150 mM Na_2HPO_4 and distilled H_2O. The two tubes are combined and incubated at room temperature for 15 min, and then added to the 10-tray cell factory containing HEK293T cells and 500 ml of culture

medium. Do not use HEK293T cells beyond passage 20 for lentiviral vector production.

16. Count the cells from three wells of the 24-well plate on the day of transduction and determine the average number of cells. This value will be utilized in the calculation of lentivirus concentration following FACS analysis.
17. Thaw aliquots of lentiviral vectors at 37 °C and prepare the appropriate volume of culture medium containing lentivirus and 8 μg/ml Polybrene according to the formula $V = (\mathrm{MOI} \times C)/T$; where V volume of lentivirus, *MOI* multiplicity of infection, C number of cells to be transduced and T lentivirus titer.
18. Transgene expression following lentiviral transduction is optimal 3–5 days posttransduction. FACS analysis and sorting of transduced MSCs should be performed during this period. FACS analysis performed <3 days post transduction may result in poor transduced cell retrieval due to low fluorescent reporter expression.

References

1. In't Anker PS, Scherjon SA, Kleijburg-Van Der Keur C, De Groot-Swings GMJS, Claas FHJ, Fibbe WE et al (2004) Isolation of mesenchymal stem cells of fetal or maternal origin from human placenta. Stem Cells 22(7):1338–1345
2. Noort WA, Kruisselbrink AB, In't Anker PS, Kruger M, Van Bezooijen RL, De Paus RA et al (2002) Mesenchymal stem cells promote engraftment of human umbilical cord blood-derived CD34+ cells in NOD/SCID mice. Exp Hematol 30(8):870–878
3. Wang HS, Hung SC, Peng ST, Huang CC, Wei HM, Guo YJ et al (2004) Mesenchymal stem cells in the Wharton's jelly of the human umbilical cord. Stem Cells 22(7):1330–1337
4. Fan CG, Tang FW, Zhang QJ, Lu SH, Liu HY, Zhao ZM et al (2005) Characterization and neural differentiation of fetal lung mesenchymal stem cells. Cell Transplant 14(5):311–321
5. De Ugarte DA, Morizono K, Elbarbary A, Alfonso Z, Zuk PA, Zhu M et al (2003) Comparison of multi-lineage cells from human adipose tissue and bone marrow. Cells Tissues Organs 174(3):101–109
6. Campagnoli C, Roberts IAG, Kumar S, Bennett PR, Bellantuono I, Fisk NM (2001) Identification of mesenchymal stem/progenitor cells in human first-trimester fetal blood, liver, and bone marrow. Blood 98(8):2396–2402
7. Van Vlasselaer P, Falla N, Snoeck H, Mathieu E (1994) Characterization and purification of osteogenic cells from murine bone marrow by two-color cell sorting using anti-Sca-1 monoclonal antibody and wheat germ agglutinin. Blood 84(3):753–763
8. Eslaminejad MB, Nadri S (2009) Murine mesenchymal stem cell isolated and expanded in low and high density culture system: surface antigen expression and osteogenic culture mineralization. In Vitro Cell Dev Biol Anim 45 (8):451–459
9. Baddoo M, Hill K, Wilkinson R, Gaupp D, Hughes C, Kopen GC et al (2003) Characterization of mesenchymal stem cells isolated from murine bone marrow by negative selection. J Cell Biochem 89(6):1235–1249
10. Soleimani M, Nadri S (2009) A protocol for isolation and culture of mesenchymal stem cells from mouse bone marrow. Nat Protoc 4 (1):102–106
11. Siclari VA, Zhu J, Akiyama K, Liu F, Zhang X, Chandra A et al (2013) Mesenchymal progenitors residing close to the bone surface are functionally distinct from those in the central bone marrow. Bone 53(2):575–586
12. Sun S, Guo Z, Xiao X, Liu B, Liu X, Tang PH et al (2003) Isolation of mouse marrow mesenchymal progenitors by a novel and reliable method. Stem Cells 21(5):527–535
13. Short BJ, Brouard N, Simmons PJ (2009) Prospective isolation of mesenchymal stem cells

from mouse compact bone. Methods Mol Biol 482:259–268
14. Gerace D, Martiniello-Wilks R, Simpson AM (2015) Persistence of luciferase expressing bone marrow-derived mesenchymal stem cells (BMSCs) in non-obese diabetic (NOD) and NOD/Scid mice (abstract). Mol Ther 23:S101
15. Gerace D, Martiniello-Wilks R, Habib R, Ren B, Nassif NT, O'Brien BA et al (2019) Ex vivo expansion of murine MSC impairs transcription factor-induced differentiation into pancreatic beta-cells. Stem Cells Int 2019:1–15
16. Geraerts M, Michiels M, Baekelandt V, Debyser Z, Gijsbers R (2005) Upscaling of lentiviral vector production by tangential flow filtration. J Gene Med 7(10):1299–1310
17. Ren B, O'Brien BA, Byrne MR, Ch'ng E, Gatt PN, Swan MA et al (2013) Long-term reversal of diabetes in non-obese diabetic mice by liver-directed gene therapy. J Gene Med 15 (1):28–41
18. Gerace D, Ren B, Hawthorne WJ, Byrne MR, Phillips PM, O'Brien BA et al (2013) Pancreatic transdifferentiation in porcine liver following lentiviral delivery of human furin-cleavable insulin. Transplant Proc 45(5):1869–1874
19. National Health and Medical Research Council (2013) Australian code for the care and use of animals for scientific purposes, 8th edition. Canberra: National Health and Medical Research Council
20. Groskreutz DJ, Sliwkowski MX, Gorman CM (1994) Genetically engineered proinsulin constitutively processed and secreted as mature, active insulin. J Biol Chem 269(8):6241–6245
21. Choi JK, Hoang N, Vilardi AM, Conrad P, Emerson SG, Gewirtz AM (2001) Hybrid HIV/MSCV LTR enhances transgene expression of lentiviral vectors in human CD34(+) hematopoietic cells. Stem Cells 19(3):236–246
22. Ren B, O'Brien BA, Swan MA, Koina ME, Nassif N, Wei MQ et al (2007) Long-term correction of diabetes in rats after lentiviral hepatic insulin gene therapy. Diabetologia 50 (9):1910–1920
23. Wang XY, Martiniello-Wilks R, Shaw JM, Ho T, Coulston N, Cooke-Yarborough C et al (2004) Preclinical evaluation of a prostate-targeted gene-directed enzyme prodrug therapy delivered by ovine atadenovirus. Gene Ther 11(21):1559–1567

Chapter 17

Bone Marrow-Derived Progenitor Cells Mediate Immune Cell Regulation

Kisha N. Sivanathan and Patrick T. Coates

Abstract

Human bone marrow (BM) derived mesenchymal stem cells (MSC) have high capacity to propagate ex vivo with superior reparative, immunosuppressive, and anti-inflammatory properties. Here we describe standardized protocols and culture conditions that enable the isolation, expansion and maintenance of a highly purified and homogenous population of human MSC. These third party-derived off-the-shelf MSC from healthy human bone marrow donors can potently inhibit mitogenically or allogeneically activated human T cells in proliferation assays. The standard operating procedures described in this chapter can be applied to researchers aiming to enhance MSC immunosuppressive properties and defining MSC mechanisms of action. Importantly, these assays can be incorporated into clinical protocols where the safety and efficacy of human BM MSC can be verified in diseases that are modulated by T cell responses.

Key words Mesenchymal stem cells, Bone marrow, T cells, Peripheral blood mononuclear cells, Bone marrow, Proliferation assays, Mixed lymphocyte reactions, Mitogen, Differentiation, Immunophenotyping

1 Introduction

The stromal tissue of the bone marrow (BM) comprise a heterogeneous population of cells including hematopoietic stem cells (HSC), endothelial progenitor cells, erythroid and myeloid precursors as well as mesenchymal stem cells (MSC) [1]. Multipotent MSC were initially described by Friedenstein and colleagues (1970) as a rare fraction of bone marrow cells (0.001–0.01%) in rodents that have in vitro clonogenicity potential (i.e., ability to form colonies defined as colony-forming unit—fibroblast; CFU-F) with self-renewal capacity and able to form bones under osteogenic defined conditions [2]. To date, there are no unique markers to accurately distinguish human BM MSC from the other constituents of the BM. *The Mesenchymal and Tissue Stem Cell Committee* however has proposed a simplistic set of generic criteria to define human BM MSC in vitro by their (1) adherence to plastic when

Mugdha V. Joglekar and Anandwardhan A. Hardikar (eds.), *Progenitor Cells: Methods and Protocols*, Methods in Molecular Biology, vol. 2029, https://doi.org/10.1007/978-1-4939-9631-5_17,

maintained in standard culture conditions; (2) trilineage differentiation potential into osteoblasts, chondroblasts, and adipocytes; and (3) cell surface expression of CD 105, CD 73, and CD 90 in >95% of the culture and negative for CD 34, CD 45, CD 14 or CD 11b, CD 79alpha or CD 19, and MHC-class-II [3].

MSC can indirectly modulate proinflammatory T cell responses by promoting the generation of immune-regulatory cells (regulatory T cells, Tregs or regulatory dendritic cells, DC) or by interacting with antigen presenting cells such as macrophages, dendritic cells (DC) and B cells. MSC may also directly suppress T cell activation, proliferation, and effector function via the secretion and expression of a number of immunosuppressive factors [4–7]. MSC therapy has been translated into the clinic to prevent and treat various inflammatory diseases including graft-versus-host disease, autoimmune disease, ischemic–reperfusion injury and allograft rejection with varying efficacies in vivo [4, 8–11]. Absence of standardization in BM MSC isolation and maintenance protocols may be a contributing factor to the conflicting results on the in vitro and in vivo therapeutic efficacy of MSC in preclinical mouse models and in patients.

Here, we provide a concise, detailed and standardized set of operating procedures based on our previous publications, to isolate high quality human bone marrow MSC. We describe in detail, the cell culture conditions and approaches that can be applied to maintain human MSC in their multipotent state with high purity and potent in vitro immunosuppressive function on human T cells. Quality control analysis techniques for MSC include CFU-F assays that identify MSC progenitor cells with greatest potential to propagate ex vivo [12–14]. Our laboratory routinely applies this assay to determine the initial quality of human BM aspirates and subsequent MSC enrichment. We next characterize MSC based on the expression of standard MSC markers (>95% positive) and determine the levels of contaminating cells (<1% positive) that co-exist in the MSC cultures. The final quality control experiment we perform on all BM MSC donors is to verify the capacity of these cells to functionally differentiate into adipocytes, osteoblasts, and chondrocytes.

Based on these protocols and culture conditions, we were able to establish and enrich a highly purified and homogenous population of human BM MSC with potent in vitro immunosuppressive capacity. Human third party-derived ("off-the-shelf") MSC from healthy bone marrow donors inhibit T cell proliferation in a dose-dependent manner. Regardless of the different human T cell sources (e.g., buffy coat T cell blood donors), our "quality-control" qualified human MSC donors showed consistent suppression of mitogen- or allogeneic- induced T cell proliferation in a mixed lymphocyte reaction (MLR) [5]. The level of T cell proliferation was measured in a methyl-tritiated thymidine assay. MLR assays are clinically relevant and are the "classical" in vitro method to test the safety and efficacy of chemical compounds or cellular therapies in a setting that partly mimic the complex immune interactions of immune cells in vivo [15].

Hence, this protocol chapter provides easy guidelines to other workers in the field aiming to isolate, expand, maintain and determine the immunosuppressive potential of human BM MSC on T cells. By standardizing in vitro MSC culture conditions and T cell immunosuppressive assays, investigators may alter therapeutic approaches—either to apply strategies to enhance MSC immunosuppressive properties or to identify novel mechanisms and candidate molecules by which MSC modulate T cell function. From a clinical perspective, these assays can be incorporated into quality control assessment of human BM MSC under good manufacturing practice (GMP) settings to test the in vitro therapeutic efficacy of autologous, allogeneic, or third party-derived MSC on recipient T cells prior to patient administration.

2 Materials

Prepare all media fresh on the day of experiment and use within 14 days. Replenish L-glutamine every 7 days.

2.1 MSC Culture Media

1. *Primary MSC medium* (*see* **Note 1**): α-MEM (Minimum Essential Medium alpha, nucleosides), 20% heat-inactivated FBS (fetal bovine serum) (*see* **Note 2**), 50 U/ml penicillin and 50 U/ml streptomycin, 2 mM L-glutamine, 100 μM L-ascorbate-2-phosphate, 1 mM sodium pyruvate.
2. 10 mM L-ascorbate-2-phosphate stock solution: dissolve 0.144 g L-ascorbate-2-phosphate in 50 ml RPMI 1640, filter-sterilize, aliquot and store at −20 °C.
3. *MSC medium*: α-MEM, 10% heat-inactivated FBS, 50 U/ml penicillin and 50 U/ml streptomycin, 2 mM L-glutamine, 100 μM L-ascorbate-2-phosphate, 1 mM sodium pyruvate.
4. *CFU-F (colony-forming unit fibroblast) medium*: α-MEM, 20% heat-inactivated FBS, 50 U/ml penicillin and 50 U/ml streptomycin, 2 mM L-glutamine, 100 μM L-ascorbate-2-phosphate, 1 mM sodium pyruvate.

2.2 MSC Differentiation Media

1. *Osteogenesis medium*: α-MEM-high glucose, 5% heat inactivated FBS (osteogenesis qualified FBS, *see* **Note 3**), 50 U/ml penicillin and 50 U/ml streptomycin, 2 mM L-glutamine, 100 μM L-ascorbate-2-phosphate, 1 mM sodium pyruvate, 10^{-7} M dexamethasone, 10 mM Hepes, 1.8 mM KH_2PO_4 (monopotassium phosphate).

2. *Adipogenesis medium*: α-MEM-high glucose, 5% heat-inactivated FBS (adipogenesis qualified FBS, *see* **Note 3**), 50 U/ml penicillin and 50 U/ml streptomycin, 2 mM L-glutamine, 100 μM L-ascorbate-2-phosphate, 1 mM sodium pyruvate, 0.5 μM isobutylmethylxanthine, 0.5 μM hydrocortisone, 60 μM indomethacin.
3. *Chondrogenesis medium (serum free)*: α-MEM-high glucose, 50 U/ml penicillin and 50 U/ml streptomycin, 2 mM L-glutamine, 100 μM L-ascorbate-2-phosphate, 1 mM sodium pyruvate, 10^{-7} M dexamethasone, 1× ITS + Premix (100×), 10 ng/ml TGF-β (transforming growth factor β), 1% BSA (bovine serum albumin).

2.3 T Cell Proliferation Assay Medium

1. *Complete RPMI 1640 medium*: RPMI 1640 (Roswell Park Memorial Institute Medium), 10% heat-inactivated FBS, 50 U/ml penicillin and 50 U/ml streptomycin, 2 mM L-glutamine, 100 μM L-ascorbate-2-phosphate, 1 mM Sodium pyruvate.

2.4 Staining Reagents

1. *Alizarin red solution*: Dissolve 0.5 g Alizarin Red in 1 ml ethanol and add 49 ml PBS into a beaker. Mix with a magnetic stirrer. When Alizarin red is fully dissolved, transfer into a bottle and mix well. Adjust to pH 4.1 to 4.3. This stock solution is stable for 1 month at RT.
2. *PBS (1× phosphate buffer solution) adjusted to pH 4.2.*
3. *Oil O Red stock solution*: Dissolve 0.5 g Oil O Red Powder in 100 ml Isopropanol. Stir well to dissolve the Oil O Red powder. This stock solution is stable for 1 year at RT.
4. *Oil O Red working solution*: Dilute 6 ml Oil O Red stock solution in 4 ml distilled water. Allow the solution to stand for 10 min and filter through No.45 Whatman filter paper. This working solution is stable for 2 h at RT.
5. *Alician Blue solution*: Dissolve 1% Alician Blue powder with 3% glacial acetic acid in water. Agitate solution for 2 h with a magnetic stirrer under gentle heat. Adjust pH to 2.5. This solution is stored at 4 °C.
6. *Toluidine blue stain (CFU-F stain)*: Dissolve 0.51 g Toluidine Blue in 30 ml 40% formaldehyde and 480 ml PBS. This solution is stored at RT.
7. *Trypan blue.*

2.5 Other Reagents and Buffers

1. *HBSS buffer*: Hanks Buffer Salt Solution (HBSS), 5% heat inactivated FBS.
2. *Cell freezing media*: 10% dimethyl sulfoxide (DMSO), 90% heat inactivated FBS.

3. *Blocking buffer*: HBSS, 1% bovine serum albumin (BSA), 5% normal human serum.
4. *FACS Fix*: 5 ml 10% formalin, 10 g glucose, 1 ml 10% sodium azide. Store at 4 °C.
5. Unconjugated or directly conjugated antibodies or isotype controls as per the need.

3 Methods

3.1 Isolation and Culture of Human Plastic Adherent Selected Mesenchymal Stem Cells

1. Obtain about 40 ml of human bone marrow aspirates from the posterior iliac crest of normal adult volunteers (subjects with informed consent; age 20–35 year) (*see* **Note 4**).
2. Dilute bone marrow in HBSS. Pool 20 ml bone marrow into a 50 ml tube and use 10 ml HBSS to rinse out all vials containing bone marrow for maximum recovery.
3. Remove any small blood clots or bone chips from the bone marrow using a 70 μM cell strainer. Determine initial viable cell counts of bone marrow aspirate on a hemocytometer using trypan blue (*see* **Note 5**).
4. Isolate human bone marrow mononuclear cells (BM MNC) by Lymphoprep™ density gradient centrifugation (*see* **Note 6**). Place 3 ml Lymphoprep™ (stored at RT and light protected) in the bottom of a 12 ml round bottom tube (*see* **Note 7**). Carefully overlay 7.5 ml bone marrow onto Lymphoprep™. Centrifuge at 400 × *g*, 30 min (low acceleration, no brake) at 22 °C.
5. To isolate the BM MNC layer from the Lymphoprep™ interface, carefully remove the top plasma layer fraction up to half above the BM MNC layer. Carefully harvest the enriched cell fraction (BM MNC) using a Pasteur pipette and transfer into a new 12 ml tube containing cold HBSS buffer.
6. Wash BM MNC twice with HBSS buffer at 400 × *g*, 7 min, 4 °C.
7. Perform cell count on a hemocytometer. Determine cell viability and yields using trypan blue. Ideally, the final BM MNC cell count should be about 10% of the initial cell count in **step 3**.
8. Seed BM MNC at 30,000 cells/cm^2 in tissue culture flasks to establish the primary plastic adherent derived MSC cell lines (20 ml primary MSC media/T-75 flask) (*see* **Note 8**). Establish MSC cultures in the primary MSC medium at 37 °C in a humidified tissue culture incubator with 5% CO_2 for 7-14 days with first medium change at day 7 (*see* **Note 9**). We define MSC at this stage of culture as Passage 0 (P0). MSC exhibit a fibroblast-like appearance in culture (Fig. 1).

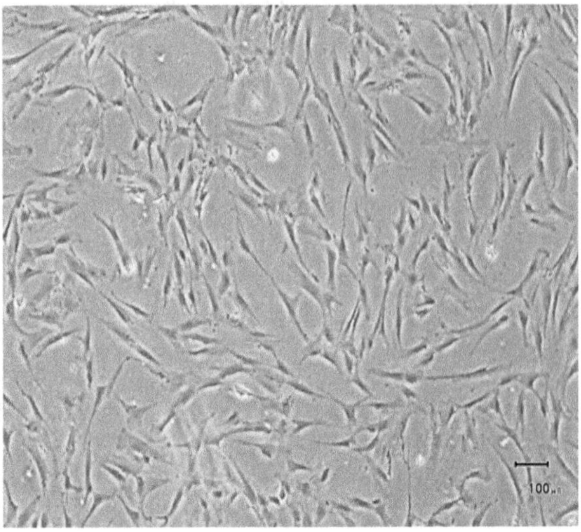

Fig. 1 BM-MSC in culture. BM-MSC after about 5–7 days of initial culture in primary MSC medium display a typical fibroblast-like morphology. The Olympus IX51 inverted microscope was used to capture image. Original magnification ×40 and scale bar is 100 μm

Table 1
Guidelines for the maintenance of MSC in culture

Flask	Surface area (cm^2)	Total cells/flask	MSC media (ml)	0.25% trypsin/EDTA (ml)
T-25	25	1.5×10^5–2.5×10^5	4	0.5
T-75	75	5×10^5–7.5×10^5	10	1
T-175	175	12×10^5–17.5×10^5	20	2

9. Detach adherent MSC when cells are at 80–90% confluency using 0.25% trypsin/EDTA (ethylenediaminetetraacetic acid, *see* Table 1) for 4 min, 37 °C in humidified tissue culture incubator with 5% CO_2. Inactivate trypsin/EDTA by adding equal volumes of HBSS buffer (*see* **Notes 10** and **11**). Tap flasks to lift off the cells that remain attached and collect detached cells with a serological pipette. Rinse flasks several times with HBSS buffer to collect all cells. Check under the microscope to ensure most of the detached cells have been collected (*see* **Note 12**). MSC are referred to as P1 following the first trypsinization step.
10. Wash cells once in HBSS buffer, centrifuging at 400 × *g*, 7 min, 4 °C.

11. Reseed MSC for expansion at a density of approximately 6500–10,000 cells/cm^2 in tissue culture flasks and maintain cells in the MSC medium (*see* Table 1). Perform medium change every 3 days. Use MSC at passages 3–8 in experiments (*see* **Note 13**).
12. MSC can be cryopreserved and stored at any stage of subculturing for later use in experiments.

3.2 Cryopreservation and Log-Phase Growth Recovery of MSC

1. Human MSC expanded in culture can be cryopreserved in 1 ml cold cell freezing media at (1–2 × 10^6 cells/cryovial) (*see* **Note 14**). Store cryovials at −80 °C and transfer to liquid nitrogen by day 3 to preserve the cell viability and growth recovery (*see* **Note 15**).
2. To recover MSC following cryopreservation, thaw MSC at RT till about half thawed and transfer cells into a 12 ml conical tube. Add cold α-MEM to MSC drop-wise (10 ml α-MEM/cryovial). Wash cells twice with HBSS buffer at 400 × *g*, 7 min, 4 °C to ensure removal of DMSO.
3. Seed cells at 10,000 cells/cm^2 for expansion in MSC medium.
4. After 24 h, remove non-adherent cells by rinsing flask with HBSS buffer (*see* **Note 16**). Maintain adherent cells in MSC medium until 70–80% confluent (*see* **Notes 17** and **18**).

3.3 Characterization of Human MSC

3.3.1 CFU-F Assay

1. Plate human BM MNC in a 6-well plate with 3 ml primary MSC medium in triplicate wells at 1 × 10^5 cells/well and 3 × 10^5 cells/well. Incubate cells at 37 °C in a humidified tissue culture incubator with 5% CO_2 for 10–14 days depending on the cell growth (*see* **Note 19**).
2. At the end of this assays (i.e., before colonies start to merge into each other), rinse plates 3 times with PBS and stain the nuclei of cells with 1 ml Toluidine blue. Incubate plates on rocker for a minimum of 2 h.
3. Rinse the Toluidine blue stain gently with deionized water until water runs clear from wells and allow the plates to dry on bench.
4. Count colonies under an inverted light microscope. A colony is defined as a cluster of ≥50 cells (Fig. 2). Exclude cobblestone shaped colonies for CFU-F counts.

3.3.2 Surface Staining and Flow Cytometry Analysis (Immunophenotype Analysis)

1. Following trypsinization of MSC, wash cells once with HBSS buffer and resuspend the single cell suspension in blocking buffer for 20 min, 4 °C.
2. Stain MSC with unconjugated or directly conjugated antibodies or isotype controls for 45 min, 4 °C (*see* Table 2, **Note 20**).
3. Wash cells twice with HBSS buffer and add secondary antibodies to appropriate tubes for 30 min, 4 °C.

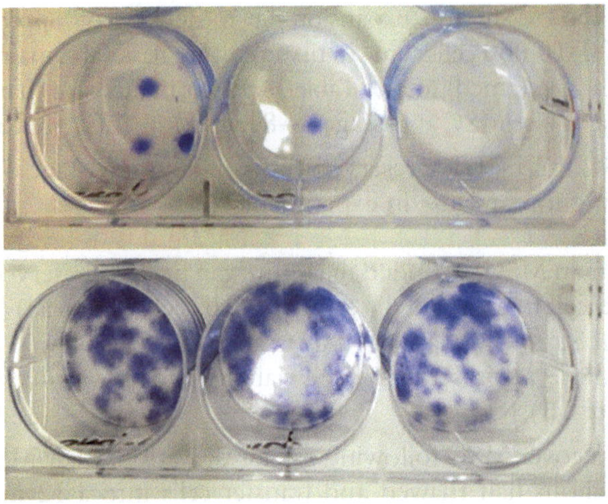

Fig. 2 BM MSC and their colony-forming unit fibroblast efficiency. The figures represent Toluidine Blue stained primary MSC from two different donors

Table 2
Human MSC antibody panel and guidelines for expected MSC culture purity

Antibody	Clone	MSC expression	% of expression in cultures
CD 44	H9H11	+	>95%
CD 73	AD2	+	>95%
CD 90	5E10	+	>95%
CD 105	266	+	>95%
CD 146	CC9	+	>95%
CD 166	3A6	+	>95%
STRO-1	STRO-1	+/−	(initial culture; variable)
STRO-4	HYB-H	+	>95%
CD 11a	HB202	−	(<1%)
CD 14	RMO52	−	(<1%)
CD 34	581	−	(<1%)
CD 45	J33	−	(<1%)
CD 40	MAB89	−	(<1%)
CD 80	L307.4	−	(<1%)
CD 83	HB15e	−	(<1%)
CD 86	2331 (FUN-1)	−	(<1%)
MHC class I	W6.32	+	>95%
MHC class II	3.54	−	(<1%)

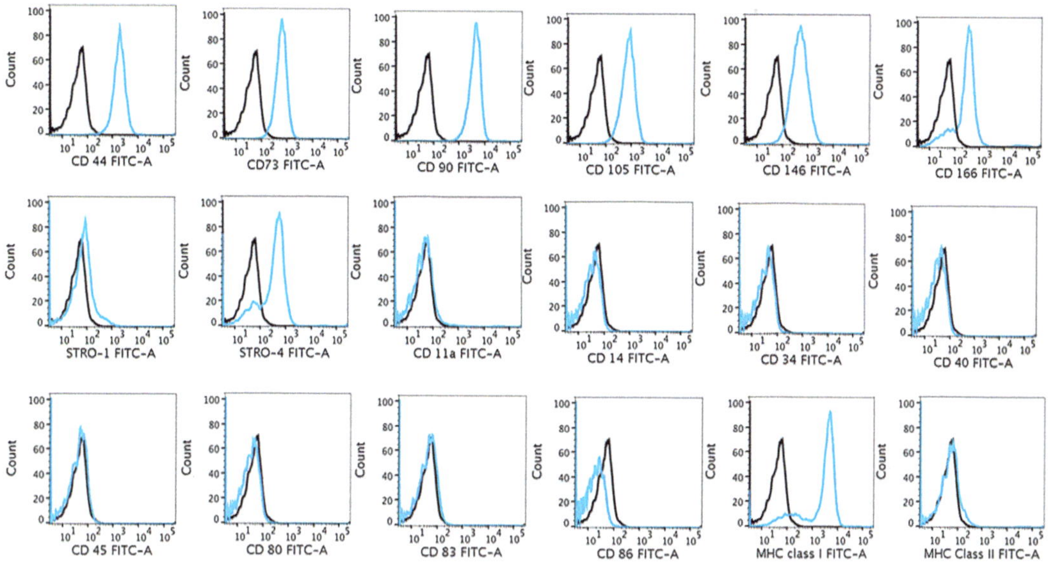

Fig. 3 MSC immunophenotype. The expression of cell surface markers on MSC was assessed by flow cytometry. Histograms show level of surface marker expression on MSC and are depicted by the isotype control (black) and MSC (blue). Data are one representative of 6 MSC donors

4. Either wash cells twice with HBSS buffer and acquire immediately on a flow cytometer or fix cells with FACS fix solution, store at 4 °C and acquire later (Fig. 3, **Note 21**).

3.3.3 Trilineage Functional Differentiation of MSC

1. *Osteogenesis assay*: Seed MSC at 2×10^5 cells/T-25 flask in 6 ml MSC medium. Allow MSC to adhere to flask overnight. Remove non-adherent cells by rinsing flasks with HBSS buffer and replace with 6 ml osteogenesis medium. Incubate cells at 37 °C in a humidified tissue culture incubator with 5% CO_2 for 2–4 weeks, with medium change performed 3× weekly. Observe changes in MSC morphology under the microscope every 3 days. At the end of differentiation, aspirate the medium and rinse flasks with PBS. Fix cells for 1 h at RT with 4 ml 4% neutral buffered formalin. Following fixation, rinse flasks 3× with pH 4.2 PBS. Stain the cells with 6 ml Alizarin Red Solution for 5 min to reveal osteocytes containing calcium deposits. Remove the stain by washing cells 3× with PBS. Add 4 ml PBS to flasks and acquire images under an inverted light microscope immediately (Fig. 4a).
2. *Adipogenesis assay* (*see* **Note 22**): Seed MSC at 2×10^5 cells/T-25 flask in 6 ml MSC medium. Allow MSC to adhere to flask. When cells reach 100% confluency, remove non-adherent cells by rinsing flaks with HBSS buffer and replace with 6 ml adipogenesis medium. Incubate cells at 37 °C in a humidified tissue culture incubator with 5% CO_2 for 2–4 weeks, with medium change performed 3× weekly. Observe changes in MSC

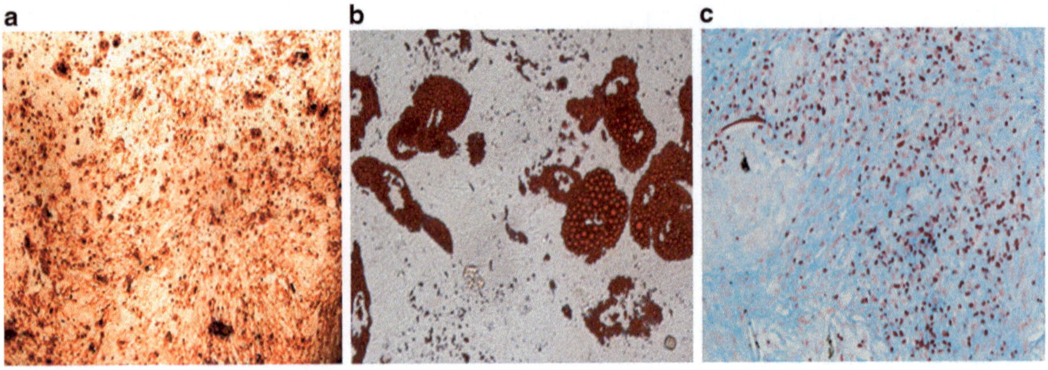

Fig. 4 MSC functional differentiation. Functional differentiation of MSC into (**a**) osteoblasts (Alizarin red stained mineral; original magnification ×200), (**b**) adipocytes (Oil red stained fat; original magnification ×200), and (**c**) chondrocytes (Alician blue stained glycosaminoglycans; original magnification ×200) for 4 weeks. The Olympus IX51 inverted microscope was used to capture images. Data are one representative of 6 MSC donors

morphology under the microscope every 3 days. At the end of differentiation, aspirate medium and rinse the adherent cell cultures exhibiting adipocytic clusters 3× gently with PBS. Fix cells for 30 min in 4% neutral buffered formalin. Following fixation, rinse flasks gently with PBS (3×) and perform another 3 rinses with distilled water. Stain cell cultures with 6 ml in Oil O Red working solution for 1 h to reveal lipid formation from adipocytes. Remove the stain by washing the cells 3× with distilled water. Add 4 ml PBS to flask and acquire images under an inverted light microscope immediately (Fig. 4b).

3. *Chondrogenesis assay*: Pellet 5×10^5 cells in 0.5 ml HBSS in a 10 ml polypropylene tubes at 400 × *g*, 10 min to remove FBS in MSC cell suspension. Gently remove the supernatant and carefully add 0.5 ml serum-deprived chondrogenesis medium without disrupting the MSC pellet. Centrifuge cells again at 400 × *g*, 5 min. Loosen the tube cap containing the pellet and incubate at 37 °C in a humidified tissue culture incubator with 5% CO_2 for 4 weeks, with medium change performed 3× weekly. At the end of differentiation, remove the chondrogenesis medium and rinse pellet once with 1 ml PBS. Fix the pellet aggregates with 4% neutral buffered formalin overnight at 4 °C. Replace fixative with 70% ethanol and paraffin embed (*see* **Note 23**). Following paraffin embedding, 5 μM cross sections are histologically stained with Alician Blue Solution (pH 2.5) to reveal proteoglycan synthesis by chondrocytes. The samples are dewaxed and rehydrated through a series of washes—2× 6 min in Solv 21C, 1× 3 min in 100% ethanol and 2× 4 min in 70% ethanol into distilled water. Immerse section in Alician Blue solution for 40 min, RT and then rinse well in deionized water. If required, samples can be counter stained

with H&E (hematoxylin and eosin) prior to dehydration and mounting. Acquire images under an inverted light microscope (Fig. 4c).

3.4 Human Blood Cell Isolation

This method is also applicable for the isolation of PBMC from whole blood. Collect fresh blood into EDTA or Lithium heparin coated tubes.

3.4.1 PBMC Isolation by Density Gradient Centrifugation

1. Add 15 ml Lymphoprep™ (stored at RT and light protected) into a 50 ml Leucosep® tube (*see* **Note 6**). Centrifuge the tube containing Lymphoprep™ separation medium for 1 min at 1000 × *g*, RT. The separation medium will be located at the bottom of the tube, below the porous barrier.
2. Dilute buffy coats to a ratio of 1:2 in 1× PBS (*see* **Note 24**).
3. Carefully add diluted buffy coats directly into the Leucosep® tubes containing the Lymphoprep™ separation medium at the bottom (*see* **Note 25**).
4. Centrifuge at 1000 × *g*, 10 min (low acceleration, no brake) at RT.
5. To isolate the peripheral blood mononuclear cells (PBMC) layer from the Lymphoprep™ interface, carefully remove the top plasma layer fraction up to half above the PBMC interface. Harvest the enriched cell monolayer (PBMC/lymphocytes) using a Pasteur pipette and transfer into a new 50 ml tube containing cold PBS.
6. Wash PBMC fraction with PBS at 300 × *g*, 10 min, 4 °C to remove platelets. Repeat the washing step twice. Perform the final wash in PBS, centrifuging at 400 × *g*, 7 min, 4 °C.
7. Perform cell counts on a hemocytometer. Determine viable cell counts using trypan blue.
8. Resuspend cells in complete RPMI 1640 medium.
9. Either cryopreserved or freshly isolated PBMC from buffy coats can be used in MSC immune cell regulation assays. If cryopreserved PBMC are used, thaw PBMC at RT and add cold RPMI dropwise (10 ml RPMI/cryovial). Wash cells twice with PBS, centrifuging at 400 × *g*, 7 min, 4 °C to ensure removal of DMSO. Resuspend cells in complete RPMI 1640 medium. Rest PBMC in non-adherent tissue culture flasks (about 1×10^6 cells/T-75 flask in 10 ml complete RPMI 1640 medium) overnight to allow PBMC to recover. Incubate cells at 37 °C in a humidified tissue culture incubator with 5% CO_2. Collect cells, rinse flask with 1% FBS/PBS and pellet cells at 400 × *g*, 7 min, 4 °C. Perform cell counts on a hemocytometer. Determine viable cell counts using trypan blue. Resuspend cells in complete RPMI medium.

3.4.2 T Cell Isolation Using Surface Markers

1. T cells can be isolated from either fresh or frozen PBMC (*see* **Note 26**). Isolate $CD3^+$ T cells from PBMC using commercially available T cell negative selection kits. $CD4^+$, $CD8^+$, naive T cells, or other T cell subsets of interest can also be isolated using commercial negative selection kits according to manufacturer's instructions (*see* **Note 27**). Following T cell isolation, pellet cells at 400 × *g*, 7 min, 4 °C and resuspend cells in complete RPMI medium. Perform cell counts on a hemocytometer and determine viable cell counts using trypan blue.
2. Alternatively, $CD3^+$ T cells can be isolated from fresh or frozen PBMC using nylon wool columns (*see* **Note 28**), Enrich T cells by removing monocyte population by panning PBMC in a T-75 flask (5×10^7 cells/ml in a total of 10 ml/flaks) for 45 min–1 h in 1% FBS/RPMI 1640 at 37 °C in a humidified tissue culture incubator with 5% CO_2. Collect the non-adherent cell fraction by rinsing flasks with PBS and centrifuge at 400 × *g*, 7 min, 4 °C. Resuspend the pellet in complete RPMI 1640 medium. Load 5 ml of non-adherent cells (maximum 1×10^8 cells/column) into sterile nylon wool columns preequilibrated at 37 °C for 30 min and saturated with RPMI 1640. Incubate the column upright at 37 °C, 30 min in a humidified tissue culture incubator with 5% CO_2. Elute T cells with 13 ml prewarmed complete RPMI 1640 medium and collect T cells into a clean tube. B cells are selectively retained in the column, allowing the enrichment of T cells to flow through the column when eluted with medium. Centrifuge the eluted T cell fraction at 400 × *g*, 7 min, 4 °C and resuspended cells in complete RPMI 1640 medium. Perform cell counts on a hemocytometer and determine viable cell counts using trypan blue. $CD3^+$ T cells isolated using traditional nylon wool columns are also high in purity (>90%). This method is a cheaper than using commercially available negative selection kits but is time consuming and laborious.

3.5 Human MSC and T Cell Immunosuppressive Assays

Use at least 3–4 MSC donors and different T cell buffy coats for each experiment. The source of MSC can either be autologous (MSC from donor T cells) or allogeneic (MSC from responder T cells) or third-party cells (MSC not related to donor or responder T cells).

1. Detach adherent MSC from tissue culture flasks at 80–90% confluency using 0.25% trypsin–EDTA for 4 min, 37 °C as described in Subheading 3.1.
2. Wash MSC once in HBSS buffer at 400 × *g*, 7 min, 4 °C.
3. Resuspend MSC in complete RPMI 1640 medium.
4. Gamma-irradiate MSC at 30 Grays prior to their use in T cell immunosuppressive assays (*see* **Note 29**).

Table 3
MSC dose to responder T cells

MSC dose (%)	Ratio of MSC: responder T cells	MSC concentration (add 50 μl/well)	Final number of MSC/well
10	1:10	2×10^5 MSC/ml	1×10^4
1	1:100	2×10^4 MSC/ml	1×10^3
0.1	1:1000	2×10^3 MSC/ml	1×10^2

5. Perform cell counts on a hemocytometer. Determine viable cell counts using trypan blue.
6. Adjust MSC to concentrations of 2×10^5 MSC/ml, 2×10^4 MSC/ml, and 2×10^3 MSC/ml in complete RPMI 1640 medium.

3.5.1 Mitogen (PHA) Induced T Cell Proliferation Assay

1. Seed 50 μl of 30 Grays irradiated MSC at the varying concentrations into 96-well flat bottom plates (Table 3). Allow MSC to adhere to wells overnight or for a minimum of 4 h prior to the addition of purified T cells.
2. Adjust purified T cells to 2×10^6 cells/ml in complete RPMI 1640 medium. Seed 50 μl of T cells into the 96-well flat bottom plates in the presence or absence of MSC. The final number of T cells per well will be 1×10^5 cells.
3. Prepare PHA at a concentration of 40 μg/ml in complete RPMI 1640 medium and add 50 μl of PHA to wells containing either T cells alone or MSC co-cultured with T cells. The final concentration of PHA in each well is 10 μg/ml.
4. Add complete RPMI 1640 medium alone to appropriate wells. The total volume in each well should be 200 μl.
5. Add 200 μl of sterile PBS to any unused wells to reduce evaporation of medium from the plate during culture.
6. Incubate cells at 37 °C in a humidified tissue culture incubator with 5% CO_2 for 5 days.
7. The baseline culture conditions and controls that are essential for this assay are:
   (a) T cells alone.
   (b) MSC alone.
   (c) PHA and T cells.
   (d) MSC, T cells and PHA.

3.5.2 Mixed Lymphocyte Reaction (MLR)

1. Seed 50 μl of 30 Grays irradiated MSC at the varying concentrations into 96-well round bottom plates (Table 3). Allow MSC to adhere to wells overnight or for a minimum of 4 h prior to the addition of PBMC or T cells (*see* **Note 30**).

2. Adjust responder cells (PBMC or purified T cells) to 2×10^6 cells/ml in complete RPMI 1640 medium and add 50 μl responder cells/well. The final number of responder cells per well is 1×10^5 cells.
3. For a one-way MLR, isolate stimulator PBMC from a different buffy coat donor to responder cells. Gamma-irradiate simulator cells at 30 Grays prior to their use in the one-way MLR experiment. Adjust stimulator PBMC to 4×10^6 cells/ml in complete RPMI 1640 medium and add 50 μl stimulator cells/well. The final number of stimulator cells per well is 2×10^5 cells.
4. For a two-way MLR, co-culture MSC with 1×10^5 responder and stimulator PBMC isolated from two different buffy coat donors.
5. Add complete RPMI 1640 medium alone to appropriate wells to a final volume of 200 μl and fill the unused wells with 200 μl sterile PBS.
6. Incubate cells at 37 °C in a humidified tissue culture incubator with 5% CO_2 for 5 days.
7. The baseline culture conditions and controls that are essential for MLR are as follows:
   (a) MSC alone.
   (b) Responder PBMC or T cells alone.
   (c) Stimulator PBMC alone.
   (d) MSC and responder PBMC or T cells.
   (e) MSC and stimulator PBMC.
   (f) MSC, responder PBMC or T cells, and stimulator PBMC.

3.5.3 Measurement of Human T Cell Proliferation by [^{3}H]-Thymidine Assay

1. On day 4 of MSC and T cell or PBMC proliferation assay or MLR, responder T cells are pulsed with 1 μCi methyl-tritiated-thymidine ([^{3}H]-thymidine, concentration 1 mCi/ml; 37 KBq/well) per well for 18 h to assess levels of T cell proliferation. Prepare [^{3}H]-thymidine from a stock at 1:20 dilution in complete RPMI 1640 medium and add 20 μl/well (*see* **Note 31**).
2. Following 18–24 h pulsing of T cells with [^{3}H]-Thymidine, harvest cells onto glass fiber filters using a cell harvester system, where cellular DNA will bound onto the filter mat. Follow the harvester operating instructions to prime the system. Align the glass filter onto the harvester platform and pre-wet the filter mat by performing at least three washes with distilled water. Place the 96-well plate of MSC and T cell or PBMC co-cultures onto the harvester and harvest the cells onto the glass fiber filter. Perform four initial washes based on the harvester

standard protocol. Then, remove adherent cells by adding 50 μl of 0.25% trypsin/EDTA to each well and incubate for 4–5 min maximum at 37 °C. Leave filter mat in place during this step. Continue harvesting cells by performing another 8 washes. Observe plates under the microscope to ensure that all cells are transferred onto the filter mat. If there are any cells adhered to the plate, perform another 4 washes or more until all cells have been completely transferred. Follow operating instructions to wash system and dispose waste (*see* **Note 31**).

3. Air-dry the filter mat on the bench overnight or place in a designated dry incubator at 60 °C for 1 h.
4. After the filter mat is completely dry, place it in a glass fiber filter-bottom 96-well microplate. Add MicroScint scintillation fluid (a radiometric detection solution) to the glass fiber filters (30 μl/well). Seal the wells using a clear adhesive film for microplates (*see* **Note 32**).
5. The incorporation of [^{3}H]-Thymidine into cellular DNA is counted using a scintillation counter (TopCount NXT Microplate Scintillation and Luminescence counter).
6. The levels of T cell proliferation are expressed by mean counts per minute (cpm) of triplicate wells (*see* **Note 33**).
7. Determine the levels T cell proliferation in the presence of MSC (Fig. 5a, b).

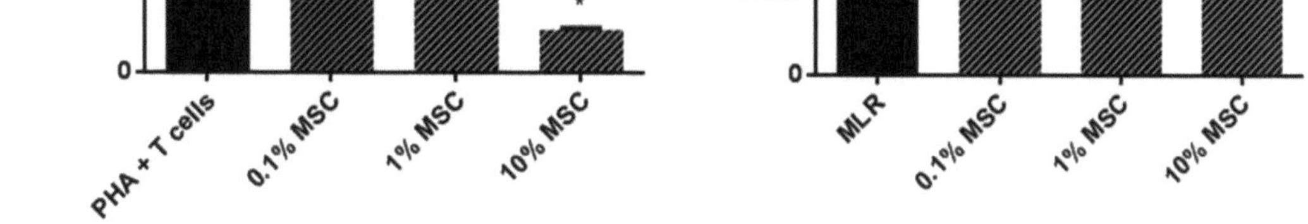

Fig. 5 MSC inhibit T cell proliferation in a dose-dependent manner. MSC were co-cultured with either (**a**) PHA activated CD3$^+$ T cells or (**b**) responder PBMC and stimulator PBMC (30 Grays irradiated) from two different buffy coat donors in a one-way MLR at 0.1% MSC, 1% MSC, and 10% MSC. T cell proliferation was measured in a [^{3}H]-Thymidine assay. Data are (**a**) one representative of 6 independent experiments and (**b**) one representative of 3 independent experiments. $*$, $p < 0.05$ versus PHA+ T cells or MLR determined by one-way ANOVA with post-Sidak multiple comparison test. Error bars depict means of triplicates $\pm$ SD. Irradiated MSC alone, T cells alone, responder PBMC alone and irradiated stimulator PBMC alone controls showed minimal or undetectable levels of cell proliferation

4 Notes

1. Use freshly prepared primary MSC medium for the establishment of primary MSC cultures.
2. FBS batches vary in quality and composition. The choice of FBS used in culture is crucial for optimal isolation, expansion, maintenance, and function of MSC. FBS can affect adherence and mitotic expansion of primary MSC cultures as well as the retention of MSC in undifferentiated states during subcultivation [14, 16]. It is therefore crucial that investigators use MSC qualified FBS, either commercially available or laboratory batch tested for MSC. FBS batch testing can be conducted in the laboratory by determining the effects of different FBS on MSC progenitor cell enrichment in CFU-F assays, MSC morphology (fibroblastic-like appearance) and the growth potential of MSC following subculturing.
3. FBS can also affect MSC trilineage differentiation potential. We therefore recommend investigators to use osteogenesis or adipogenesis qualified FBS, either commercially available or laboratory batch tested for MSC differentiation. If FBS is incompatible, the differentiated cells will detach prematurely during the differentiation protocol. Successful induction of adipogenesis and osteogenesis can be confirmed by Oil O Red or Alizarin Red staining, respectively.
4. 40 ml bone marrow aspirate can be used to generate a large bank of human plastic-adherent MSC for subsequent use in experiments.
5. Bone chips can give rise to normal bone cells (osteoblasts) when cultured in appropriate osteoblast media.
6. Ficoll-Paque™ can also be used as an alternative to Lymphoprep™ without the need to change this protocol.
7. We recommend using multiple 12 ml tubes instead of a 50 ml tube for highest recovery of BM MNC from the layer.
8. If investigators are generating primary MSC cultures from purified BM MNC fractions (e.g., STRO-1 positive selected BM MNC to establish primary MSC cultures [17]), isolated BM MNC can be seeded at a lower seeding density—that is, 10,000 cells/cm^2 in 20 ml primary MSC media/T-75 flask.
9. BM MNC can also be cryopreserved and stored in liquid nitrogen for later isolation or establishment of primary MSC cultures. We however recommend the establishment of primary MSC cultures from freshly isolated bone marrow aspirates for highest MSC yields.

10. It is important that MSC are not overconfluent prior to trypsinization as this can affect their subsequent proliferative capacity when maintained in culture.
11. Trypsin/EDTA should be inactivated using HBSS buffer (supplemented with FBS) within 4–5 min maximum. Prolonged exposure of MSC to trypsin can be toxic to cells.
12. Other contaminating bone marrow cells co-exist in early MSC cultures. These cells remain more tightly adhered to tissue culture flasks following trypsinization. Avoid collecting these contaminating cells.
13. We recommend using MSC for experiments between P3 and P8. At P3 of culture, almost all other contaminating hematopoietic cells from the bone marrow are eliminated by subculturing. MSC at a later passage, beyond P8, undergo senescence and have altered immunomodulatory secretion profiles as compared to the cells at early passages that might affect their immunosuppressive properties [16].
14. It is important not to exceed this freezing density as it can affect cell viability and growth phase recovery of MSC.
15. Transfer cells into liquid nitrogen the next day (optimal) or by day 3 as this can affect cell viability or MSC growth recovery.
16. Rinsing the non-adherent cells following cryo-recovery of MSC can promote MSC growth in cultures. MSC can be ready for passaging in 3–5 days.
17. We have been able to maintain MSC up to greater than passage 10 with good and consistent population doublings between cell passages. We, however, recommend not to use cells beyond passage 8 in functional T cell immunosuppressive assays.
18. Allow cryopreserved MSC to undergo one log-phase growth recovery prior to using them in any functional assays.
19. It is important to perform CFU-F assay on human bone marrow aspirates as this is a key indicator of human bone marrow quality. CFU-F assays can identify MSC progenitor cells with greatest propagation potential. Human bone marrow quality is determined by the bone marrow donor and (or) the extraction procedure. CFU-F assay can therefore help researchers identify poor quality marrow to be excluded in subsequent functional assays.
20. Table 2 is a list of key MSC markers to determine MSC purity. We recommend these cell surface markers for complete human MSC immunophenotyping. These are the list of antibody clones used in our previous publications [5, 6].
21. Stro-1 is a progenitor cell marker that is downregulated in MSC upon subculturing.

22. It is important to induce adipogenesis when MSC reach 100% confluency, otherwise cells will detach prematurely during the differentiation protocol. Treat flaks gently at all stages of differentiation, when performing media changes and during the staining steps. Adipocytes can easily detach from the flask.
23. Be careful not to disrupt the pellet during medium change and fixation. The pellet can be small and easily lost while processing, prior to paraffin embedding.
24. Buffy coat is the fraction of anticoagulated blood that contains most leukocytes and platelets.
25. PBMC can also be isolated using standard 50 ml conical tubes. For this, add 10 ml buffy coats into a 50 ml tube. Dilute buffy coats with 25 ml of PBS. Underlay diluted buffy coats with 15 ml Lymphoprep™ and centrifuge at 600 × *g*, 20 min, 22 °C (low acceleration, no brake).
26. If investigators are working with frozen PBMC, we recommend that T cell enrichment is performed on the day of MSC and T cell co-culture. We also suggest that PBMC instead of enriched T cells are cryopreserved. The recovery and yield of cryopreserved PBMC are generally low (50–90%) and is variable between buffy coat donors.
27. It is important to use negative selection kits to isolate T cells for MSC and T cell immunosuppressive assays. The use of immunomagnetic positive selection kits can pre-activate T cells.
28. Nylon wool columns can be prepared in advance. Pack about 2 g of teased nylon wool firmly, evenly but not too tight into a 10 ml syringe to the 10 ml mark (remove the plunger) with sterile tweezers. Place columns in a paper bag and autoclave. When pre-equilibrating the columns during experiments, wrap the top and bottom of the column with clean Parafilm. Rinse the bottom end of the column with 70% ethanol and allow to dry before loading non-adherent cells into the column and when eluting T cells.
29. Gamma-irradiation inhibits MSC proliferation without affecting MSC viability and their immunosuppressive potential [18].
30. We recommend the use of 96-well round bottom plates for MLR experiments as cells can be in close proximity and stimulator cells can induce more robust responder cell proliferation as compared to seeding cells in a 96-well flat bottom plate.
31. Wear personal protective equipment when handling [^{3}H]-thymidine (two pairs of gloves, laboratory gown and safety glasses). Use designated pipettes to add [^{3}H]-thymidine to wells and dispose all tips and plates containing [^{3}H]-thymidine, appropriately into medical waste. Institutional guidelines for disposal of radioactive waste must be followed.

32. The glass fiber filter-bottom 96-well microplate can be reused. Soak microplates overnight with 5% Decon Neutracon (prepared in distilled water) to chemically decontaminate the plates. Dispose the waste in a fume hood sink with running water for 5 min. Rinse microplates with distilled water and leave to dry overnight before reusing.
33. Radiolabeled thymidine (methyl-tritiated thymidine; [^{3}H]-Thymidine) is used in cell proliferation assays as it is incorporated into DNA of dividing cells. The level of [^{3}H]-Thymidine is measured using a radiometric detection liquid scintillation counter. This measure is proportional to the amount of cell proliferation.

Acknowledgments

We thank all bone marrow donors and the Royal Adelaide Hospital, Adelaide, South Australia as well as the Australian Red Cross Blood Service for all the buffy coats. We would like to specially thank and acknowledge Svjetlana Kireta and Julie Johnston for their expertise, invaluable input throughout their long-term service in our laboratory at the University of Adelaide and Royal Adelaide Hospital and for proof-reading the manuscript. We thank the Australian Government and the National Health and Medical Council (NHMRC) for K.N.S CJ Martin Biomedical Postdoctoral Early Career Fellowship.

References

1. Mendez-Ferrer S, Michurina TV, Ferraro F et al (2010) Mesenchymal and haematopoietic stem cells form a unique bone marrow niche. Nature 466:829–834
2. Friedenstein AJ, Chailakhjan RK, Lalykina KS (1970) The development of fibroblast colonies in monolayer cultures of guinea-pig bone marrow and spleen cells. Cell Tissue Kinet 3:393–403
3. Dominici M, Le Blanc K, Mueller I et al (2006) Minimal criteria for defining multipotent mesenchymal stromal cells. The International Society for Cellular Therapy position statement. Cytotherapy 8:315–317
4. Sivanathan KN, Gronthos S, Rojas-Canales D et al (2014) Interferon-gamma modification of mesenchymal stem cells: implications of autologous and allogeneic mesenchymal stem cell therapy in allotransplantation. Stem Cell Rev 10:351–375
5. Sivanathan KN, Rojas-Canales DM, Hope CM et al (2015) Interleukin-17A-induced human mesenchymal stem cells are superior modulators of immunological function. Stem Cells 33:2850–2863
6. Sivanathan KN, Rojas-Canales D, Grey ST et al (2017) Transcriptome profiling of IL-17A preactivated mesenchymal stem cells: a comparative study to unmodified and IFN-gamma modified mesenchymal stem cells. Stem Cells Int 2017:1025820
7. Sivanathan KN, Coates PT (2018) IL-17A-induced mesenchymal stem cells have promising therapeutic value for clinical translation. Kidney Int 93:771–773
8. Togel FE, Westenfelder C (2012) Kidney protection and regeneration following acute injury: progress through stem cell therapy. Am J Kidney Dis 60:1012–1022
9. Hickson LJ, Eirin A, Lerman LO (2016) Challenges and opportunities for stem cell therapy in patients with chronic kidney disease. Kidney Int 89:767–778
10. Casiraghi F, Perico N, Cortinovis M et al (2016) Mesenchymal stromal cells in renal

transplantation: opportunities and challenges. Nat Rev Nephrol 12:241–253

11. Squillaro T, Peluso G, Galderisi U (2016) Clinical trials with mesenchymal stem cells: an update. Cell Transplant 25:829–848
12. Digirolamo CM, Stokes D, Colter D et al (1999) Propagation and senescence of human marrow stromal cells in culture: a simple colony-forming assay identifies samples with the greatest potential to propagate and differentiate. Br J Haematol 107:275–281
13. Pochampally R (2008) Colony forming unit assays for MSCs. Methods Mol Biol 449:83–91
14. Sivanathan KN, Gronthos S, Grey ST et al (2017) Immunodepletion and hypoxia preconditioning of mouse compact bone cells as a novel protocol to isolate highly immunosuppressive mesenchymal stem cells. Stem Cells Dev 26:512–527
15. Lindemann M (2014) Ex vivo assessment of cellular immune function - applications in patient care and clinical studies. Tissue Antigens 84:439–449
16. Liu Y, Li YQ, Wang HY et al (2015) Effect of serum choice on replicative senescence in mesenchymal stromal cells. Cytotherapy 17:874–884
17. Psaltis PJ, Paton S, See F et al (2010) Enrichment for STRO-1 expression enhances the cardiovascular paracrine activity of human bone marrow-derived mesenchymal cell populations. J Cell Physiol 223:530–540
18. De Andrade AV, Riewaldt J, Wehner R et al (2014) Gamma irradiation preserves immunosuppressive potential and inhibits clonogenic capacity of human bone marrow-derived mesenchymal stromal cells. J Cell Mol Med 18:1184–1193

Chapter 18

Flow Cytometry and Cell Sorting Using Hematopoietic Progenitor Cells

Sangeetha Vadakke-Madathil, Lalita S. Limaye, Vaijayanti P. Kale, and Hina W. Chaudhry

Abstract

Flow cytometry is a widely used laser-based technology for rapid analysis of the expression of cell surface antigens and intracellular molecules in various cell types including hematopoietic stem/progenitor cells (HSPCs). Multiparametric analysis of individual cells within a short time frame makes this tool attractive and indispensable in the field of stem cell research. This is accomplished by harnessing the specific light scattering ability of the cell type, which determines its size and internal complexity. In addition, use of fluorescently conjugated antibodies allows the detection of a specific surface or intracellular antigen present at that particular stage. Fluorescent Activated Cell Sorting (FACS) is used to separate a subset of cells from a heterogeneous cell population based on fluorescent labeling. Here we describe the general principles of flow cytometry and detailed methods for the isolation of HSPCs using flow cytometry as a tool.

Key words Flow cytometry, LSK population, CD34$^+$ progenitor cells, Cell sorting

1 Introduction

Flow cytometry as a technique dates back to the late 1960s [1] and the current term "flow cytometry" was later ascribed to it that suggests "measuring properties of a cell (cyto = cell) in a flowing system" (metry = measurement). The basic principle of a flow cytometer relies on the emission of a beam of laser light into a stream of fluid [2]. This component of a flow cytometer that involves the phenomenon of hydrodynamic focusing is termed ***fluidics***. The second part, ***optics***, refers to the lasers that illuminate the particles followed by their collection by the detectors. The placement of the detectors is such that they deflect the light beam in line with the beam based on the cell characteristics (Forward Scatter-FSC) and some perpendicular to the beam (Side Scatter-SSC). FSC reflects the relative size of the particle/cell and hence can reveal the heterogeneity of the given population. SSC, on the other hand, reflects the granularity or internal complexity of the

Mugdha V. Joglekar and Anandwardhan A. Hardikar (eds.), *Progenitor Cells: Methods and Protocols*, Methods in Molecular Biology, vol. 2029, https://doi.org/10.1007/978-1-4939-9631-5_18,

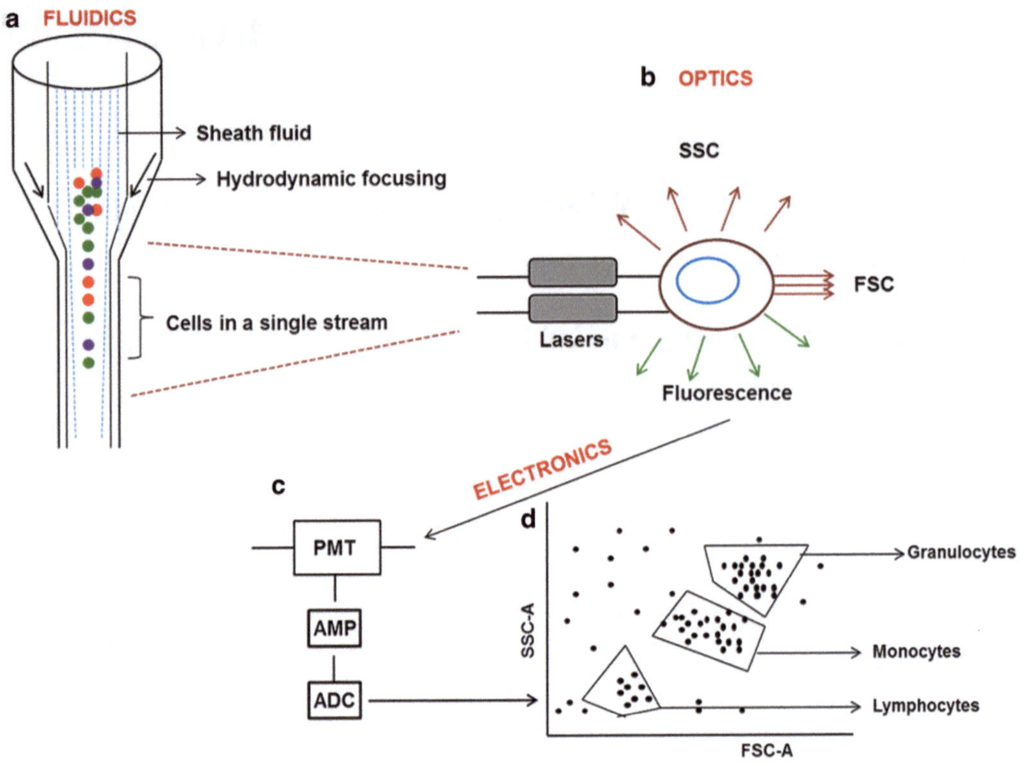

Fig. 1 Basic principle of flow cytometer—a schematic representation: (**a**) The hydrodynamic focusing streamlines cell population into a single stream (fluidics) so that one cell passes through the laser (**b**—optics) at a time. The light signals are then converted into electric signals (**c**—electronics) and can be processed using software to obtain the characteristic dot plot profile (**d**) of the cell population of interest. The FSC shows relative size of the cells and can be gated to different subsets as shown. SSC-A reflects the degree of internal complexity or granularity of the cells (e.g., granulocytes in the top and monocytes in the middle and lymphocytes in the bottom)

cells [2]. Additionally, fluorescent activated cell sorting (FACS) can be carried out to isolate a specific cell population of interest provided they are stained with fluorochrome-tagged antibodies. The third component ***electronics*** converts the light signals into electric signals that can be processed by a computer to facilitate the post-acquisition process. A simplified overview of this process is presented in Fig. 1.

Flow cytometer-based separation of cells is very crucial for cell biology; hematopoietic stem cell research, in particular, has greatly benefited from the cutting-edge technologies associated with the evolution of flow cytometers. Bone marrow niches serve as a pool of adult hematopoiesis that involves differentiation of multipotent cells into committed progenitors with restricted renewal at each step of differentiation [3]. Hematopoietic stem/progenitor cells (HSPCs) are the best-characterized adult stem cells to date, mainly due to the availability of cell surface markers with established

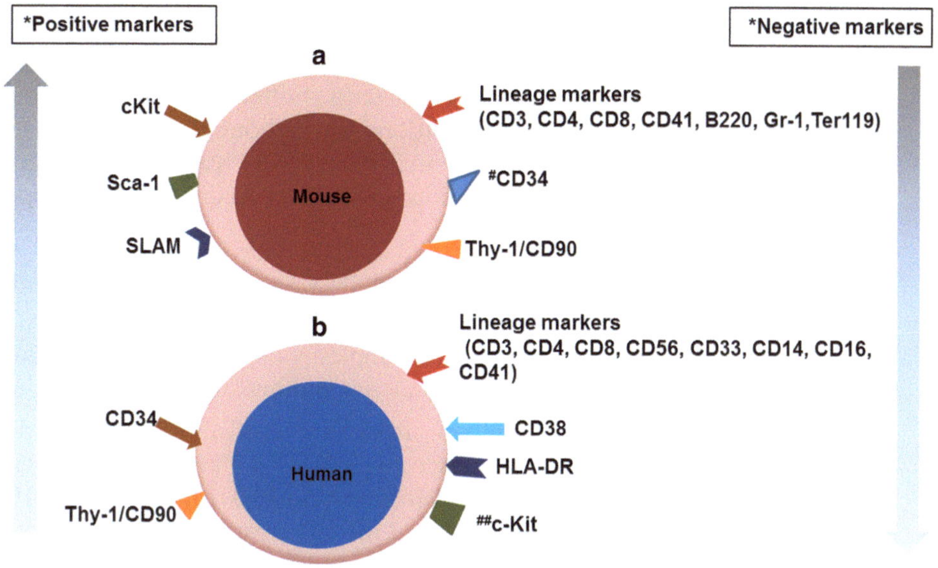

Fig. 2 Phenotypic characterization of hematopoietic progenitors: Hematopoietic progenitors can be isolated by the virtue of the presence and absence of surface antigenic markers. Figure 2 depicts a schematic representation of positive and negative markers that can be used to separate (**a**) murine and (**b**) human HSPCs by flow cytometry based sorting. ∗ The list of markers is not exhaustive. # Murine HSPCs can be $CD34^{+/-}$ based on short-term (SH-HSC, $CD34^{+}$) and long-term (LT-HSC, $CD34^{-}$) repopulating ability, ## Human HSPCs are $c\text{-}Kit^{low/-}$

fluorescent-tagged antibodies, enabling the separation of specific stem cells and their committed progenies. This characterization has ultimately facilitated the studies to device strategies for the efficient expansion of HSPCs in culture and the functional analysis thereafter [4–7]. The isolation of HSPCs is based on a group of cluster of differentiation (CD) antigens that is present on the cell surface combined with a simultaneous state of absence of the lineage-committed markers (Fig. 2). In general, the markers associated with human hematopoietic stem cells are $CD34^{+}CD38^{-}CD45RA^{-}CD90/Thy1^{+}c\text{-}Kit^{low/-}$ and Lin^{-} [8]. Mouse HSPCs are considered $CD34^{low/-}c\text{-}Kit^{+}$ $Sca\text{-}1^{+}$ CD90/ $Thy1^{low}CD38^{+/-}$ and Lin^{-} [9]. Most of the studies in murine HSPCs also utilize signaling lymphocyte activation markers (SLAM family) CD150 and CD48 ($CD150^{+}CD48^{-}$) in combination with c-Kit and Sca-1 markers [10, 11]. However, primitive long-term HSPCs are found to be overlapping between these two categories; hence, many studies utilize a battery of markers, or a combination of them to isolate highly purified HSPCs. The methods in this chapter can be adapted for isolating further subsets of progenitors as well. Below we describe the stepwise protocol to isolate human umbilical cord blood $Lin^{-}CD34^{+}CD38^{-}$ progenitors and murine bone marrow-derived $Lin^{-}c\text{-}Kit^{+}$ $Sca\text{-}1^{+}$ (LSK) progenitors using flow cytometry.

2 Materials

As a rule of thumb, the reagents, buffers, and antibodies are best handled in sterile conditions throughout the procedure for cell sorting. Except for the antibodies, other reagents can be stored at room temperature during the procedure, unless specified otherwise.

1. Umbilical cord blood samples: Collected within 0–24 h post-delivery and stored at 4 °C prior to use (*see* **Note 1**).
2. Mouse bone marrow samples: Collect the bone marrow from tibia and femurs (detailed in Subheading 3).
3. Complete medium: Iscove's Modified Dulbecco's Medium with 10% fetal calf serum (IMDM+10% FCS) and 2× penicillin and streptomycin (Penstrep). Add 10 mL of FCS into 100 mL of IMDM followed by 2 mL of Penstrep (stock usually of 100× concentration). Filter the medium through 0.22-μm sterile filter prior to use.
4. Phosphate buffered saline (PBS): 80 g NaCl, 2 g KCl, 14.4 g Na_2HPO_4, 2.4 g KH_2PO_4. To prepare a 1-liter 10× stock, dissolve the salts in 800 mL of sterile milli-Q water, adjust the pH to 7.4, and make the volume up to 1000 mL. A working stock can be made by diluting the 10× stock (1:10) with milli-Q water, filter-sterilize using 0.22 μ filters.
5. FACS buffer: 0.1% bovine serum albumin (BSA) in PBS (*see* **Note 2**): Dissolve 0.1 g of BSA in 100 mL of PBS. Filter-sterilize using 0.22 μ sterile filters.
6. Hydroxyethyl starch (HES).
7. Ficoll-Paque™ PLUS solution (density 1.007 g/mL).
8. Heparin Solution (Preservative free): Heparin sodium salt (20,000 IU) dissolved in 5 mL of plain IMDM and filtered through a 0.22 μ filter. The sterile solution should be stored at 4 °C.
9. RBC lysis buffer: 0.17 M Tris base, 0.16 M NH_4Cl. For 100 mL stock, 10 mL of Tris base should be combined with 90 mL of NH_4Cl, and pH should be adjusted to 7.2.
10. Turk's solution: 0.01 g crystal violet stain, 3 mL glacial acetic acid, 100 mL milli-Q water. All components are mixed and the solution is stored at room temperature.
11. Trypan blue dye: Dissolve 0.4 g trypan blue in 100 mL of Hanks' Balanced Salt Solution (HBSS) and store at room temperature.
12. Fluorochrome-tagged antibodies: Anti-human lineage cocktail FITC, anti-human CD34-PE (clone 8G12), anti-human

CD38-APC (clone HB7), anti-human Fc receptor blocker, anti-mouse CD16/32 Fc blocker, anti-mouse lineage cocktail FITC, anti-mouse c-Kit PE (clone CB8), anti-mouse Sca-1 APC (clone-E13.161.7), and respective Isotype control (mIgG1 tagged FITC/PE/APC) antibodies. The given fluorochrome conjugates are for representation purposes and can be replaced with other available fluorochromes as per the requirement.

13. Sterile conical tubes (15 mL, 50 mL).
14. 1.5 mL clear sterile microfuge tubes.
15. Polystyrene or polypropylene round bottom FACS tubes.
16. 25 gauge needle.
17. Sterilized surgical scissors.
18. 70% ethanol: Mix 70 mL absolute ethanol and 30 mL milli-Q water.
19. Special equipment: High-speed cell sorter (e.g., BD influx or BD FACS Aria).

3 Methods

3.1 Collection of Cord Blood (CB) Samples

1. CB samples are obtained after full-term deliveries and collected with preinformed consent from the mother and should conform to the ethical compliance of the respective research institute
2. Collect the samples in sterile containers containing preservative-free heparin (40 IU of heparin/mL of blood in plain IMDM) and process within 24 h to isolate mononuclear cells (MNCs).

3.2 Isolation of Mononuclear Cells from Cord Blood

1. In a laminar hood cabinet, dilute larger volume of CB samples (>40 mL) at ratio 1:1 using plain IMDM. Mix gently but thoroughly with the diluent.
2. Mix the diluted sample with HES or Ficoll-Hypaque (*see* **Note 3**).
3. For HES based separation, add 6% HES (4 mL of diluted blood: 1 mL of HES) and dispense up to 25 mL into a 50 mL falcon tube (*see* **Note 3**).
4. Keep the tubes in a slanting position in an incubator for 1 h (37 °C, 5% CO_2) to allow the sedimentation of RBCs.
5. After the incubation, collect the upper leukocyte-rich layer carefully (*see* **Note 3**) into 15 mL conical tubes and centrifuge at 800 × *g* for 5 min at room temperature. Discard the supernatant and wash the pellet using plain IMDM.

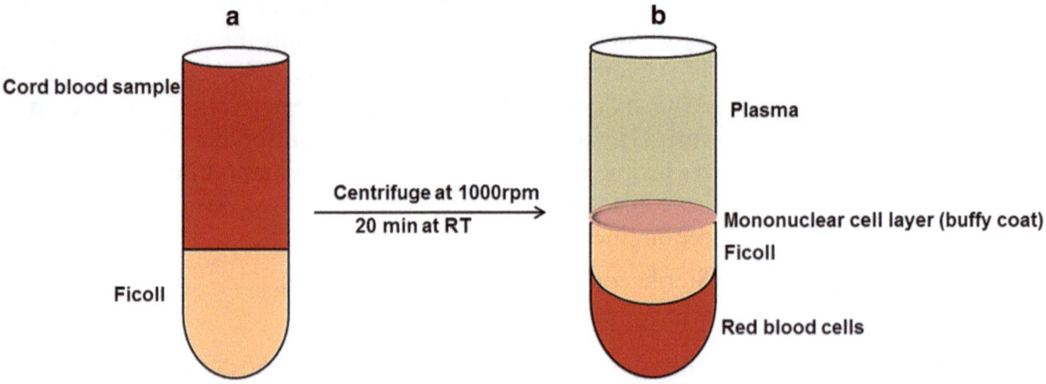

Fig. 3 Ficoll-Hypaque-based separation of MNCs from cord blood samples: (**a**) 8 mL of diluted blood sample (top red) is slowly layered over 2.5 mL of Ficoll solution and centrifuged at room temperature (RT). (**b**) Phase separation after centrifugation showing the interphase (buffy coat) enriched in mononuclear cells. Plasma separates as the top layer, and RBCs sediment at the bottom of the tube

6. For Ficoll-based separation (especially for smaller samples <40 mL, *see* **Note 3**) samples can be diluted with plain IMDM in a 1:1 ratio. Then slowly layer 8 mL of the diluted blood onto 2.5 mL of Ficoll-Hypaque (density 1.007 g/mL) that is predispensed in a 15 mL conical tubes for density gradient separation of MNCs (*see* **Note 4** and Fig. 3).
7. Centrifuge the tubes at 400 × *g* for 20 min at room temperature in a swinging bucket rotor.
8. Discard the plasma and platelets in the top clear layer and carefully aspirate the buffy interface containing MNCs (maximum 1–2 mL from a tube), leaving behind a layer of Ficoll and sedimented RBCs and mature granulocytes. Transfer these MNCs into a fresh 15 mL conical tube containing 5 mL of plain IMDM. Centrifuge at 400 × *g* for 5 min at room temperature and wash the pellet with plain IMDM to remove traces of Ficoll.
9. If the blood samples are freshly collected and processed, RBCs will be efficiently eliminated when subjected to the HES/Ficoll method. Therefore, RBC lysis step is not essential. However for specific requirements that need complete removal of RBCs or in case RBCs are not efficiently removed, follow RBC lysis steps.

3.3 Red Blood Cell Lysis

1. Wash the isolated MNCs using 1× PBS and then resuspend in 5 mL of 1× RBC lysis buffer and incubate in the dark, at room temperature for 15 min.
2. Centrifuge the sample tubes at 400 × *g* or 5 min at room temperature and discard the supernatant containing lysed RBCs. Mix the pellet gently and wash twice using 1× PBS to remove the residual lysis buffer and RBCs.

3.4 Processing of Mouse Bone Marrow Samples

1. Dissect the hind limb long bone from the mice after euthanizing the animal. Adhere to the requirements of handling and euthanizing animals of the ethics committee of respective research Institute. The tibias and femurs are separated and the tissues are cleaned using scissors.
2. After cleaning the bones, make a small cut (or pierce) on both ends. Insert the needle and flush the bones into 5 mL of complete medium using a 25 G needle and syringe (*see* **Note 5**). The bones will turn transparent as the bone marrow is being flushed out.
3. Mix the cells gently using a pipette and filter using 40 μm sterile filter into a fresh tube in order to avoid clumping.
4. Centrifuge the cell suspension at 400 × *g* for 3 min at room temperature.
5. Wash once with 1× PBS and repeat the centrifugation.
6. Proceed to Subheading 3.3 if required.

3.5 Sample Preparation and Sorting

Resuspend the cells (human cord blood or mouse bone marrow MNCs) in sterile FACS buffer and carry out nucleated cell count using Turk's solution/trypan blue and a hemocytometer. The samples are now ready to be stained using a panel of fluorochrome-tagged antibodies prior to cell sorting.

1. Enumerate the nucleated cells using Turk's solution (*see* **Note 6**) and aliquot 5×10^5–1×10^6 MNCs into each tube. Additionally a viability count using trypan blue can be done to get a viable cell count of the starting population. Addition of vital dyes like propidium iodide/7-AAD/Hoechst is recommended after the incubation with antibodies, in order to gate the live cells during sort setup.
2. Add Fc blocker antibody at a concentration of 1–2 μg per 1×10^6 MNCs in to each tube and incubate for 10–15 min at 4 °C.
3. Do not wash the cell suspension, and add lineage depletion cocktail and respective isotype controls (*see* **Note 7** *for controls and compensation setup*) to murine and human MNC samples.
4. Add an optimized concentration of antibodies (including anti-human lineage cocktail FITC, anti-human CD34-PE, and anti-human CD38-APC) to the human MNC cell suspension and incubate at 4 °C for 40 min in dark, with intermittent flicking of the tubes to maximize the antibody contact with the cells.
5. Similarly, into murine bone marrow MNCs, add lineage cocktail antibody (FITC-tagged) followed by c-Kit-PE and Sca-1 APC antibodies and incubate as in **step 3** above. Fluorescent tags can vary according to the experimental need.

6. After incubation, centrifuge the samples at 500 × *g* for 3 min at 4 °C and wash twice using FACS buffer. Filter through a 40 μm cell strainer, dispense into FACS tubes, and keep at 4 °C until sorter is set up.
7. Once the staining of the cells is completed, the performance of the sorter is assessed and is made optimal for the current sorting. For example, for BD FACS Aria, CS&T beads calibration is performed followed by setting up gating strategies using unstained cells, positive control, Fluorescent Minus One (FMO) control, and isotype controls (*see* **Note 8**) using BD FacsDiva software.
8. If there is no template yet, open a new experiment and once the gating strategies are defined (FSC-A, SSC-A, and fluorescent filters, respectively), collection tubes, which are precoated with heat-inactivated FCS (*see* **Note 9**) and containing sterile complete medium (>3 mL), are placed in the sample collection port.
9. For cord blood MNCs, the dot plots and the gated population hierarchy are described below (a schematic representation shown in Fig. 4).
10. Plot 1 FSC-A vs. SSC-A. Gate live cells using free-hand gating tool avoiding debris—this is named as Gate P1 (Fig. 4a). Alternatively, addition of a viability dye like propidium iodide/7-AAD or Hoechst is highly recommended.
11. Plot 2 is FSC-H vs. FSC-W and Plot 3 is SSC-H vs. SSC-W. Apply Gate P1(viability gate) to plots 2 and 3.
12. Now gate the lower population in Plot 2 (Fig. 4b) avoiding cell aggregates or doublets as Gate P2. Derive a similar gate also in Plot 3 (note that this is derived from gate P2). This population is now Gate P3 (Fig. 4c). This gating is important to avoid clumps/aggregates and to get single cells.
13. Plot 4 is SSC vs. Lin-FITC, and plot 5 is CD34-PE vs. CD38-APC.
14. In the next step apply Gate P3 into plot 4 and plot 5.
15. Now in plot 4, gate the cells that are negative for FITC fluorescence. This is now P4 (Lin^- population)—Fig. 4d.
16. Apply Gate P4 on plot 5 and select populations that are PE^+ ($CD34^+$) and APC^-($CD38^-$) and this will be the final sorting gate P5—Fig. 4e.
17. Sort all populations within Gate P5 ($Lin^-CD34^+ CD38^-$) into the collection tube.
18. Centrifuge the sorted fraction containing $Lin^-CD34^+CD38^-$ progenitors and use for further experiments accordingly.

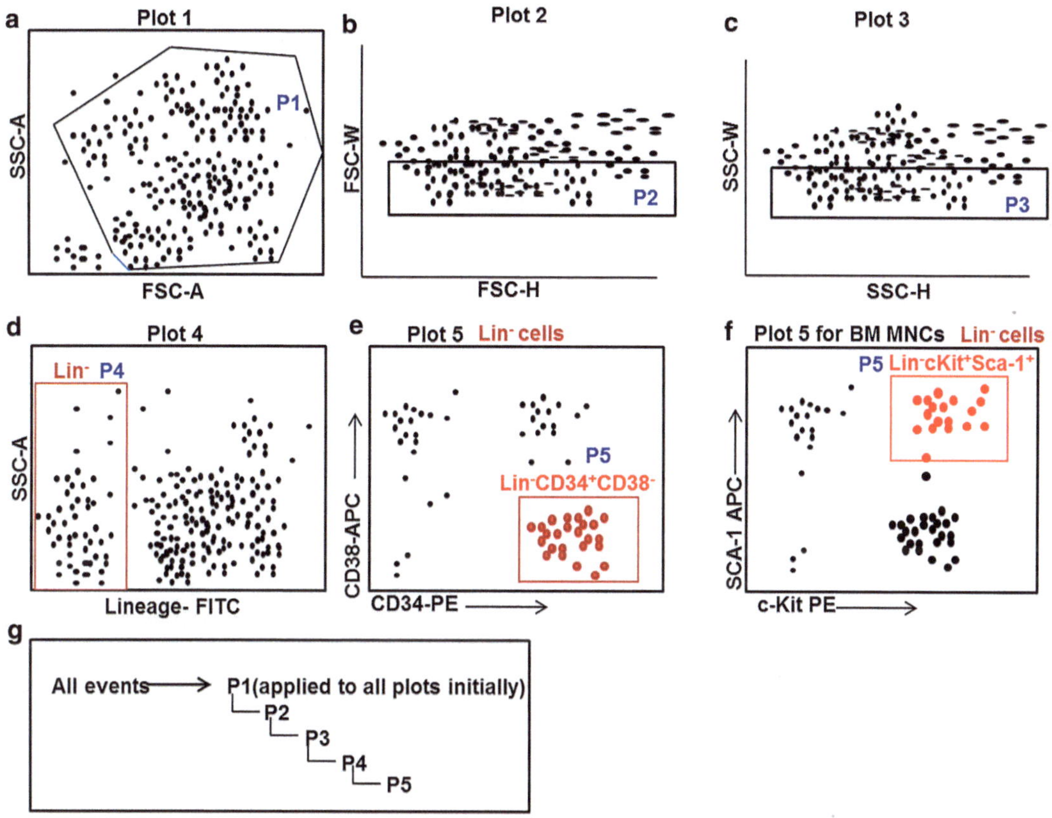

Fig. 4 Schematic representation of gating strategy for sorting: (**a**) Plot 1 is FSC-A vs. SSC-A and gated for live cells in Gate P1 (**b**) FSC-W vs. FSC-H derived from P1. P2 represents the cells without aggregates and should be applied to (**c**) plot 3, which is SSC-H vs. SSC-W. (**d**) Plot 4 is derived from P3 gated cells and represents SSC-A vs. the major population, which is lineage-FITC. Gate P4 is further selected based on the absence of fluorescence (Lin^- cells). Gate P4 is applied to Plot 5, and the sorting Gate P5 is defined in CB MNCs (**e**) and in murine BM MNCs (**f**) as shown. (**g**) The population hierarchy for the derivation of the cells is shown

19. For murine bone marrow MNCs, the four dot plots and the gated population hierarchy should be identical as in **steps 10–15** except for the threshold adjustments required in the FSC and SSC to delineate the cell population and the differences in the CD markers used here for demonstration.
20. Apply Gate P4 on plot 5 and select populations that are double positive PE^+ (c-Kit+) and APC^+(Sca-1^+) and this will be the sorting gate P5—Fig. 4f.
21. Sort all populations within gate P5 (Lin^-c-Kit^+ $Sca1^+$) into the collection tube.
22. Centrifuge the sorted fraction containing LSK progenitors and use for further experiments accordingly.

4 Notes

1. Cord blood samples are best when processed within 24 h of collection. Proper mixing of the collected samples with heparin solution is critical to prevent clotting. Samples should be mixed gently, but thoroughly with the heparin solution and should be shipped on ice packs. Care should be taken to frequently change the surgical gloves and masks in between the processing steps to avoid contamination.
2. FACS buffer composition is 0.1% BSA dissolved in 1× PBS. Addition of BSA helps to reduce nonspecific binding and maintain cell survival.
3. Hydroxyethyl starch or HES has been widely used to sediment RBCs [12, 13]. Contact with HES increases the specific gravity of RBCs, resulting in their efficient separation and leaving behind an MNC-enriched straw-colored supernatant. However, if the blood sample is hemolyzed during collection, in spite of sedimented RBCs, the supernatant will be transparent red in color and such samples should be discarded. Alternatively, both HES and Ficoll-Hypaque can be used for MNC separation irrespective of cord blood volume.
4. It is very important to layer the blood very carefully without mixing up with the Ficoll solution in the tube and is best accomplished when the tube is held in an angle. Swinging rotor centrifuge (with "brake-off" option) is critical. After layering of the blood, the Ficoll and the blood sample should be visible as two separate layers as shown in Fig. 3a. After centrifugation, the RBCs and the granulocytes that have higher density tend to settle in the bottom. Proper dilution is necessary to get a clear separation of MNCs from the blood sample at the bottom versus the plasma in the top layer. The MNCs will appear as a buffy coat (most cases in a ring shape) in the interphase.
5. Care should be taken while dissecting the femur to include the femur head and not to cut it off to obtain a good yield of progenitor cells. Cleaning of the bones can be made easier by removing the tissues with delicate task wipes like Kimwipes followed by rinsing the bones in sterile PBS or plain IMDM. Additionally, the best method to flush out bone marrow is to prefill the syringe (half volume) with plain IMDM and flush the bones multiple times until the bone marrow is completely flushed out.
6. Crystal violet-based method is used to enumerate the nucleated cells. The MNCs are diluted in a 1:10 ratio with crystal violet (Turk's) solution and counted using a hemocytometer. For trypan blue dye-exclusion method, dilute the MNCs and

0.4% trypan blue in a 1:1 ratio. The viable cells will exclude the dye while the dead cells will take up the blue dye when observed under a hemocytometer.

7. It is a good option to use a strong fluorochrome conjugate (like PE or PE cy5) for rare population of interest. To avoid autofluorescence, an aliquot of unstained cells can serve as a control sample (unstained sample). However, when performing multicolor flow cytometry, single stained samples are important to determine the spectral overlap between fluorophores (especially PE and FITC). Additionally, fluorescence minus one controls (FMO) are important to set gates. This setup includes cells with all fluorophores added, except one (e.g., FITC-FMO, PE-FMO, and APC-FMO). Tubes of isotype control to eliminate any background staining/nonspecific staining of the antibodies are recommended. Biological controls include negative controls (cells that are not of hematopoietic origin and hence do not express the antigens of interest). Antibody concentration has to be defined empirically prior to staining for sorting.
8. An experienced operator is usually available in most research institutions to train or help set up the flow cytometer and to assist in sorting. It is crucial to maintain the sterility and the calibration of the sorter in a systematic manner.
9. Precoating of the collection tube using sterile BSA or FBS overnight at 4 °C (ensure that the entire wall of the tube is filled with serum) prevents the cells from sticking to the wall of the tube during collection, thus preventing cell loss. It is also important to filter the cells (40 μm sterile filter) just before sorting and to maintain the FACS tubes on ice in order to prevent cell clumping.

References

1. Robinson JP, Roederer M (2015) HISTORY OF SCIENCE. Flow cytometry strikes gold. Science 350(6262):739–740
2. Brown M, Wittwer C (2000) Flow cytometry: principles and clinical applications in hematology. Clin Chem 46(8 Pt 2):1221–1229
3. Morrison SJ, Scadden DT (2014) The bone marrow niche for haematopoietic stem cells. Nature 505(7483):327–334
4. Sangeetha VM, Kadekar D, Kale VP, Limaye LS (2012) Pharmacological inhibition of caspase and calpain proteases: a novel strategy to enhance the homing responses of cord blood HSPCs during expansion. PLoS One 7(1): e29383
5. Sangeetha VM, Kale VP, Limaye LS (2010) Expansion of cord blood CD34 cells in presence of zVADfmk and zLLYfmk improved their in vitro functionality and in vivo engraftment in NOD/SCID mouse. PLoS One 5(8): e12221
6. Ko KH, Nordon R, O'Brien TA, Symonds G, Dolnikov A (2017) Ex vivo expansion of hematopoietic stem cells to improve engraftment in stem cell transplantation. Methods Mol Biol 1524:301–311
7. Guo B, Huang X, Lee MR, Lee SA, Broxmeyer HE (2018) Antagonism of PPAR-gamma signaling expands human hematopoietic stem and progenitor cells by enhancing glycolysis. Nat Med 24(3):360–367

8. Bhatia M, Wang JC, Kapp U, Bonnet D, Dick JE (1997) Purification of primitive human hematopoietic cells capable of repopulating immune-deficient mice. Proc Natl Acad Sci U S A 94(10):5320–5325
9. Zhao Y, Lin Y, Zhan Y, Yang G, Louie J, Harrison DE, Anderson WF (2000) Murine hematopoietic stem cell characterization and its regulation in BM transplantation. Blood 96 (9):3016–3022
10. Yilmaz OH, Kiel MJ, Morrison SJ (2006) SLAM family markers are conserved among hematopoietic stem cells from old and reconstituted mice and markedly increase their purity. Blood 107(3):924–930. https://doi.org/10.1182/blood-2005-05-2140
11. Kiel MJ, Yilmaz OH, Iwashita T, Terhorst C, Morrison SJ (2005) SLAM family receptors distinguish hematopoietic stem and progenitor cells and reveal endothelial niches for stem cells. Cell 121(7):1109–1121
12. Solves P, Mirabet V, Planelles D, Blasco I, Perales A, Carbonell-Uberos F, Soler MA, Roig R (2005) Red blood cell depletion with a semiautomated system or hydroxyethyl starch sedimentation for routine cord blood banking: a comparative study. Transfusion 45 (6):867–873
13. Lee JH, Leitman SF, Klein HG (1995) A controlled comparison of the efficacy of hetastarch and pentastarch in granulocyte collections by centrifugal leukapheresis. Blood 86 (12):4662–4666

Chapter 19

Isolation of Epidermal Progenitor Cells from Rat Tympanic Membrane

Lawrence J. Liew, Allen Y. Wang, and Rodney J. Dilley

Abstract

The eardrum is an important structural component for hearing, but it is delicate and subject to traumatic injury and disease. Healing mechanisms are activated after injury but sometimes healing fails and chronic perforations develop, requiring surgical intervention. To model the wound healing responses we established a simple method for isolating keratinocytes and progenitors from individual eardrums. The central region of the eardrum contains epidermal proliferative centers that produce keratinocytes which migrate to cover the eardrum surface. We dissected out the central region and explanted it to the plastic membrane of a culture well insert. Epidermal cells grew from the explant onto the surface of the insert membrane. The cells could be serially harvested and passaged for continuous culture and characterization. Magnetic immunoseparation methods were used to enrich for epithelial cells with stem cell-like characteristics. Proliferation and migration in vitro was demonstrated, and the cells were shown suitable for tissue engineering applications.

Key words Epidermal stem cells, Tympanic membrane, Explant culture, Magnetic sorting, Tissue engineering

1 Introduction

The tympanic membrane (TM) is a thin, translucent membrane in the middle ear which plays important roles in sound transmission and protecting the middle ear from infection. Injury activates robust wound healing mechanisms but in some patients chronic perforations occur, requiring surgical grafting intervention to restore membrane integrity. Studies to understand TM wound repair and to optimize graft materials are ongoing.

The TM is a conical-shaped fibrocellular sheet nearly 1 cm in diameter and 50–100 μm thick. It has a trilaminar structure, covered by keratinized stratified squamous epithelium on the lateral (ear canal) side, a fibrous middle layer, and a mucosal epithelium on the medial (middle ear) side [1, 2]. Pars tensa is the thin membrane which vibrates in response to sound and forms the major structural

Mugdha V. Joglekar and Anandwardhan A. Hardikar (eds.), *Progenitor Cells: Methods and Protocols*, Methods in Molecular Biology, vol. 2029, https://doi.org/10.1007/978-1-4939-9631-5_19, © Springer Science+Business Media, LLC, part of Springer Nature 2019

component of the TM. A thickened ligament-like structure, the annulus, lies at the periphery of the TM, suspending it from the bony ear canal. At the center (the umbo) of the TM the membrane is attached to the malleus bone medially, and there is an epidermal proliferative center recognized for substantial keratinocyte production [3–6].

Putative epidermal stem cells expressing markers such as integrin α6/β1 and cytokeratin 19 have been found in both the umbo and annulus regions of TM [4, 6, 7]; physical trauma appears to play a role in the activation of these proliferative areas, and a role in wound healing has been proposed [3, 6, 8]. The study of progenitor cells of the TM is important as it provides insight into the development of TM-related pathologies such as chronic perforations and cholesteatoma. Cultivation of human TM-derived keratinocytes has been reported [9] and proven to be useful in evaluating scaffold materials [10, 11]. However, it remains unclear whether these terminally differentiated cells are suitable models to understand the biology of TM.

We describe here a simple method to explant TM umbo to a supporting membrane where migrating keratinocytes can be isolated from the culture surface and enriched for progenitor cells.

2 Materials

2.1 Instruments and Supplies

1. 12 cm Adson tissue forceps.
2. 13 cm curved scissors.
3. 17 cm Mayo curved scissors.
4. Oval ear speculum.
5. #21 blade steel scalpel.
6. Microscope.
7. 100 mm diameter cell culture dish.
8. 50 mL conical tubes.
9. Tissue culture flask T-75.
10. 35 mm diameter cell culture dish.
11. 5 mL syringe.
12. 21 gauge needle.
13. Benchtop centrifuge.
14. Tissue culture plate, 12-well flat bottom.
15. Culture well inserts.
16. Hemocytometer.

2.2 Animal Surgery Supplies

1. Isoflurane.
2. Lethabarb (pentobarbitone).
3. Inhalation anesthesia system.
4. Absorbent surgical pad.
5. Oto-microscope.

2.3 Cell Culture Reagents

1. Cell culture setting, including cell culture incubator (37 °C with 95% air/5% CO_2, humid atmosphere), cell culture hood, sterile pipettes, and lab ware.
2. Keratinocyte serum-free medium (KSFM).
3. Fetal bovine serum (FBS).
4. Dulbecco's modified Eagle's medium/Ham's F-12 Nutrient Medium 1:1.
5. Antibiotic–antimycotic (A/A).
6. Dispase II solution (stock solution). *4% Dispase II solution can be prepared in PBS or culture medium by diluting the stock 1:25.*
7. Growth medium: KSFM +10% FBS + 5% A/A.
8. Cell dissociation enzyme (such as TrypLE™ Express).
9. Phosphate buffered saline, pH 7.4.
10. Dimethyl sulfoxide (DMSO).
11. Cryopreservation container.

2.4 MACS Reagents and Equipment

1. Rabbit anti-rat CK 19 antibody.
2. Wash Buffer: Hank's Balanced Salt Solution +0.2% bovine serum albumin (BSA).
3. Anti-rabbit IgG microbeads (Miltenyi Biotec).
4. Mini MACS starter kit with MS column (Miltenyi Biotec).

3 Methods

3.1 Rat Primary Cell Isolation

Tympanic membrane cells can be isolated from adult rats. We prefer using animals not more than 12 weeks old as older animals tend to have stronger bones that are difficult to dissect.

1. Under suitable general anesthesia regime (e.g., gaseous induction: 4% isoflurane 4 L/min O_2; maintenance: 2% isoflurane 2 L/min O_2), examine all animals using ear speculum and microscope to exclude any with ear canal or middle ear infection before surgery.
2. Euthanize animal via intra-abdominal injection of Lethabarb according to approved protocol and lay animal on an absorbent surgical pad.

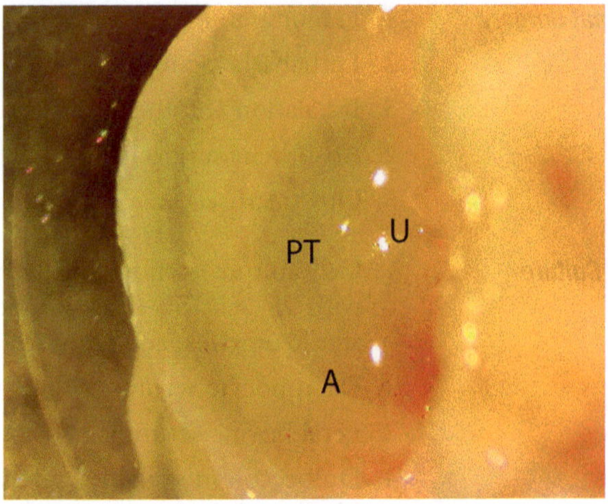

Fig. 1 A tissue block of temporal bone bulla containing the tympanic membrane with annulus (A), pars tensa (PT), and umbo region (U)

3. Make an incision across the nape of the neck and remove the external ears at the osseocartilaginous junction.
4. Decapitate the animal through the neck in axial plane.
5. Incise through the superior scalp/skull along the sagittal plane.
6. Dissect with fine tip dissecting scissors the whole tympanic bulla from the temporal bone.
7. Crack the bony bulla from the medial side and carefully trim off excess tissue to create a block of bulla tissue containing the tympanic membrane with annulus and handle of malleus (Fig. 1).
8. Place the tissue in KSFM at 4 °C for transport or short-term storage (<24 h) (*see* **Note 1**).
9. Treat tissue with 4% Dispase II solution overnight at 4 °C (*see* **Note 2**).
10. Aspirate Dispase II solution and rinse tissues gently with PBS.
11. Carefully peel the TM from the bulla using a pair of forceps (*see* **Note 3**) (Fig. 2).
12. Dissect the umbo region of the TM using a pair of forceps and scalpel blade.
13. Make a 1 mm incision in the membrane (0.4 μm pore size) of a culture well insert and feed the malleus bone attached to the umbo tissue into the incision (*see* **Note 4**). The epidermal surface of the umbo faces vertically and is positioned close to the upper surface of the culture well insert supporting membrane (Fig. 3).
14. Place the culture well insert into a 12-well plate and top up with 1 mL of growth medium.
15. Replenish with growth medium twice weekly.

Fig. 2 The tympanic membrane after peeling from the bulla bone after Dispase II digestion

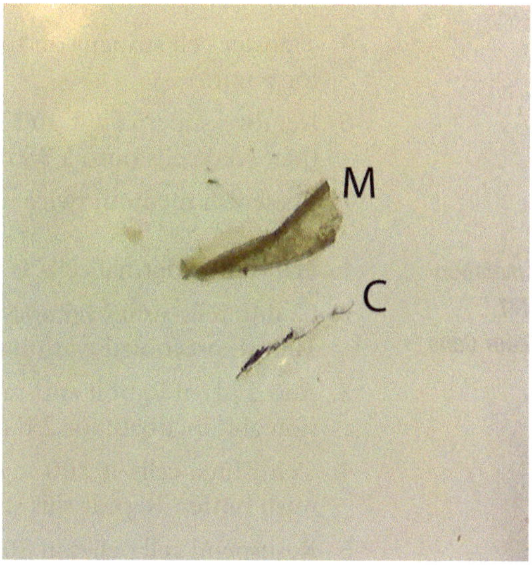

Fig. 3 The malleus bone (M) attached to the umbo tissue is placed on the culture well insert membrane and a 1 mm incision is cut in the membrane (C)

3.2 Subculture of TM Primary Cells

1. Subculture cells after 10 days in culture or when the cells approach the periphery of the culture well insert (Fig. 4).
2. Aspirate growth medium and rinse twice with PBS.
3. Add 1 mL of prewarmed TrypLE™ Express and incubate at 37 °C for 15–20 min until cells are detached.
4. Add 2 mL of growth medium to neutralize TrypLE™ Express.

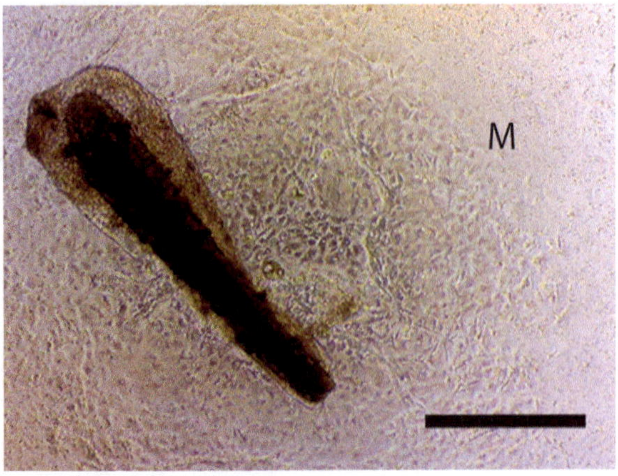

Fig. 4 Phase-contrast micrograph showing a tympanic membrane explant in place on the surface of the culture well insert membrane. The cells have migrated about 1 mm from the explant onto the membrane surface. The migrating front (M) can be monitored as it approaches the edge of the culture surface before subculturing. Scale bar = 500 μm

5. Transfer cell suspension to a conical tube and spin at 300 × *g* for 5 min.
6. Remove supernatant and resuspend cells in growth medium, then seed cells onto a 35 mm cell culture dish.
7. Replenish medium twice weekly until ~80% confluent.

3.3 MACS Isolation of Putative TM Epidermal Stem Cells

1. Harvest epidermal cells as previously described.
2. Count cells using hemocytometer and resuspend 1×10^7 in 100 μL precooled wash buffer.
3. Add 2 μL of Rabbit anti-rat CK 19 antibody to the cell suspension and incubate for 2 h on ice (*see* **Note 5**).
4. Centrifuge cells at 200 × *g* for 5 min and resuspend cells with wash buffer. Repeat this step one more time.
5. Resuspend cell pellet in 80 μL of wash buffer and add 20 μL of anti-rabbit IgG microbeads. Incubate for 40 min on ice.
6. Centrifuge cells at 200 × *g* for 5 min and resuspend cells with wash buffer. Repeat this step one more time.
7. Resuspend cell pellet in 500 μL of wash buffer.
8. Place a new MS column onto a magnetic separator and rinse column with 500 μL of wash buffer.
9. Apply cell suspension to the column. Collect unlabeled cells which pass through the column.
10. Wash column three times with wash buffer.

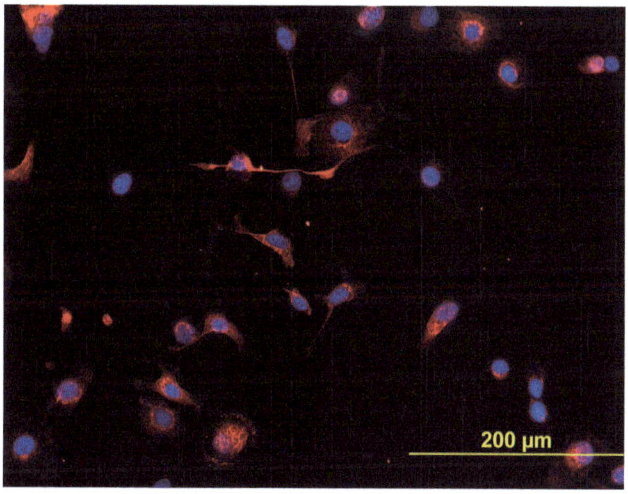

Fig. 5 Fluorescence micrograph showing cytokeratin 19 immunostaining of tympanic membrane epidermal cells after enrichment by immunomagnetic separation. Scale bar = 200 μm

11. Remove column from the magnetic separator, immediately flush out the magnetically labeled cells with 1 mL of wash buffer using the plunger supplied.
12. The magnetically labeled cells can be further expanded in culture (*see* **Note 6**) or fixed for immunofluorescent staining (Fig. 5).

3.4 Cryopreservation of Epidermal Cells

1. Resuspend 5×10^5 cells in growth medium containing 10% DMSO and add 1 mL of the suspension to cryopreservation vials.
2. Cryopreserve cells in a freezing container at −80 °C.
3. Transfer cells to liquid nitrogen for long-term storage (*see* **Note 7**).

4 Notes

1. Rat TM should be fixed in 10% formalin immediately (no need to keep in culture medium) if cell isolation is not needed.
2. 5 mL of 4% Dispase II solution is usually sufficient for two rat TM.
3. Tissue explants from the flat part of the rat TM (pars tensa) can be adhered directly to collagen I-coated cell culture dishes for a more traditional explant culture.

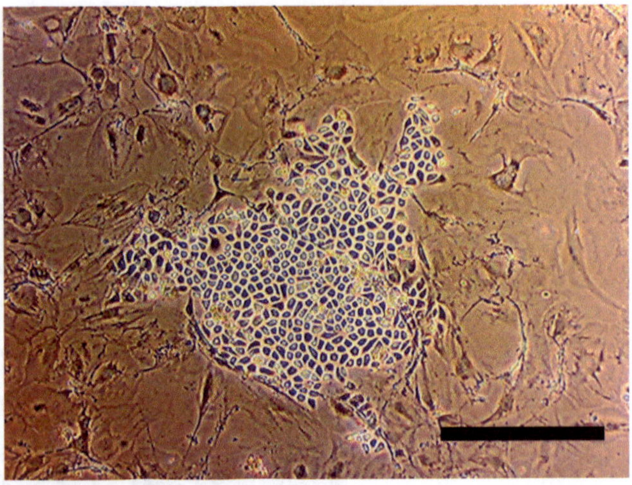

Fig. 6 Epidermal stem cells colonies grow well in coculture with feeder cells (mouse embryonic fibroblasts; Mitomycin-c inactivated). Scale bar = 500 μm

4. Umbo grafts are partially inserted through the incision to prevent them from floating off when medium was added. Grafts are unable to adhere to membrane surface.
5. CK 19 can be replaced by other putative stem cell markers.
6. Enriched epidermal stem cells grow best with feeder cells such as mouse embryonic fibroblasts (Mitomycin-c inactivated) (Fig. 6)
7. Cells have low viability if stored long-term at −80 °C.

Acknowledgments

This work was in part funded by the Garnett Passe and Rodney Williams Memorial Foundation and National Health and Medical Research Council of Australia. We thank KARL STORZ Endoscopy Australia Pty. Ltd. for the supply of some specialist surgical instruments.

References

1. Lim DJ (1968) Tympanic membrane. Electron microscopic observation. I: pars tensa. Acta Otolaryngol 66:181–198
2. Lim DJ (1968) Tympanic membrane. II. Pars flaccida. Acta Otolaryngol 66:515–532
3. Koba R (1995) Epidermal cell migration and healing of the tympanic membrane: an immunohistochemical study of cell proliferation using bromodeoxyuridine labeling. Ann Otol Rhinol Laryngol 104:218–225
4. Knutsson J, von Unge M, Rask-Andersen H (2011) Localization of progenitor/stem cells in the human tympanic membrane. Audiol Neurootol 16:263–269
5. Wang WQ, Wang ZM, Chi FL (2004) Spontaneous healing of various tympanic membrane perforations in the rat. Acta Otolaryngol 124:1141–1144
6. Liew LJ, Chen LQ, Wang AY, von Unge M, Atlas M, Dilley RJ (2018) Tympanic membrane

derived stem cell-like cultures for tissue regeneration. Stem Cells Dev 27:649–657

7. Kim SW, Kim J, Seonwoo H, Jang K-J, Kim YJ, Lim HJ, Lim K-T, Tian C, Chung JH, Choung Y-H (2015) Latent progenitor cells as potential regulators for tympanic membrane regeneration. Sci Rep 5:11542
8. Wang AY, Shen Y, Liew LJ, Wang JT, von Unge M, Atlas MD, Dilley RJ (2015) Searching for a rat model of chronic tympanic membrane perforation: healing delayed by mitomycin C/dexamethasone but not paper implantation or iterative myringotomy. Int J Pediatr Otorhinolaryngol 79:1240–1247
9. Redmond SL, Levin B, Heel KA, Atlas MD, Marano RJ (2011) Phenotypic and genotypic profile of human tympanic membrane derived cultured cells. J Mol Histol 42:15–25
10. Levin B, Redmond SL, Rajkhowa R, Eikelboom RH, Atlas MD, Marano RJ (2013) Utilising silk fibroin membranes as scaffolds for the growth of tympanic membrane keratinocytes, and application to myringoplasty surgery. J Laryngol Otol 127(Suppl 1): S13–S20
11. Danti S, Mota C, D'alessandro D, Trombi L, Ricci C, Redmond SL, De Vito A, Pini R, Dilley RJ, Moroni L (2015) Tissue engineering of the tympanic membrane using electrospun PEOT/PBT copolymer scaffolds: a morphological in vitro study. Hearing Balance Commun 13:133–147

Chapter 20

Measurement of Store-Operated Calcium Entry in Human Neural Cells: From Precursors to Differentiated Neurons

Renjitha Gopurappilly, Bipan Kumar Deb, Pragnya Chakraborty, and Gaiti Hasan

Abstract

Calcium imaging in an ex-vivo setup is used to understand the calcium status of isolated cells or tissue. In this chapter we explain the use of the ratiometric chemical indicator Fura-2 which can be loaded into isolated cells in the form of lipophilic acetomethyl (AM) esters. Fura-2 is a combination of calcium chelator and fluorophore, and can be used with dual wavelength excitation (340/380 nm) for quantitative calcium concentrations. The cells can then be viewed using a fluorescence microscope and captured by a CCD camera. We specifically discuss the technique involved in understanding the endoplasmic reticulum (ER)-driven store-operated calcium entry (SOCE) in human neural precursors (NPCs) and spontaneously differentiated neurons derived from a pluripotent human embryonic stem cell (hESC) line. The derivation of neural precursors from stem cells and their subsequent spontaneous neural differentiation is also explained. The method can be used for various non-excitable and excitable cell types including neurons, be it freshly isolated, from frozen vials, or derived from different stem cell lines.

Key words Fura-2 AM, Ratiometric calcium indicators, Thapsigargin, SOCE, Progenitors, Calcium mapping, Calcium indicators, Stem cells, Neural differentiation

1 Introduction

Store-operated calcium entry (SOCE) is based on the interaction of the calcium sensor STIM proteins in the ER [1, 2] and Orai, the pore-forming subunit of the calcium-release activated calcium (CRAC) channels in the plasma membrane [3–5]. SOCE can be monitored by imaging of cultured cells after the depletion of ER stores with thapsigargin in very low external calcium and then adding back physiological levels of calcium [6]. Significant roles for intracellular calcium stores and SOCE have been suggested in neurogenesis and neural development [7]. SOCE has been investigated in mouse [8] and human neural precursors [9] to understand

Renjitha Gopurappilly, Bipan Kumar Deb, and Pragnya Chakraborty contributed equally to this work.

Mugdha V. Joglekar and Anandwardhan A. Hardikar (eds.), *Progenitor Cells: Methods and Protocols*, Methods in Molecular Biology, vol. 2029, https://doi.org/10.1007/978-1-4939-9631-5_20, © Springer Science+Business Media, LLC, part of Springer Nature 2019

its role in neural development. SOCE has also been characterized in fly and rodent neurons [6, 10–12]. The development of highly sensitive buffers and fluorescent calcium indicators derived from the parent compound 1,2-bis (o-aminophenoxy) ethane-*N,N,N′,N′*-tetraacetic acid (BAPTA) [13] has made calcium imaging a valuable technique to understand calcium handling at a single-cell level. Among the first generation of fluorescent calcium indicators, Fura-2 [14] is considered superior as it shows much stronger fluorescence, wavelength shifts upon calcium binding, and better selectivity against magnesium and heavy metals. Fura-2 is excited at 340 and 380 nm of light, and emits at 510 nm of light. The use of the ratio automatically cancels out confounding variables, such as variable dye concentration and cell thickness. We have described the derivation of neural precursors (Fig. 1), imaging of SOCE in them with Fura-2 (Fig. 2) and also spontaneously differentiated neurons and their SOCE dynamics (Figs. 3 and 4).

2 Materials

2.1 Equipment and Supplies

1. Biosafety cabinet.
2. Centrifuge.
3. Stereomicroscope.
4. Microscope fitted with high speed imaging and time-lapse acquisition mode in software (e.g., Andor iXON 897E EMCCD camera and AndoriQ 2.4.2 imaging software).
5. 37°C, 5% CO_2 incubator
6. 37°C water bath
7. Cryogenic handling gloves and eye protection.
8. Cryovials.
9. Isopropanol freezing containers (Mr. Frosty).
10. Refrigerator.
11. Liquid nitrogen storage tank.
12. −20°C and −80°C freezers
13. Coverslips, 6-well plates, 100 mm non-TC treated bacteriological dishes.
14. 5 mL and 10 mL sterile serological pipettes, 2 mL aspirating pipette.
15. 15 mL conical tubes, 50 mL conical tubes.
16. 1.5 mL eppendorf tubes.
17. 70% ethanol.
18. Marker pens.

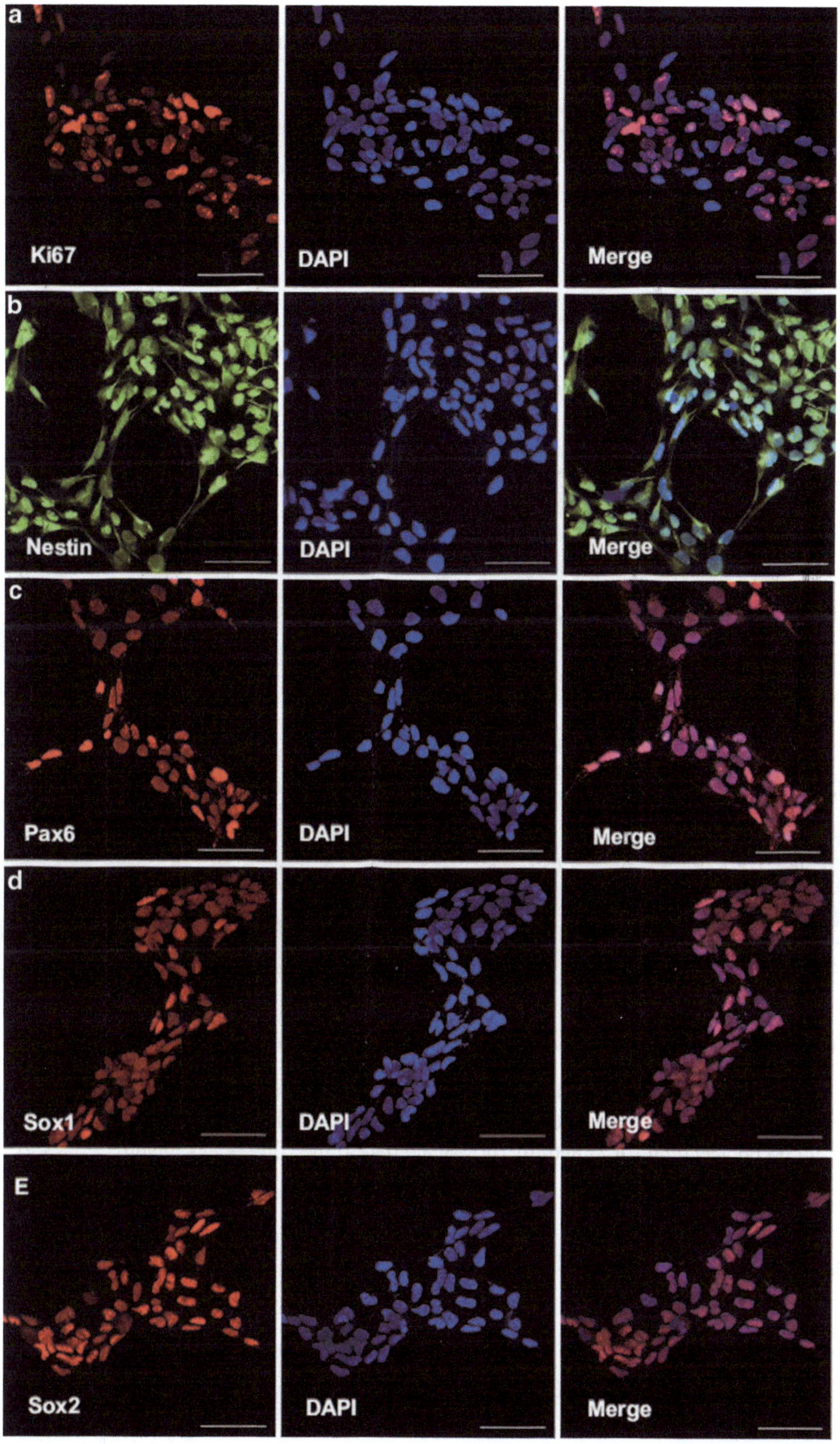

Fig. 1 Characterization of NPCs derived from hESC. Immunostaining of NPCs with (**a**) Ki67, a proliferation marker; (**b**) Nestin (**c**) Pax6 (**d**) Sox1 and (**e**) Sox2, which are classical NPC markers. Scale bar 50 μM

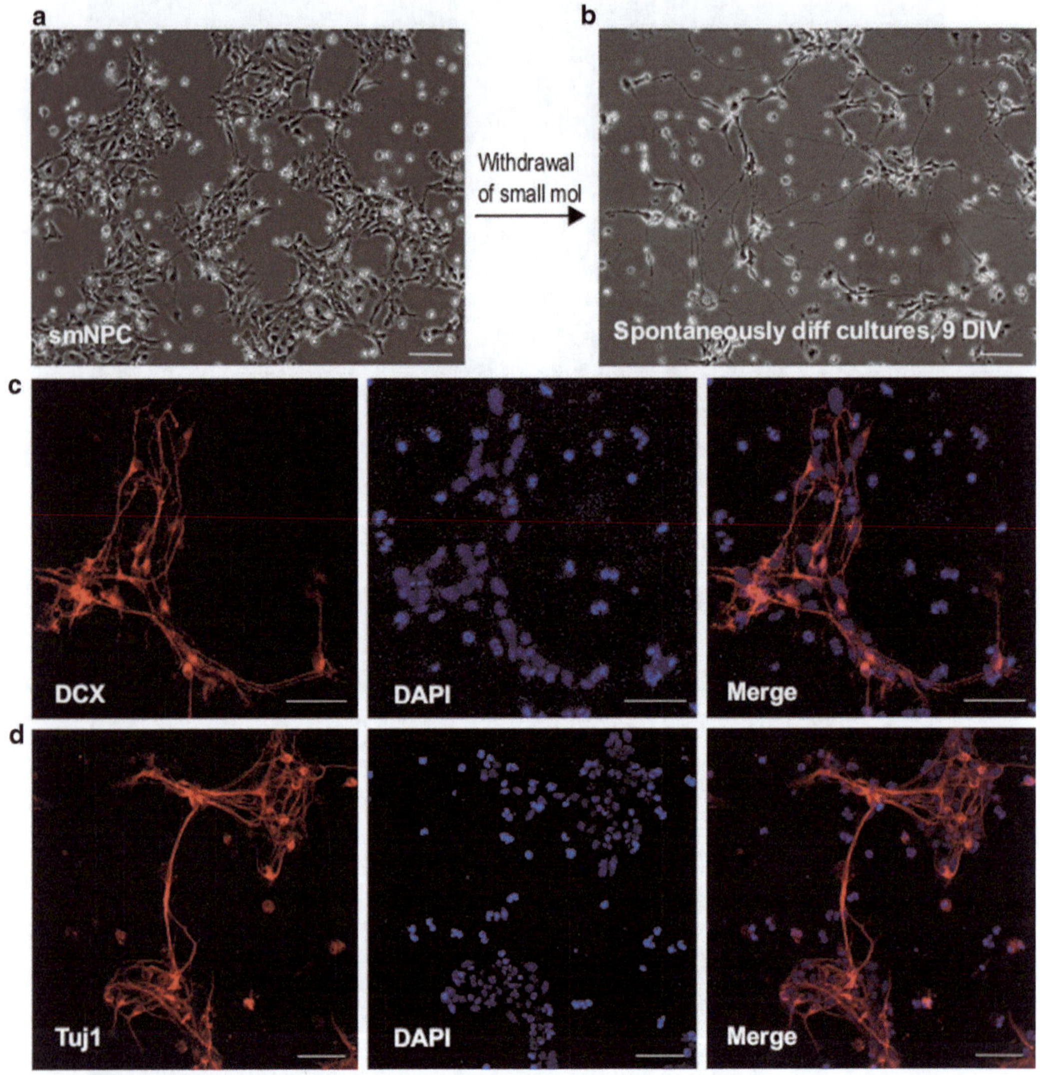

Fig. 2 Spontaneous differentiation of NPCs into neuronal cells. (**a**) NPCs in culture (**b**) Spontaneous differentiation of NPCs into neuronal cells after small-molecule withdrawal. Immunostaining of the differentiated neurons with (**c**) doublecortin (DCX) and (**d**) Tuj1. Scale bar 50 μM; DIV—days in vitro

2.2 Reagents

1. PBS.
2. hESC qualified Matrigel.
3. Growth factor reduced Matrigel.
4. PDL (poly-D-lysine).
5. mTeSR1 medium (StemCell Technologies, 05850).
6. DMEM/F-12 Medium.
7. Neurobasal medium.

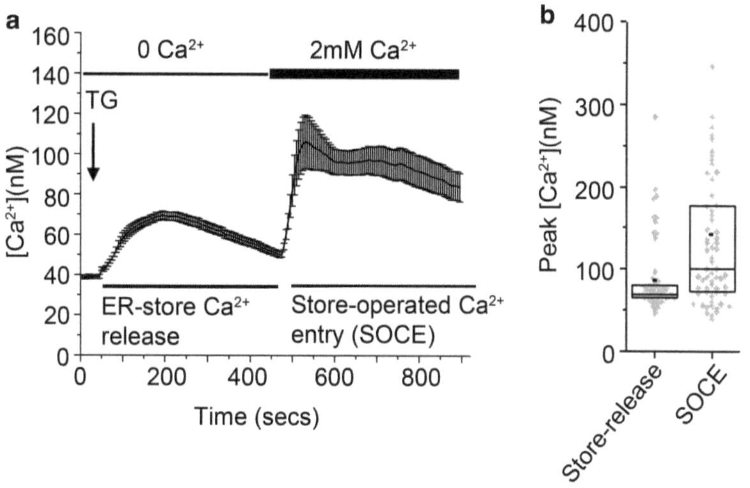

Fig. 3 SOCE measurements in human NPCs. (**a**) Ca^{2+} changes during ER-store depletion using thapsigargin, TG (10 μM) and SOCE after Ca^{2+} add-back. The trace shows the mean $[Ca^{2+}] \pm$ SEM from >50 cells. (**b**) Box plot quantifying the peak store-release and SOCE in hNPCs

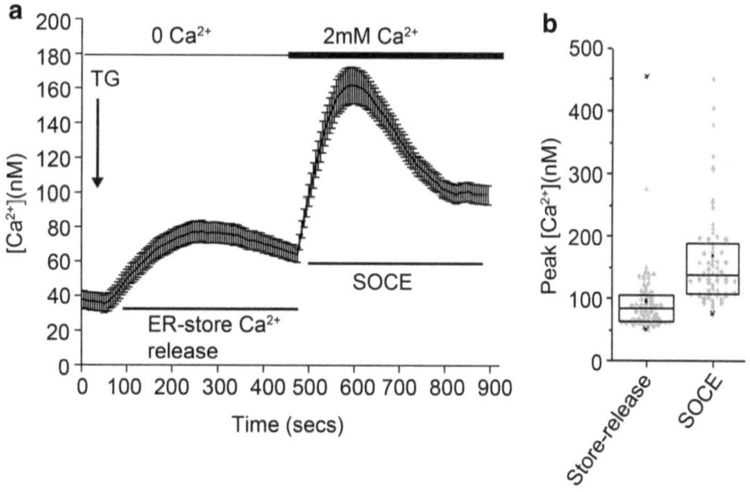

Fig. 4 SOCE-measurements in neurons spontaneously differentiated from human NPCs. (**a**) Ca^{2+} changes during ER-store depletion using TG (10 μM) and SOCE after Ca^{2+} add-back in neurons differentiated from hNPCs for 10 days. The trace shows the mean $[Ca^{2+}] \pm$ SEM from >50 cells. (**b**) Box plot quantifying the peak store-release and SOCE in the neurons

8. N2 supplement.
9. B27 supplement without vitamin A.
10. B27 with vitamin A (for differentiation).
11. Penicillin/Streptomycin/glutamine.
12. MEM nonessential amino acids.
13. β-mercaptoethanol.
14. Accutase.
15. KOSR (Knockout Serum Replacement).
16. Fetal bovine serum (Hyclone).
17. Dimethyl sulfoxide (DMSO) 10 mL ampoules.
18. Small molecules- SB43152, CHIR 99021, dorsomorphin, purmorphamine.
19. Rock Inhibitor (Y-27632 dihydrochloride).
20. Ascorbic acid.
21. Fura-2-AM.
22. Pluronic F-127.
23. 0 Ca^{2+} HBSS.
24. Thapsigargin (TG).
25. Calcium chloride, sodium chloride, potassium chloride, magnesium chloride.
26. HEPES.
27. Glucose.
28. EGTA.
29. Ionomycin.

2.3 Solutions

1. *Matrigel (applicable for hESC-qualified and growth factor-reduced versions)*

   Keep the Matrigel™ frozen in −80°C until you are ready to aliquot it. It gels rapidly at room temperature. 100 μL aliquots are generally made (can vary depending on lab requirements). Thaw the Matrigel overnight in 4°C refrigerator in ice. Tips, 1.5 mL tubes and serological pipettes are kept in the −20°C or −80°C freezer to keep them cold. Mix Matrigel using 5 mL serological pipette, aliquot keeping the bottle on an ice using cold pipette tips in cold 1.5 mL eppendorf tubes in sterile biosafety cabinet. Place each aliquot on ice and transfer them to −80°C till further use. Once thawed, an aliquot cannot be reused (*see* **Notes 1** and **2**).

2. *mTeSR™ Medium*

   mTeSR comes as a −20°C supplement and a basal media to be stored at 4°C. Thaw the Supplement (50×) at 2–8°C overnight. To prepare 500 mL of complete mTeSR medium, add the 100 mL supplement to the 400 mL basal medium. Aliquot it in 50 mL and store it in −20°C till further use.

3. *EB media*

   To make 25 mL, filter-sterilize the following: 19.5 mL DMEM/F-12 medium, 5 mL Knockout Serum Replacer, 250 μL 200 mM L-glutamine, 250 μL MEM nonessential amino acids, 45 μL β-mercaptoethanol.

4. *EB media for neural induction*

   EB media which contains 10 μM SB-431542, 1 μM dorsomorphin 3 μM CHIR 99021, and 0.5 μM purmorphamine, PMA.

5. *N2B27 media/Differentiation media*

   To make 50 mL, filter-sterilize the following: 25 mL DMEM/F-12 medium, 25 mL Neurobasal media (commercially available), 250 μL N2 supplement, 250 μL 200 mM L-glutamine, 500 μL B27 supplement without vitamin A (when used for differentiation, replace with B27 with vitamin A), 1× penicillin/streptomycin.

6. *N2B27 media for neural induction*

   N2B27 media which contains 10 μM SB-431542, 1 μM dorsomorphin 3 μM CHIR 99021, and 0.5 μM purmorphamine, PMA (*see* **Note 3**).

7. *NPC expansion media*

   To filter-sterilized 50 mL N2B27 media, add 3 μM CHIR 99021, 0.5 μM purmorphamine, PMA and 150 μM Ascorbic Acid, AA (Volumes of small molecules vary depending on reconstitution).

8. *Cryopreservation Media*

   hESC: To make 10 mL of Cryopreservation Medium, add 2 mL defined FBS to 8 mL mTeSR medium, filter-sterilize and then add 1 mL sterile DMSO. Keep it on ice until use.

   NPC: To make 10 mL of Cryopreservation Medium, add 9 mL filter-sterilized NPC medium to 1 mL sterile DMSO. Keep it on ice until use.

9. *0 Ca^{2+} HBSS for Calcium imaging.*

   20 mM HEPES, 137 mM sodium chloride, 5 mM potassium chloride, 2 mM magnesium chloride, pH = 7.3, supplemented with 0.5 mM EGTA and 10 mM Glucose.

3 Methods

3.1 Neural Precursor Cell (NPC) Derivation

3.1.1 Matrigel Coating of Plates

1. Thaw Matrigel aliquot on ice at 4°C.
2. Transfer 10 mL cold, sterile DMEM/F-12 medium to a 15 mL conical tube in sterile biosafety cabinet, add 1 mL of cold DMEM/F-12 medium from the tube to the 100 μL Matrigel aliquot and transfer it to 15 mL conical tube. Dissolve thoroughly using a 5 mL or 10 mL serological pipette.
3. Add 1 mL of Matrigel solution to each well of a 6-well plate (or as required). Transfer the remaining Matrigel solution to 4°C (and can be used for 7 days). Incubate the Matrigel-coated plate for 2 h at 37°C (or 4°C overnight).

3.1.2 Thawing hESC on Matrigel

1. Remove the frozen vial of hESC from the liquid nitrogen storage tank.
2. Immerse the cryovial in a 37°C water bath; swirl gently till a small ice crystal remains.
3. Spray the vial with a 70% ethanol and transfer the vial to the sterile biosafety cabinet.
4. Add 1 mL of prewarmed mTeSR medium to the vial and transfer the cells gently into a sterile 15 mL conical tube.
5. Add another 3 mL of mTeSR medium dropwise to cells in the 15 mL conical tube. The tube can be flicked or tapped gently while adding the medium for proper mixing and minimal osmotic shock.
6. Centrifuge the cells at 200 × *g* for 2 min. Aspirate the supernatant with a sterile aspirating pipette. Resuspend the cell pellet in 2 mL of mTeSR medium for every well that will receive cells with a cut 1 mL tip to avoid disintegration into single cells (*see* **Note 4**).
7. Add 1 mM Rock Inhibitor to the medium only for the first day (*see* **Note 5**).
8. Aspirate the Matrigel from each well, transfer hESC cell clumps to the wells and place the plate into the incubator. It is extremely important to move the plate back and forth and side to side to evenly distribute the cells.
9. After 12–24 h, remove the spent medium using a sterile aspirating pipette, add 2 mL of mTeSR medium dropwise to the each well, and place the plate back into a 37°C incubator.
10. Repeat until ready to passage or freeze.

3.1.3 Passaging and Freezing of hESC Cultured on Matrigel

1. Split hESC when the plate is confluent or when the colonies are becoming differentiated. The split ratio is decided based on the above factors. We maintain a split ratio of 1:3–1:5 for our lines when the cells are 80–90% confluent.

2. Transfer the plate into the hood equipped with a stereomicroscope and remove differentiated colonies with a sterile micropipette tip.
3. Then transfer the plate to the biosafety cabinet and aspirate the spent medium that contains dead cells and differentiated colonies, rinse with 1 mL warmed DMEM-F12.
4. Add 1 mL room temperature Accutase solution to each well to be passaged.
5. Observe under the microscope till the edges of the colonies curl up, this should not take more than 2–3 min.
6. Aspirate the Accutase solution without disturbing the attached hESC colonies.
7. Add 2 mL of mTeSR medium to each well, gently make zigzag motions in all directions using a sterile micropipette tip on the adherent surface (use a sterile cell scraper if necessary). Make sure that the colonies are not dispersing into single cells (*see* **Note 6**).
8. Once the cell clumps start floating in the medium, transfer them to a 15 mL sterile conical tube after all wells are scraped and the cells are removed from the surface of the well.
9. The cells can be spun down at 200 × g for 2 min or allowed to settle down in the tube. According to the split ratio (*see* **Note 7**), resuspend in required volume of mTeSR medium and add 2 mL each onto Matrigel-coated plates with 1 mM Rock Inhibitor (as described in Subheading 3.1.3).
10. Incubate cells overnight to allow colonies to attach.
11. For freezing the hESC, 1 mL cryopreservation medium should be added dropwise to the colony clumps. Pipet up and down to mix.
12. Transfer the cell suspension to cryovials appropriately labeled, tighten caps and place cryovials in an isopropanol containing freezing container.
13. Place the freezing containers in the −80°C freezer overnight and transfer cell vials to liquid nitrogen the following day.

3.1.4 Derivation of Small Molecules-Induced Neural Precursors (hNPCs) from hESC

1. Once hESC colonies are 70–80% confluent, passage them as described. Transfer all the floating iPSC colonies to a 15 mL tube and add at least twice the volume of neutralization medium. (mTeSR medium can be used for neutralization.)
2. Keep the tube upright in a tube rack undisturbed for 5 min to allow colonies to settle. Centrifugation might result in smaller colony clumps which reduce the efficiency of embryoid body (EB) formation (*see* **Note 6**).

3. Aspirate medium from the top without disturbing the colonies in the bottom of the tube.

   Resuspend in EB medium (not mTeSR) and distribute them to desired numbers of 100 mm low-adherent bacteriological dishes (~2 wells of confluent 6-well plates/dish).
4. We followed the Reinhardt protocol [15] for NPC derivation. EB medium (without FGF2) is supplemented with 10 μM SB-431542, 1 μM dorsomorphin for neural induction through dual smad inhibition, and 3 μM CHIR 99021, a GSK 3β inhibitor and 0.5 μM purmorphamine, PMA (canonical Wnt stimulator). Medium is replaced on day 2 with DMEM-F12/ Neurobasal 50:50 with 1:200 N2 supplement, 1:100 B27 supplement lacking vitamin A with 1× penicillin/streptomycin/glutamine with the same small molecules (*see* **Notes 8–10**).
5. On day 4, SB-431542 and dorsomorphin is withdrawn and 150 μM ascorbic acid, AA is added to the medium (This is the expansion media with CHIR, PMA, and AA.).
6. On day 6, the EBs, which show intensive neuroepithelial outgrowths, are triturated with a 1000 μL pipette into smaller pieces (use Accutase if necessary) and plated on Matrigel-coated 6-well dishes with ~20 EBs per well in NPC expansion medium.
7. For coating, Matrigel (growth factor reduced version and not the hESC qualified one) is diluted to a final dilution of 1:100 in DMEM-F12.
8. The first split can be performed at a 1:5 ratio on day 4 after plating using Accutase as described earlier. The only difference from hESC is that cells are allowed to form a single cell suspension before neutralization (7–10 min) and centrifuged at 250 × *g* for 3–5 min.
9. After 5–6 splits, cultures are homogenous with NPC-like cells. Media is changed every other day. The cells can be tested for NPC marker expression (Ki67 for proliferation and Nestin, Pax6, Sox1, Sox2 as neuroepithelial markers) by immunocytochemistry to confirm their identity (Fig. 1) (*see* **Notes 11** and **12**).
10. The NPCs could be maintained in culture till 25 passages as described above. Alternatively propagation as neurospheres (*see* **Note 13**) can also be performed in bacteriological dishes in the same expansion media.

3.1.5 Freezing of hNPCs

1. Add room temperature Accutase solution to each well/plate to be passaged. Volume of Accutase depends on the cell culture surface area.

2. Once the cells detach from the surface, triturate a little to break the clumps, add any basal media to dilute the Accutase solution, and transfer to a 15 mL sterile conical tube.
3. Centrifuge at 250 × *g* for 3–5 min.
4. Decant the supernatant and tap the pellet gently.
5. Add required volume of NPC cryopreservation medium. Pipet up and down to mix. Generally 1–2 × 10^6 cells are frozen per vial.
6. Transfer the cell suspension to cryovials appropriately labeled, tighten caps, and place cryovials in an isopropanol containing freezing container.
7. Place the freezing containers in the −80°C freezer overnight and transfer cell vials to liquid nitrogen the following day.

3.1.6 Spontaneous (Undirected) Neuronal Differentiation of NPCs

1. Plate cells at a density of 10,000–15,000 viable cells/cm^2 in a 12-well plate or PDL-coated coverslips. Alternatively, growth factor reduced Matrigel or PLO/Laminin can also be used for biocoating (*see* **Notes 14–16**).
2. Grow cells in DMEM-F12/Neurobasal 50:50 with 1:200 N2 supplement, 1:100 B27 supplement with vitamin A with 1× penicillin/streptomycin/glutamine. Small molecules are not added to the media to allow differentiation of NPCs.
3. Change media every other day. Gently add media through the walls as differentiated cells are more fragile and tend to lift up.
4. Cells can be digested to single cells in case they become confluent and replated on fresh coated plates and further differentiated. Expect some cell death.
5. Cells can be kept in the differentiation media for up to 4 weeks. We are showing 9-day differentiated neurons, immunopositive for DCX and Tuj1 as an example (Fig. 2). Astrocytes and oligodendrocytes also come up in culture in lesser numbers. 50 μM dbcAMP [16, 17] can be added for better survival (optional) 0.1 mM DAPT can also be used to enhance neuronal differentiation [18].

3.2 Ca^{2+} Imaging from hNPCs and Spontaneously Differentiated Neurons

1. For plating cells for Ca^{2+} imaging, coat coverslips with 0.2 mg/mL poly-D-lysine (PDL) for 2 h at 37°C or overnight at 4°C.
2. Wash the PDL-coated coverslips 3 times with sterile phosphate buffered saline (PBS) and air-dry inside a laminar hood for 30 mins before plating the cells.
3. Plate hNPCs onto these PDL-coated coverslips at a confluency of ~80%. These hNPCs can be used for Ca^{2+} imaging 24–48 h after plating. In case of the neurons, allow the hNPCs plated onto PDL-coated coverslips to spontaneously differentiate into

neurons for a period of 10 days (or more) after withdrawal of small molecules [15] necessary for maintaining the neural precursor or stem cell state (*see* **Note 17**).

4. On the day of Ca^{2+} imaging wash plated hNPCs or neurons 3 times with the culture medium after which load them with 2.5 μM of the ratiometric Ca^{2+}-indicator Fura-2-AM in the culture medium supplemented with 0.002% Pluronic (*see* **Note 18**) in dark for 45 min.
5. After loading, wash cells 3 times with the culture medium (*see* **Notes 19** and **20**), after which keep them in the culture medium until the start of imaging.
6. Remove the culture medium and cover cells with 100 μL of "0 Ca^{2+}" HBSS. Adjust focus to find a field containing preferably 40–50 single cells.
7. Record fluorescence emission intensities from each cell after exciting the Fura-2 in the cells at 340 nm and 380 nm (F340 and F380) alternatively with a time delay of 1 s.
8. Capture F340 and F380 for each field of cells every 2 s during the entire course of imaging (Figs. 3 and 4).
9. After the start of imaging, capture F340/F380 for 25 time points at 2 s interval to monitor resting fluctuations in cytosolic Ca^{2+} levels in hNPCs (Fig. 3a) or neurons (Fig. 4a) bathed in very low extracellular Ca^{2+} ("0 Ca HBSS") (*see* **Note 21**).
10. Add the sarcoendoplasmic reticulum ATPase (SERCA)-blocker thapsigargin (TG) [19] to the cells to deplete the ER Ca^{2+} store. The increase in [Ca^{2+}] observed after TG addition (Figs. 3a and 4a) represents the Ca^{2+} released from the ER into cytosol during ER-store release. Monitor ER-store release and subsequent Ca^{2+} clearance from the cytosol for 450 s post TG-addition.
11. For monitoring SOCE, add back 2 mM Ca^{2+} to the extracellular medium and record F340/F380 for the next 450 s.
12. Calculate ratio of F340 and F380 (F340/F380) for each time point and calibrate into [Ca^{2+}] by using the Grynkiewicz equation [14] as follows:

$$\left[Ca^{2+}\right]\ (\text{nM}) = K_{\text{d}} \times \beta \times (R - R_{\text{min}})/(R_{\text{max}} - R),$$

where K_d for Fura-2 in human cells = 225 nM, β refers to scaling factor and R refers to F340/F380 ratio at a particular time point. R_{min} refers to the minimum F340/F380 obtained after addition of 10 mM EGTA to maximally chelate most of free cytosolic Ca^{2+}. R_{max} refers to the maximum F340/F380 obtained after addition of ionomycin (10 μM) in presence of 10 mM extracellular Ca^{2+}. This results in saturation of the Fura-2 with Ca^{2+} and hence gives the maximum possible value of R.

13. Calculate the average $[Ca^{2+}]$ for each time point for all the cells imaged and plot the mean $[Ca^{2+}]$ on the *y*-axis with the time (in seconds) on the *x*-axis to get an average Ca^{2+}-response curve for store-release and SOCE in hNPCs (Fig. 3a) and spontaneously differentiated neurons (Fig. 4a).
14. Calculate the peak cytosolic $[Ca^{2+}]$ during ER-store release and SOCE for each cell and plot them as box plots as shown in Figs. 3b and 4b. Such box plots with the distribution of the raw data superimposed on them can be plotted for different genotypes and treatment conditions and used for computation of statistical significance.

4 Notes

1. Matrigel should be handled carefully. As it gels at room temperature, thawing at 4°C refrigerator in ice is optimal.
2. Lot-to-lot variability in concentration is common in Matrigel. Dilution can be optimized in the lab. 1:100 is a reference.
3. Small molecules and growth factors are generally not passed through filters to prevent adsorption and product loss.
4. Cut the tip of a 1 mL tip while passaging hESC to ensure gentle dispersion of bigger colonies into small clumps, not single cells.
5. Rock Inhibitor should be stored in small aliquots to avoid repeated freeze–thaw cycles.
6. hESC passaging in clumps increases survival and EB formation. Optimize the enzyme incubation time in the lab. Accutase can be replaced with dispase or collagenase for hESC passaging.
7. Split ratio of hESCs can be variable depending on the cell line used.
8. The protocol defined here for NPC derivation can be applied across different ESC and iPSC lines.
9. Feeder-dependent hESC cultures can also be used to derive NPCs.
10. NPC or neural stem cell derivation can be achieved through various published protocols.
11. It is advisable to check the karyotype of hESCs and NPCs when maintained in culture routinely.
12. Mycoplasma testing in cultures should be done at least once in 3 months.
13. NPCs can also be grown as Neurospheres in bacteriological dishes and dispersed into single cells using Accutase before plating for differentiation or imaging.
14. For differentiation into neurons PDL can be replaced with PLO/Laminin or growth factor-reduced Matrigel.

15. Differentiated neurons should be handled with care, the plates should be never dry, and media should be added dropwise or through the walls of the culture plate.
16. Seeding densities can be changed for directed differentiation into different neuronal subtypes or neuroglia.
17. While plating for calcium imaging, ensure that most cells are plated as single cells or a cluster of cells with clearly demarcated boundaries.
18. Pluronic F-127 can be warmed at ~40°C to resolubilize a solidified aliquot.
19. Bring the culture medium (stored preferably at 4°C) to room temperature before washing cells or dye loading to prevent cold shock to the cells.
20. When washing the cells before or after dye loading, do not add the culture medium directly on top of the cells as this can dislodge some cells or put unnecessary mechanical strain on them. Instead add or remove medium from the borders of the coverslip.
21. Do not add "0 Ca^{2+} HBSS" to cells until the microscope and the imaging software has been set up for starting the imaging. In this way, the cells would stay in the "0 Ca^{2+} HBSS" for only about 2–3 mins, the time taken to select a suitable field of cells and not more.

Acknowledgments

We are grateful to Central Imaging and Flow Cytometry Facility, National Centre for Biological Sciences (NCBS), TIFR for the use of microscopes and Stem Cell Facility for the infrastructure for hESC and NPC work. This work was funded by a grant from the Department of Biotechnology, Ministry of Science and Technology, Govt. of India (BT/PR6371/COE/34/19/2013) and core funds from NCBS, TIFR, to GH. Financial support from the DBT-RA program in Biotechnology and Life Sciences, Govt. of India to RG and INSPIRE fellowship under DST, Govt. of India to PC are gratefully acknowledged.

References

1. Liou J, Kim ML, Won DH et al (2005) STIM is a Ca2+ sensor essential for Ca2+−store-depletion-triggered Ca2+ influx. Curr Biol 15:1235–1241. https://doi.org/10.1016/j.cub.2005.05.055
2. Roos J, DiGregorio PJ, Yeromin AV et al (2005) STIM1, an essential and conserved component of store-operated Ca 2+ channel function. J Cell Biol 169:435–445. https://doi.org/10.1083/jcb.200502019
3. Feske S, Gwack Y, Prakriya M et al (2006) A mutation in Orai1 causes immune deficiency by abrogating CRAC channel function. Nature 441:179–185. https://doi.org/10.1038/nature04702

4. Vig M, Peinelt C, Beck A et al (2006) CRACM1 is a plasma membrane protein essential for store-operated Ca2+ entry. Science 312:1220–1223. https://doi.org/10.1126/science.1127883
5. Prakriya M, Feske S, Gwack Y et al (2006) Orai1 is an essential pore subunit of the CRAC channel. Nature 443:230–233. https://doi.org/10.1038/nature05122
6. Venkiteswaran G, Hasan G (2009) Intracellular Ca2+ signaling and store-operated Ca2+ entry are required in Drosophila neurons for flight. Proc Natl Acad Sci 106:10326–10331. https://doi.org/10.1073/pnas.0902982106
7. Toth AB, Shum AK, Prakriya M (2016) Regulation of neurogenesis by calcium signaling. Cell Calcium 59:124–134
8. Somasundaram A, Shum AK, McBride HJ et al (2014) Store-operated CRAC channels regulate gene expression and proliferation in neural progenitor cells. J Neurosci 34:9107–9123. https://doi.org/10.1523/JNEUROSCI.0263-14.2014
9. Gopurappilly R, Deb BK, Chakraborty P, Hasan G (2018) Stable STIM1 knockdown in self-renewing human neural precursors promotes premature neural differentiation. Front Mol Neurosci 11:178. https://doi.org/10.3389/fnmol.2018.00178
10. Hartmann J, Karl RM, Alexander RPD et al (2014) STIM1 controls neuronal Ca2+signaling, mGluR1-dependent synaptic transmission, and cerebellar motor behavior. Neuron 82:635–644. https://doi.org/10.1016/j.neuron.2014.03.027
11. Lalonde J, Saia G, Gill G (2014) Store-operated calcium entry promotes the degradation of the transcription factor Sp4 in resting neurons. Sci Signal 7:ra51–ra51. https://doi.org/10.1126/scisignal.2005242
12. Pathak T, Agrawal T, Richhariya S et al (2015) Store-operated calcium entry through Orai is required for transcriptional maturation of the flight circuit in *Drosophila*. J Neurosci 35:13784–13799. https://doi.org/10.1523/JNEUROSCI.1680-15.2015
13. Tsien RY (1980) New calcium indicators and buffers with high selectivity against magnesium and protons: design, synthesis, and properties of prototype structures. Biochemistry 19:2396–2404. https://doi.org/10.1021/bi00552a018
14. Grynkiewicz G, Poenie M, Tsien RY (1985) A new generation of Ca2+ indicators with greatly improved fluorescence properties. J Biol Chem 260:3440–3450
15. Reinhardt P, Glatza M, Hemmer K et al (2013) Derivation and expansion using only small molecules of human neural progenitors for neurodegenerative disease modeling. PLoS One 8:e59252. https://doi.org/10.1371/journal.pone.0059252
16. Mena MA, Casarejos MJ, Bonin A et al (1995) Effects of dibutyryl cyclic AMP and retinoic acid on the differentiation of dopamine neurons: prevention of cell death by dibutyryl cyclic AMP. J Neurochem 65:2612–2620. https://doi.org/10.1046/j.1471-4159.1995.65062612.x
17. Tojima T, Kobayashi S, Ito E (2003) Dual role of cyclic AMP-dependent protein kinase in neuritogenesis and synaptogenesis during neuronal differentiation. J Neurosci Res 74:829–837. https://doi.org/10.1002/jnr.10754
18. Crawford TQ, Roelink H (2007) The Notch response inhibitor DAPT enhances neuronal differentiation in embryonic stem cell-derived embryoid bodies independently of Sonic Hedgehog signaling. Dev Dyn 236:886–892. https://doi.org/10.1002/dvdy.21083
19. Lytton J, Westlin M, Hanley MR (1991) Thapsigargin inhibits the sarcoplasmic or endoplasmic reticulum Ca-ATPase family of calcium pumps. J Biol Chem 266:17067–17071

Chapter 21

Embryoid Body Differentiation of Mouse Embryonic Stem Cells into Neurectoderm and Neural Progenitors

Rachel A. Shparberg, Hannah J. Glover, and Michael B. Morris

Abstract

Mouse embryonic stem cells (mESCs) are pluripotent cells capable of differentiating in vitro to form the ~200 types of cells of the developing embryo and adult, including cells of the nervous system. This makes mESCs a useful tool for studying the molecular mechanisms of mammalian embryonic development. Many protocols involving the use of growth factors and small molecules to differentiate mESCs into neural progenitors and neurons currently exist. However, there is a paucity of protocols available that recapitulate the developmental process. Our laboratory has developed a protocol to recapitulate mammalian neural lineage development by differentiating mESCs to mature neurons via intermediate cell populations observed during in vivo embryo development. This protocol uses the amino acid L-proline to direct the differentiation of mESCs, grown as embryoid bodies, into *Sox1*$^+$ neurectoderm, followed by differentiation to form Nestin$^+$, BLBP$^+$, and NeuN$^+$ neural cell types.

Key words Definitive ectoderm, Directed differentiation, Early primitive ectoderm-like cells, L-Proline, Mouse embryonic stem cells, Neurectoderm, Neural progenitor, Neuron

1 Introduction

Early embryonic development begins with the fertilization of the oocyte to form a totipotent zygote. Following a series of cleavage divisions, the blastocyst is formed between 3.5 and 4.5 days post-coitum (dpc) in the mouse. The blastocyst consists of two cell populations from which the entire developing embryo and its supporting structures are derived. Of these two populations, the *Oct4*$^+$/*Rex1*$^+$ pluripotent inner cell mass (ICM) is responsible for forming the cells of the embryo and some support structures. By 5.5 dpc, the ICM differentiates to a second pluripotent cell population of *Dnmt3b*$^+$/*Fgf5*$^+$/*Oct4*$^+$/*Rex1*$^-$ primitive ectoderm. Following gastrulation at 6.5 dpc, the trilaminar embryo—consisting of the three multipotent germ layers of definitive mesoderm, definitive endoderm, and definitive ectoderm—is formed by 7.0 dpc [1–3]. The neural lineage develops from the definitive ectoderm,

Mugdha V. Joglekar and Anandwardhan A. Hardikar (eds.), *Progenitor Cells: Methods and Protocols*, Methods in Molecular Biology, vol. 2029, https://doi.org/10.1007/978-1-4939-9631-5_21, © Springer Science+Business Media, LLC, part of Springer Nature 2019

which gives rise to the *Sox1*⁺ neural plate, the precursor tissue of the nervous system, by 7.5 dpc [4]. The primary neural progenitor population of *BLBP*⁺ radial glial cells develops from the neuroepithelium and give rise to *NeuN*⁺ neurons from 10.0 to 14.5 dpc [5, 6].

mESCs are derived from the 3.5 to 4.5 dpc blastocyst and, like the ICM, can differentiate to any cell types of the embryo or adult by manipulation of the in vitro environment. When cultured in the presence of LIF and BMP4, mESCs can be maintained as pluripotent, self-renewing cells that express markers of the ICM including *Rex1* and markers of pluripotency including *Oct4* and *Nanog* [7, 8]. When grown in suspension culture under conditions which permit spontaneous differentiation, mESCs produce embryoid bodies (EBs) with an outer layer of extraembryonic primitive endoderm cells and an inner population of epiblast cells. Structurally, these EBs are comparable to the 4.5 dpc mouse embryo [9–11]. mESCs can also be directed to differentiate toward lineages of interest by the addition of growth factors, inhibitors and small molecules to promote lineage- and cell-specific differentiation.

Amino acids have been shown to influence development (e.g., by promoting blastocyst formation and hatching) and regulate pluripotency and differentiation of mESCs [12–24]. L-Proline is a conditionally essential amino acid that promotes the differentiation of mESCs to *Rex1*⁻/*Fgf5*⁺ primitive ectoderm-like (EPL) cells under otherwise self-renewing conditions [12, 13, 25–28]. Similarly, when mESCs are cultured in HEPG2 conditioned medium (which contains a high concentration of L-Proline), they differentiate to *Sox1*⁺ and *Pax6*⁺ neurectoderm [29, 32].

Our laboratory has developed a simplified protocol to recapitulate mouse nervous system development using L-Proline to direct the differentiation of mESCs down the neural lineage (Fig. 1). This protocol recapitulates the stepwise differentiation to intermediate cell populations seen during in vivo embryonic nervous system development. mESCs sequentially differentiate to *Dnmt3b*⁺/*Fgf5*⁺ EPL cells, *Penk1*⁺ definitive ectoderm, *Sox1*⁺ neurectoderm, Nestin⁺/BLBP⁺ neural progenitors, and mature NeuN⁺ neural cells as confirmed by a combination of marker expression analysis by qPCR, flow cytometry, immunofluorescence microscopy and phenotype analysis by time-lapse and phase-contrast imaging.

2 Materials

2.1 Cell Culture Reagents

1. 0.1% porcine gelatin prepared in Ca^{2+}-free PBS.
2. 0.25% trypsin–EDTA.
3. β-mercaptoethanol (β-Me): 0.1 M stock in Milli-Q water by adding 3.5 μL β-Me in 500 μL Milli-Q water. This stock should be stored at 4 °C for 1–2 weeks maximum (*see* **Note 1**).

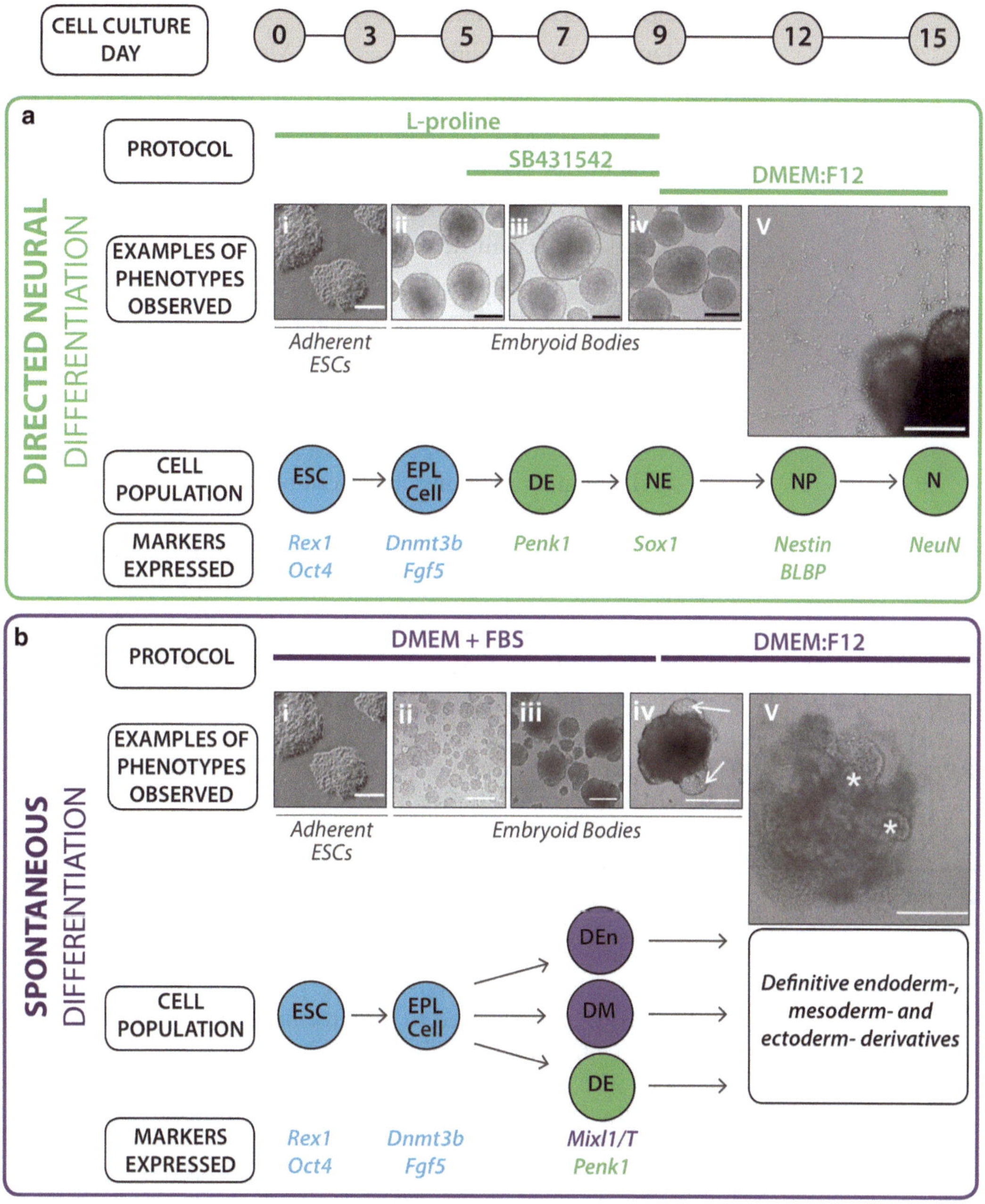

Fig. 1 Differentiation protocol used to recapitulate mammalian nervous system development using mESCs. mESCs were cultured in suspension to form EBs using (**A**) the *Directed Neural Differentiation* protocol containing 400 μM L-Proline for 9 days (to drive the differentiation of mESCs to neurectoderm (NE) via EPL-cell and definitive ectoderm (DE) populations) plus the addition of the nodal inhibitor SB431542 from Day 6 (to prevent off-pathway differentiation to definitive mesoderm (DM) and definitive endoderm (DEn) cells). EBs were also cultured using (**B**) the *Spontaneous Differentiation* protocol (without L-Proline or SB431542) for 9 days (as a control to allow for spontaneous differentiation into cells derived from all three germ layers). On Day 10, EBs from both sets of culture conditions were seeded in *Serum-Free Medium* (DMEM:F12) and allowed to differentiate for a further 6 days. On Day 15, EBs were scored for the presence of neural progenitors (NP),

4. Bovine serum albumin (BSA).
5. Dulbecco's Modified Eagle's Medium (high glucose) (DMEM).
6. F12 Nutrient Mix (1×).
7. Fetal bovine serum (FBS) (*see* **Note 2**).
8. Insulin–transferrin–sodium selenite (ITSS) (100×).
9. Knock-out Serum Replacer (KOSR).
10. L-proline: 0.1 M stock in Milli-Q water (*see* **Note 3**).
11. Mouse Leukemia Inhibitory Factor (LIF): 10^6 U/mL working stock by reconstituting 100 μg LIF in 1 mL 0.1% BSA/PBS to make 10^7 U/mL. Prepare the working stock of 10^6 U/mL by making a tenfold dilution of 10^7 U/mL stock in 1% BSA/PBS. Stocks are stored at −20 °C if not required immediately for up to 12 months. Working stocks can be stored at 4 °C for up to a month (*see* **Note 4**).
12. Penicillin/Streptomycin (Pen/Strep) (100×): 5000 U/mL penicillin G sodium salt and 5 mg/mL streptomycin sulfate.
13. Phosphate Buffered Saline, Ca^{2+}-free (PBS).
14. SB431542: 2.6 mM stock in DMSO (*see* **Note 5**).
15. Trypan Blue: 0.4 g Trypan Blue in PBS, pH 7.2–7.3.

2.2 Culture Media

The culture media used in these protocols are outlined in Tables 1 and 2, and all cell culturing is performed under sterile conditions.

3 Methods

The cell lines used with this protocol are the D3 and 46C cell lines. The D3 cell line is a wild-type cell line [30] while the 46C cell line is a *Sox1*-GFP cell line, used to monitor *Sox1*-driven GFP expression during in vitro differentiation [4, 31].

For the continuous passage method described below, the mESC line chosen should be able to grow without the need for a mouse embryonic fibroblast feeder layer.

For these experiments, incubation always occurs at 37 °C in a humidified incubator (5% CO_2).

Fig. 1 (continued) neurons (N) and beating patches of cells, indicative of cardiomyocytes (∗). (**A**i and **B**i) Representative image of an adherent D3 mESC colony (20× objective). (**A**ii-iv and **B**ii-iv) Representative images of suspension culture EBs at Days 5, 7, and 9, respectively (10× objective except **B**iv: 20× objective). Note that the EBs cultured using the *Directed Neural Differentiation* protocol are smoother and larger compared to those cultured by *Spontaneous Differentiation* (uneven, rough borders often associated with cyst formation (white arrows)). (**A**v and **B**v) Seeded-down EBs at Day 15 (10× objective). Scale bar = 200 μm

Table 1
Culture conditions for self-renewal of mESCs

Plate type	Medium constituents	Volume of medium	Seeding density
60 mm tissue-culture treated (for continuous passage)	*mESC Complete Medium*	5 mL	5×10^4–5×10^5 cells
100 mm tissue-culture treated (for continuous passage)	90% DMEM 10% FBS 0.1 mM β-Me 1× Pen/Strep 1000 U/mL LIF	10 mL	2×10^5–1.5×10^6 cells

This table provides the medium composition for continuous culturing of mESCs as undifferentiated stem cells, as well as suitable plate types, medium volumes and seeding densities. Note that the tissue-culture plates are first gelatinized

Table 2
Culture conditions used for embryoid body differentiation experiments

Plate type	Seeding density	Condition	Embryoid body day	Composition
100 mm suspension culture (not tissue-culture treated)	5×10^5 cells	Directed differentiation to neurectoderm	Day 0–5	*Day 0–5 Directed Differentiation Medium* 80% DMEM, 10% FBS, 10% KOSR, 0.1 mM β-Me, 1× Pen/Strep, 400 μM L-proline
			Day 6–9	*Day 6–9 Directed Differentiation Medium* 80% DMEM, 10% FBS, 10% KOSR, 0.1 mM β-Me, 1× Pen/Strep, 400 μM L-proline ± 10 μM SB431542
		Spontaneous differentiation controls	Day 0–9	*Spontaneous Differentiation Medium* 90% DMEM, 10% FBS, 0.1 mM β-Me, 1× Pen/Strep
48-well tissue-culture treated	3 EBs per well (*see* **Notes 17** and **18**)	All conditions	Day 10–15	*Day 10–15 Serum-Free Medium* 50% DMEM, 50% F12, 1× ITSS, 1× Pen/Strep

This table provides the differentiation media compositions as well as suitable plate types, optimum seeding densities and timelines for changing media during the differentiation process. *Directed Differentiation Medium* is used to drive the differentiation of mESCs to neurectoderm via EPL cells, definitive ectoderm and neurectoderm populations. Alternatively, a *Spontaneous Differentiation Medium* (without L-proline or SB) is used as a control to allow for spontaneous differentiation into cells derived from the three germ layers. *Serum-Free Medium* is used after Day 10 for both sets of culture conditions. β-mercaptoethanol (β-Me), fetal bovine serum (FBS), knock-out serum replacer (KOSR), penicillin–streptomycin (Pen/Strep), insulin + transferrin + sodium selenite (ITSS)

3.1 Passaging mESCs for Routine Maintenance

1. Before passaging mESCs, observe the cells under the brightfield microscope to ensure that the colonies are characteristic of mESCs (round and dome-shaped) and have reached an appropriate confluence: A maximum confluence of 70% is recommended as the growth and differentiation properties of overconfluent mESCs can be compromised. There should be little or no cell debris and no floating cells.
2. Prepare a plate for continuous passage of mESCs by adding ~2 mL 0.1% gelatin to a 60 mm tissue-culture treated dish (~3 mL for a 100 mm dish) and set aside for at least 10 min (*see* **Note 6**).
3. Prepare *mESC Complete Medium* as per Table 1.
4. Aspirate gelatin from the plates prepared in **step 2**, add the appropriate volume of *mESC Complete Medium* (Table 1) and set aside.
5. To the plate containing the adherent mESCs, aspirate the medium and wash twice with PBS (*see* **Note 7**). Add 0.25% trypsin–EDTA: For a 100 mm plate, add 1 mL and for a 60 mm plate, add 350 μL. Repeatedly tilt the plate to ensure the entire bottom is in contact with the trypsin. After 15–20 s deactivate the trypsin: For a 100 mm plate, add 1 mL *mESC Complete Medium* and for a 60 mm plate add 650 μL.
6. Triturate with a 1 mL pipette tip to dislodge the mESCs (*see* **Note 8**).
7. Collect the cells in a microcentrifuge tube and triturate 10–15 times to obtain a single-cell suspension.
8. Pulse-spin the cells for 10 s, remove the supernatant and wash with ~1 mL PBS. Repeat this step to ensure the trypsin has been removed.
9. Carefully aspirate the PBS and suspend the cells in 1 mL fresh *mESC Complete Medium*.
10. Perform a cell count by haemocytometer after adding 10 μL single-cell suspension to 40 μL Trypan Blue (*see* **Note 9**).
11. For continuous passage, add the appropriate number of cells to the culture plates prepared in **step 4** (Table 1) (*see* **Note 10**).
12. Distribute the cells evenly in the medium by using a figure-eight motion (both backward and forward direction), then place the plate into the incubator. Check the cells every day to assess growth, colony morphology, and confluency. The number of passages for these cells should be limited to ~35 to limit genetic mutations.

3.2 Embryoid Body Differentiation Protocol

3.2.1 Day 0: Preparing Suspension Cultures and Sample Collection

1. Prepare the differentiation culture plates by adding 10 mL *Day 0–5 Directed Differentiation Medium* and *Spontaneous Differentiation Medium* (Table 2) onto the appropriately labeled suspension culture plates (do not precoat these plates with gelatin) (*see* **Note 11**).
2. Passage the cells and prepare a single-cell suspension as described above (Subheading 3.1; **steps 1–9**). Save cells for Day 0 analysis. The number will depend on the assays you wish to perform (qPCR, flow cytometry, immunofluorescence, etc.).
3. Perform a cell count (Subheading 3.1; **step 10**) and add 5×10^5 cells to each differentiation plate. Evenly distribute the cells in the plates using the figure eight motions, then place the plates into the incubator for 3 days (*see* **Note 11**).

3.2.2 Day 3: Replacing the Medium and Sample Collection

Expected Cell Populations: EPL Cells (Fig. 1)

1. Swirl the EBs to the center of the plate and observe them under the brightfield microscope. By Day 3, they should be smooth and round, and large enough to see macroscopically (*see* **Note 12**).
2. Use a pipette to collect EBs for analysis (for qPCR, flow cytometry, immunofluorescence, etc.) (*see* **Note 13**).
3. Aspirate the medium from the edges of the plate avoiding as much as possible aspirating the EBs in the center.
4. Replace the medium with *Day 0–5 Directed Differentiation Medium* or *Spontaneous Differentiation Medium* (Table 2). Evenly distribute the EBs in the plates using the figure-eight motions and incubate the plates for another 2 days (*see* **Note 12**).

3.2.3 Day 5: Changing the Medium and Sample Collection

Expected Cell Populations: EPL Cells, Definitive Ectoderm and Neurectoderm (Fig. 1)

1. Repeat **steps 1–4** of Subheading 3.2.2 at the end of Day 5 but this time using *Day 6–9 Directed Differentiation Medium* and *Spontaneous Differentiation Medium* (Table 2) in **step 4** (*see* **Notes 14** and **15**).

3.2.4 Day 7: Replacing the Medium and Sample Collection

Expected Cell Populations: Definitive Ectoderm and Neurectoderm (Fig. 1)

1. Repeat **steps 1–4** of Subheading 3.2.2 at the end of Day 7 using *Day 6–9 Directed Differentiation Medium* and *Spontaneous Differentiation Medium* (*see* **Notes 14** and **15**).

3.2.5 Day 8: Adherent EB Plate Preparation

1. Prepare 48-well plates by coating each well with 0.1% gelatin and set aside in the incubator overnight (*see* **Note 16**).

3.2.6 Day 9: Plating EBs into Adherent Conditions

Expected Cell Populations: Neurectoderm (Fig. 1)

1. Repeat **steps 1** and **2** of Subheading 3.2.2 for Day 9 EBs.
2. Aspirate gelatin from the 48-well plate and add 0.5 mL *Day 10–15 Serum-Free Medium* (Table 2) to each well.
3. To each well add ~3 EBs (*see* **Notes 17** and **18**).
4. After covering the plate with the lid, lift the plate up to the light and observe from the bottom that the correct number of EBs has been added and that they are well spread out (*see* **Note 18**). Adjust numbers (e.g., by moving EBs between wells) as necessary.
5. Incubate for 3 days.

3.2.7 Day 12: Replacing the Medium

Expected Cell Populations: Neural Progenitors and Neurons (Figs. 1 and 2)

1. Observe EBs under the brightfield microscope: By this time, neural progenitors will appear as neural rosettes or isolated "plump," usually bipolar, cells with short neural extensions, and neurons can be identified by the presence of long, often branched neurites and axonal bundles (Fig. 1).
2. Replace the medium with fresh *Day 10–15 Serum-Free Medium* (Table 2) (*see* **Note 19**).

3.2.8 Day 15

Expected Cell Populations: Neural Progenitors and Neurons (Fig. 1)

1. Observe EBs under the brightfield microscope (*see* **Note 20**).
2. As a semiquantitative measure of neural-cell production, score the EBs for the presence of neural phenotypes including neural progenitors and neurons (Fig. 2). The percentage of EBs with neural phenotypes can be compared to EBs with nonneural phenotypes to estimate the efficiency of neural cell production. Also score for the presence of mesoderm-derived clusters of beating cardiomyocytes (*see* **Note 20**).
3. These cells can be used for different downstream analyses/ quantitation such as qPCR, flow cytometry, immunofluorescence microscopy, and time-lapse imaging.

4 Notes

1. β-Me is an organic liquid, has a strong sulfuric odor, and is toxic if inhaled or comes in contact with skin. Work quickly with this reagent, especially in the pure form.

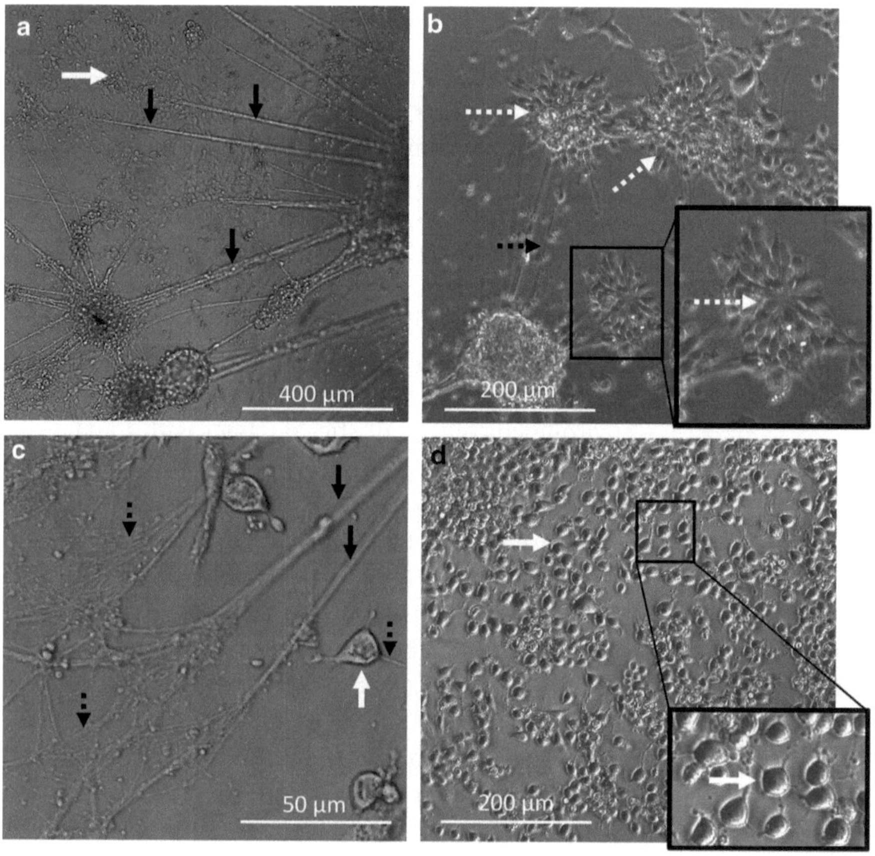

Fig. 2 L-proline-mediated differentiation of mESCs to neural cell types. At the end of the differentiation protocol (Day 15), the differentiated cell population can be scored for neural phenotypes including branched/unbranched neurite projections (broken black arrows) and axonal bundles (black arrows), neural progenitors (solid white arrows) and neural rosettes (broken white arrows). Insets have been used to zoom in on the structures of interest. Note that the neural progenitors are often bipolar with short, often branched neurite projections but can also be multisided with these projections extending from the vertices. Representative images were taken with a phase-contrast microscope using a 10× objective, scale bar = 400 μm; 20×, scale bar = 200 μm; 40×, scale bar = 50 μm

2. The FBS used for these experiments should be suitable for mESC culture. To test this, mESCs should be cultured in *mESC Complete Medium* for 3 passages and tested for Oct4 expression using flow cytometry. The FBS is suitable for mESC culture if >80% cells are Oct4 positive.

3. 0.1 M L-proline is prepared in 0.5–1 mL aliquots and stored at −20 °C if not required for experiments immediately. During experiments, working stocks of 0.1 M L-proline aliquots can be thawed and stored at 4 °C.

4. Freeze-thaw cycles should be avoided. To test the potency of LIF, an alkaline phosphatase stain can be performed, using a kit from, for example, Sigma (Kit 86-C). To do this, grow mESCs

in *mESC Complete Medium* but with different dilutions of LIF (down to 0 U/mL) for 3 passages (12- or 24-well gelatinized plates are convenient), then perform the staining. Observe the following features which indicate self-renewing mESCs: (a) intense, uniform staining. (b) round dome-shaped colonies.

5. A 2.6 mM SB431542 stock is prepared and aliquoted into small volumes (~100 μL) and stored at −20 °C if not required immediately. Working stocks can be stored at 4 °C for a few days. DMSO freezes at 19 °C so bring solutions to room temperature and mix well before adding to medium.

6. Plates for adherent culture can be prepared with gelatin solution and left for 1–2 weeks at room temperature in a draft-free environment (e.g., a plastic container with a lid).

7. Trypsin is deactivated by Ca^{2+}, which is present in culture medium. Hence *two* washes with Ca^{2+}-free PBS are essential. Do not overtrypsinize mESCs as it greatly compromises their viability and seeding efficiency.

8. Ensure cells are dislodged across the whole plate, paying careful attention to the junction between the bottom and side of the plate.

9. Perform cell counts within a few minutes as mESCs lose viability rather quickly in PBS and replating the cells should preferably occur within 15 min. Dead cells in rapidly counted cells should not exceed 5%; anything higher than this strongly suggests something wrong with the cells or the various reagents being used. If a large number of cells is expected following the single-cell suspension, the dilution of the volume of suspended cells to Trypan Blue can be increased (e.g., 10× dilution).

10. The number of cells plated for continuous passage will depend on cell proliferation rates of each batch of cells, the plate size, and how soon the cells will be required after seeding. For example, if 2×10^5 cells are plated on a 60 mm plate, they should be ~50% confluent (~2×10^6 cells) after 2 days.

11. EB culture is a form of suspension culture. As the cells differentiate and the EBs get larger, they can be prone to adhering to the plate: Test your suspension culture plates for EB adhesion. If EBs do stick, try using sterile petri dishes.

12. EBs cultured under neural-inducing conditions (i.e., with L-proline) should remain smooth and round, while EBs cultured under conditions of spontaneous differentiation will have partial or completely roughened surfaces, consistent with the presence of extraembryonic endoderm. If the EBs are overcrowded, and/or not sufficiently dispersed, some aggregation will occur, especially around days 4–7. Overcrowding can be avoided by removing excess EBs at any time during the protocol (collecting samples for analysis is a convenient way to do

this). If excessive aggregation does occur, this can be partially corrected by swirling the EBs into the center of the plate and *very gently* triturating 2–3 times with a 1 mL pipette tip. Excessive or vigorous trituration results in considerable debris, and very "torn" EBs.

13. EBs can be observed easily macroscopically from Day 2, so they can be pipetted and manipulated with the naked eye.
14. The addition of SB431542 is optional: qPCR and flow cytometry results show statistically that *Sox1*$^+$ neurectoderm is produced to the same degree in the absence of SB431542, though the trend is that its presence results in increased production. Changes to the protocol can be made based on gene and protein expression analysis; that is, if *Sox1*$^+$ neurectoderm is produced prior to Day 5 for a particular cell line, then serum-free treatment can be introduced earlier to optimize neural progenitor production. This may be necessary for the optimization of the protocol for different cell lines.
15. Add the aliquot of SB431542 away from the EBs (i.e., toward the edge of the plate), then mix the medium well, to prevent cell toxicity.
16. If EBs are required to be seeded on glass coverslips (for microscopy), ensure that the glass coverslips are sterile (we recommend storing the glass coverslips in 70% ethanol and then drying them under UV). Following sterilization, the coverslips should carefully be loaded into the wells and then gelatinized for at least 48 h.
17. To reduce damaging the EBs, use a P1000 pipette tip (larger diameter hole) and draw up as little volume as possible.
18. Ensure a consistent number of EBs are loaded into each well as this may affect differentiation. The EBs tend to collect near the center of these small wells. Gently shaking the plate and again observing the distribution of the EBs under the light can be helpful.
19. Use a pipette to gently remove the medium from each well. Do not use an aspirator as this may dislodge the EBs and associated cells.
20. Practice in recognizing neural phenotypes (including neural progenitors and neurons under 10×, 20×, and 40× objectives) is advised. The expected yield of EBs with neural phenotypes by Day 15 when cultured in *Directed Differentiation Medium* (i.e., with L-proline) is between 50 and 70% (based on morphology assessment using brightfield microscopy). The expected yield of EBs with neural phenotypes for EBs cultured in *Spontaneous Differentiation Medium* (i.e., without L-proline) is ~10%, and the expected yield for beating cardiomyocytes under this condition is between 2% and 10%.

References

1. Nagy A, Gertsenstein M, Vintersten K, Behringer R (2003) In: Nagy A (ed) Manipulating the mouse embryo: a laboratory manual, 3rd edn. Cold Spring Harbor Laboratory Press, New York, 764 p
2. Rivera-Pérez JA, Magnuson T (2005) Primitive streak formation in mice is preceded by localized activation of Brachyury and Wnt3. Dev Biol 288(2):363–371
3. Williams M, Burdsal C, Periasamy A, Lewandoski M, Sutherland A (2012) Mouse primitive streak forms in situ by initiation of epithelial to mesenchymal transition without migration of a cell population. Dev Dyn 241 (2):270–283
4. Aubert J, Stavridis MP, Tweedie S, O'Reilly M, Vierlinger K, Li M et al (2003) Screening for mammalian neural genes via fluorescence-activated cell sorter purification of neural precursors from Sox1-gfp knock-in mice. Proc Natl Acad Sci 100. Suppl:11836–11841
5. Anthony TE, Klein C, Fishell G, Heintz N (2004) Radial glia serve as neuronal progenitors in all regions of the central nervous system. Neuron 41(6):881–890
6. Kriegstein AR, Gotz M (2003) Radial glia diversity: a matter of cell fate. Glia 43(1):37–43
7. Ying QL, Nichols J, Chambers I, Smith A (2003) BMP induction of Id proteins suppresses differentiation and sustains embryonic stem cell self-renewal in collaboration with STAT3. Cell 115(3):281–292
8. Hirai H, Karian P, Kikyo N (2011) Regulation of embryonic stem cell self-renewal and pluripotency by leukaemia inhibitory factor. Biochem J 438(1):11–23
9. Coucouvanis E, Martin GR (1995) Signals for death and survival: a two-step mechanism for cavitation in the vertebrate embryo. Cell 83 (2):279–287
10. Coucouvanis E, Martin GR (1999) BMP signaling plays a role in visceral endoderm differentiation and cavitation in the early mouse embryo. Development 126(3):535–546
11. Murray P, Edgar D (2000) Regulation of programmed cell death by basement membranes in embryonic development. J Cell Biol 150 (5):1215–1221
12. Washington JM, Rathjen J, Felquer F, Lonic A, Bettess MD, Hamra N et al (2010) L-Proline induces differentiation of ES cells: a novel role for an amino acid in the regulation of pluripotent cells in culture. Am J Physiol Cell Physiol 298(5):982–992
13. Tan BSN, Kwek J, Wong CKE, Saner NJ, Yap C, Felquer F et al (2016) Src family kinases and p38 mitogen-activated protein kinases regulate pluripotent cell differentiation in culture. Public Libr Sci One 11(10):e0163244
14. Rathjen J, Washington JM, Bettess MD, Rathjen PD (2003) Identification of a biological activity that supports maintenance and proliferation of pluripotent cells from the primitive ectoderm of the mouse. Biol Reprod 69 (6):1863–1871
15. Lane M, Gardner DK (1993) Nonessential amino acids and glutamine decrease the time of the first three cleavage divisions and increase compaction of mouse zygotes in vitro. J Assist Reprod Genet 14(7):398–403
16. Gardner DK, Lane M (1993) Amino acids and ammonium regulate mouse embryo development in culture. Biol Reprod 48(2):377–385
17. Gardner DK (1998) Changes in requirements and utilization of nutrients during mammalian preimplantation embryo development and their significance in embryo culture. Theriogenology 49(1):83–102
18. Lane M, Gardner DK (1997) Differential regulation of mouse embryo development and viability by amino acids. J Reprod Fertil 109 (1):153–164
19. Gardner DK, Lane M (1996) Alleviation of the "2-cell block" and development to the blastocyst of CF1 mouse embryos: role of amino acids, EDTA and physical parameters. Hum Reprod 11(12):2703–2712
20. Chen G, Wang J (2014) Threonine metabolism and embryonic stem cell self-renewal. Curr Opin Clin Nutr Metab Care 17(1):80–85
21. Ryu JM, Han HJ (2011) L-threonine regulates G1/S phase transition of mouse embryonic stem cells via PI3K/Akt, MAPKs, and mTORC pathways. J Biol Chem 286 (27):23667–23678
22. Carey BW, Finley LWS, Cross JR, Allis CD, Thompson CB (2014) Intracellular α-ketoglutarate maintains the pluripotency of embryonic stem cells. Nature 518 (7539):413–416
23. Rezk Y, Huff C, Rizk B (2004) Effect of glutamine on preimplantation mouse embryo development in vitro. Am J Obstet Gynecol 190 (5):1450–1454
24. Ryu JM, Lee SH, Seong JK, Han HJ (2015) Glutamine contributes to maintenance of mouse embryonic stem cell self-renewal through PKC-dependent downregulation of

HDAC1 and DNMT1/3a. Cell Cycle 14 (20):3292–3305

25. Comes S, Gagliardi M, Laprano N, Fico A, Cimmino A, Palamidessi A et al (2013) L-Proline induces a mesenchymal-like invasive program in embryonic stem cells by remodeling H3K9 and H3K36 methylation. Stem Cell Reports 1(4):307–321
26. D'Aniello C, Fico A, Casalino L, Guardiola O, Di Napoli G, Cermola F et al (2015) A novel autoregulatory loop between the Gcn2-Atf4 pathway and L-Proline metabolism controls stem cell identity. Cell Death Differ 22 (7):1094–1105
27. Casalino L, Comes S, Lambazzi G, De Stefano B, Filosa S, De Falco S et al (2011) Control of embryonic stem cell metastability by L-proline catabolism. J Mol Cell Biol 3 (2):108–122
28. Rathjen J, Lake JA, Bettess MD, Washington JM, Chapman G, Rathjen PD (1999) Formation of a primitive ectoderm like cell population, EPL cells, from ES cells in response to biologically derived factors. J Cell Sci 112 (5):601–612
29. Rathjen J, Haines BP, Hudson KM, Nesci A, Dunn S, Rathjen PD (2002) Directed differentiation of pluripotent cells to neural lineages: homogeneous formation and differentiation of a neurectoderm population. Development 129 (11):2649–2661
30. Doetschman TC, Eistetter H, Katz M, Schmidt W, Kemler R (1985) The in vitro development of blastocyst-derived embryonic stem cell lines: formation of visceral yolk sac, blood islands and myocardium. J Embryol Exp Morphol 87(1):27–45
31. Ying QL, Stavridis M, Griffiths D, Li M, Smith A (2003) Conversion of embryonic stem cells into neuroectodermal precursors in adherent monoculture. Nat Biotechnol 21(2):183–186
32. Harvey NT, Hughes JN, Lonic A, Yap C, Long C, Rathjen PD, Rathjen J (2010) Response to BMP4 signalling during ES cell differentiation defines intermediates of the ectoderm lineage. J Cell Sci 123(10):1796–1804

Index

A

Adipose tissue .... vii, 1, 117–121, 133, 197
Air-driven droplet generator 95
Alcian blue staining 118, 121, 200, 205
Alginate 94, 95, 97–99, 101
Alizarin red staining 4, 8–10, 12, 118, 121, 122, 200, 205–207, 218, 223, 224, 230
Allogeneic 216, 217, 226
Amino acids 96, 176, 262, 263, 274
Anemia 137
Anesthesia 34
Animals .... viii, 10, 27, 28, 33, 34, 94, 98, 101, 104, 129, 132, 133, 137, 150, 198, 202, 241, 249, 250
Antibodies 3, 7, 8, 12, 26, 33, 35, 65, 67, 69, 70, 72, 73, 107, 110, 119, 120, 131, 133, 134, 139–141, 149, 152, 153, 158, 164, 169–171, 173, 177, 178, 197, 204, 205, 219, 220, 231, 236–238, 241, 245, 249, 252
Arteries 5, 11, 161, 162, 165
Aseptic 26, 39, 106, 118
Atomic force microscopy (AFM) 16, 18, 21
Autologous 217, 226

B

β-cells 25–35, 38, 121
Biopsy 54, 162, 164, 165
Bladder 103–105, 111, 113
Blocking buffer 66, 72, 219, 220
Blood cells
  erythrocytes 126–128, 133–135
  granulocytes 126, 127, 149, 236, 240, 244
  lymphocytes 224, 236
  macrophages 127, 148, 156, 157, 216
  platelets 126–128, 224, 232, 240
Bone marrow
  human 136, 216, 219, 231
  mouse 132–133, 135, 238, 241

C

Cancer 15, 34, 50, 82, 104
Cardiac progenitor cells (CPCs) 161–173
Cell counting .... vii, 41
Cell sorting .... vii, 70, 107, 129, 131, 151, 153–154, 178, 197, 201, 205, 208, 235–245
Cell viability 140, 175, 212, 219, 231
Clones 65, 148, 149, 154–158, 231, 239
Collagenase 26, 28, 29, 33, 52, 54, 55, 59, 65, 69, 76, 118–121, 162, 165, 177, 268
Colony-forming assay 63–79
Compensation 70, 76, 141, 153, 159, 169, 170, 173, 212, 241
Cryopreservation 9, 57, 220, 249, 253, 263, 265, 267
Cryoprotectant 9
Cytospin 154, 156, 157

D

Definitive ectoderm (DE) 273–275, 277, 279
Definitive endoderm (DEn) 94, 100, 104–110, 113, 114, 273, 275
Density gradient separation 149–151, 240
Diabetes 25, 38, 83, 93, 94, 96–98, 118

E

Electron microscopy 17, 18, 20
Embryonic development 273
Embryonic stem cells (ES cells) .... vii, 66, 103–114, 273–283
Encapsulation .... viii, 82, 86, 89, 93–101
Epidermal stem cells 248, 252–254
Erythroid 125–143, 148, 154, 156, 157, 215
Ethynyldeoxyuridine (EdU) 26, 27, 31–34
Euthanization 28, 34, 69, 133, 202, 241, 249
Explant culture 2–3, 5, 6, 253
Expression profiling 135
Extracellular vesicles (EVs)
  Exosomes 16
  microvesicles 16
  oncosomes 16

Mugdha V. Joglekar and Anandwardhan A. Hardikar (eds.), *Progenitor Cells: Methods and Protocols*, Methods in Molecular Biology, vol. 2029, https://doi.org/10.1007/978-1-4939-9631-5, © Springer Science+Business Media, LLC, part of Springer Nature 2019

F

Flow cytometry .... vii, 3, 7, 8, 16, 109, 110, 118–120, 129, 131–134, 137, 140, 141, 152–153, 164, 169, 220, 223, 235–245, 274, 279–281, 283
Fluidics 140, 235, 236
Fluorescence microscope 171, 177–179
Fourier transformed infra-red (FTIR) spectroscopy 86, 87
Fura-2-AM 262, 268

G

Gamma-irradiation 232
Gelatin 106–108, 110, 113, 176, 178–180, 274, 278–280, 282
Gel electrophoresis 85, 86, 185, 187–189, 191
Gene delivery .... vii, 81–90
Gene transcripts 37–47
Germ layers .... vii, 94, 104, 105, 273, 275, 277
Glucose 52, 63, 67, 68, 74, 76, 78, 84, 96, 98, 118, 217–219, 262, 263, 276
Green fluorescent protein (GFP) 83–87, 89, 129, 208–210, 276
Gut progenitor cells 49–61

H

Hemocytometer 3, 5, 7, 41, 65, 71, 74, 149, 151, 166, 241, 244, 245
[^{3}H]-thymidine 228, 229, 232, 233
Hydrogel 94
Hydroxyethyl starch (HES) 238–240, 244

I

Immunoassay 202, 208, 210
Immunoisolation 94
Immunophenotyping 3, 220, 223, 231
Immunostaining .... vii, 64, 253, 259, 260
Insulin 25, 26, 32–34, 38, 63, 67, 73, 83, 93, 97, 98, 133, 140, 198, 199, 202, 207, 208, 210–211, 276, 277
Interstitial cystitis/painful bladder syndrome (IC/PBS) 104
Intestinal cell culture 50, 51
Intestinal crypts 49, 51
In vitro differentiation
- adipogenic 8, 12
- chondrogenic 119, 120
- neuronal 260, 267
- osteogenic 8, 12

Islet of Langerhans
- human 27, 31
- mouse 25–35
- rat 25–35

Isoelectric focusing (IEF) 185–187, 189, 191–192
Isoflurane 68, 75, 97, 98, 101, 249
Isotype control 3, 7, 12, 65, 70, 110, 119, 141, 219, 220, 239, 241, 242, 245

K

Keratinocytes 96, 248, 249
Kidney capsule 64, 68, 79, 106

L

Lentivirus 65, 71, 197, 200–201, 207–209, 213
Lipoaspirate 118–120
Liquid nitrogen 7, 57, 166, 191, 220, 230, 231, 253, 258, 264, 265, 267
L-proline 274–277, 281–283

M

Magnetic sorting 252, 253
Mass spectrometry 185–194
Matrigel 26, 27, 31, 32, 34, 50–52, 55–57, 60, 61, 64, 66, 71, 77, 78, 105, 106, 260, 261, 264–268
Mesenchymal stem cells (MSCs) 1–3, 5–12, 117, 162, 166, 197–213, 215–232
Metal organic frameworks .... vii, 81–90
Mitochondria 175, 179, 182
Mitogen 216, 227
Mitosis 176
Mixed lymphocyte reaction (MLR) 216, 227–229, 232
Mounting medium 33, 164, 171
Multiplicity of infection (MOI) 71, 208, 213

N

NanoDrop spectrophotometer 191
Nanoparticle tracking .... vii, 17

O

Oil Red O staining 9, 10, 12, 120, 121, 230
Organoid 50

P

Paired daughter cell assay 150, 154–158
Passaging .... vii, 40, 55–58, 97, 108, 166, 231, 264–265, 268, 278

Peripheral blood mononuclear cells (PBMC)............224, 226–229, 232
Peritoneum................29, 98
Placenta................21, 197
Plasmids................83–85, 87, 89, 198, 200, 207, 209, 212
Polymerase chain reaction (PCR)
  digital................37–47
  droplet digital................37–47
  reverse-transcriptase (RT)................vii, 107–111, 113
Progenitor cells................vii, viii, 1–12, 26, 37–47, 49–61, 63–79, 81–90, 93–101, 117–121, 125–143, 147–159, 161–173, 187, 191, 197–213, 215–233, 235–245, 247–254
Progeny................viii
Proliferation................viii, 1, 25, 50, 172, 176, 216, 218, 227–229, 259, 266, 282
Proteomics................185–194

Q

Quenching medium................26, 29, 30

R

Ratiometric calcium indicators................268
RBC lysis buffer................238, 240
Reactive oxygen species................176, 179–180
Regenerative medicine................vii, 1, 104
RNA isolation................vii, 38, 39, 41, 42, 46

S

Self-renewal................50, 64, 126, 147, 148, 158
Serum-containing medium (SCM)................38, 40, 84, 87
Single-cell assays................viii, 147–159
Single-cell sorting................153–154
Small angle X-ray scattering (SAXS)................83, 86, 87
Sodium dodecyl sulfate–polyacrylamide gel electrophoresis (SDS-PAGE)................73, 185, 187, 189, 192
Spheroids................50, 51, 53, 55–58, 60, 61
Spleen................129, 130, 133, 134, 137, 138, 140
Store-operated calcium entry (SOCE)................257–270
Stromal vascular fraction (SVF)................118, 120, 140
Surface antigens................107, 128, 138, 140, 143, 237

T

Tangential flow filtration (TFF)................197, 201, 207, 209
T cell proliferation................viii, 216, 218, 227–229
Terasaki plates................149, 150, 153–155
Thapsigargin (TG)................257, 261, 262, 268
Thawing................59, 264, 268
Tissue culture medium................7, 87, 224, 226, 277
Transduction................vii, 64, 71, 83, 201, 208–210, 213
Transfection................51, 82–85, 87, 89, 90, 198, 201
Transplantation................viii, 1, 2, 38, 64, 68, 74–76, 81, 93–101, 147, 158, 161
TRIzol................39, 41, 46, 107, 111
Trypan blue................3, 5, 65, 71, 74, 218, 219, 224, 226, 227, 238, 241, 244, 276, 278, 282
Trypsin................2, 10, 26, 38, 40, 46, 52, 53, 57, 59, 60, 68, 74, 79, 84, 87, 119, 121, 132, 136, 137, 166, 168, 177, 190, 191, 193, 220, 226, 229, 231, 274, 278, 282
2D electrophoresis................185–194
Tympanic membrane................vii, 247–254

U

Ultracentrifugation................16–18, 198, 209
Umbilical cord................1–12, 117, 197, 237, 238
Umbilical cord blood................2, 237, 238
Urethra................103, 104
Uroepithelium................103, 266
Uroplakins................103, 105, 112
Urothelium................103–114

W

Western blotting................vii, 64, 67, 73
Wharton's jelly (WJ)................2–8, 10, 197

MIX
Papier aus verantwortungsvollen Quellen
Paper from responsible sources
FSC® C105338

If you have any concerns about our products,
you can contact us on
ProductSafety@springernature.com

In case Publisher is established outside the EU,
the EU authorized representative is:
Springer Nature Customer Service Center GmbH
Europaplatz 3, 69115 Heidelberg, Germany

Printed by Libri Plureos GmbH
in Hamburg, Germany